FORGOTTEN SCIENCE

Strange Ideas from the Scrapheap of History

S. D. TUCKER

Whether mythic or scientific, the view of the world that man builds is always largely a product of his imagination.

François Jacob[1]

First published 2016
This edition published 2018

Amberley Publishing
The Hill, Stroud
Gloucestershire, GL5 4EP

www.amberley-books.com

British Library Cataloguing in Publication Data.
A catalogue record for this book is available from the British Library.

ISBN 978 1 4456 8657 8 (print)
ISBN 978 1 4456 4838 5 (ebook)

Typesetting and Origination by Amberley Publishing.
Printed in the UK.

Contents

Introduction: Today's Science, Tomorrow's Superstition

Truth emerges more readily from error than from confusion.

Francis Bacon[1]

Dogs, they say, are a man's best friend. Perhaps they would feel less well disposed towards their human masters, however, if they were aware of a bizarre series of experiments performed upon canine-kind from the 1650s onwards. The men behind these weird trials were not lunatics; nor were they criminals, sadists or simpletons. They were, for the most part, Fellows of the Royal Society, England's most prestigious new scientific institution, and included such august and celebrated personages as Robert Boyle (1627–91), often described as 'the father of chemistry', and Sir Christopher Wren (1632–1723), the anatomist, architect and designer of St Paul's Cathedral. These were brilliant men, giants of their age, towering intellects and figures of great learning – and yet they still once seriously thought it might be a good idea to fill a dog up with soup instead of blood to see what would happen.

The idea now sounds insane, but at the time would have seemed slightly less so. In 1628, William Harvey (1578–1657), court physician to Charles I, had published his *De Motu Cordis*, in which he detailed one of the most significant medical discoveries of all time, his theory of the circulation of the blood, an idea he had arrived at through a series of experiments performed upon the circulatory systems of various living creatures, including dogs. Beginning in 1616, Harvey had given a series of lectures in which he had revealed that the heart pumped a fixed amount of blood to circulate through the body's veins and arteries in a continuous fashion, thus disproving the previously accepted ideas of the ancient Roman physician Galen (*c.* 129–*c.* 216). Galen believed that blood was created inside the liver and became enriched with nutrition by the ingestion of food, a source of energy which was then distributed through the body via the veins. Meanwhile, a second separate stream of blood, distributed via the heart and the arteries, transported air throughout the body. During this process, old blood was constantly being consumed, and new blood being created in the liver. His experiments on animals, though, led Harvey to realise that if this was really true then the liver would have to be producing prodigious amounts of the red stuff each and every day, far more than it would be possible for the body to absorb. Instead, Harvey saw that blood in animals – and thus by implication in humans – flowed through an essentially circular system, passing out from the heart through the arteries, and then back to the heart again through the veins.[2]

By the 1650s, Harvey's findings were common knowledge among well-educated men like Boyle and Wren, but still somewhat uncertain was the precise reason why blood circulated in the first place. We now know that blood cells' main function is to carry oxygen around the body, but at the time this was not quite so obvious. Harvey himself admitted he didn't know why blood needed to be pumped so quickly throughout the body; he just knew that it did.[3] The only way for interested parties like Boyle and Wren to find out for sure what exactly blood could and could not do was to perform some practical experiments upon living beings. As such, in 1656 Wren got hold of a dog, dissolved a quantity of opium in some wine, and then introduced this heroin-like substance into the animal's bloodstream. At first the dog seemed reluctant to aid science, and had to be held still while a syringe was inserted into a ligatured vein. Very quickly the drugs worked their magic and the animal became intoxicated, putting up less of a struggle. Wren's conclusion was that this proved Harvey's theory admirably, the opium circulating throughout the dog's whole body via its bloodstream with great rapidity. Eventually, after being whipped around a nearby garden to snap him out of it, the dog returned to normal, and Wren was able to draw some basic conclusions about how it was that poisons, intoxicants and medicines might be made to work more effectively by injecting them straight into the bloodstream rather than administering them orally. If, today, you have a fear of visiting a doctor for a needle, then this is where the procedure effectively began. As for the junkie dog, meanwhile – after Wren made his findings public, it became famous and was stolen by someone wanting a celebrity pet![4]

Wren's experiments, then, as later written up by his colleague Boyle, gave some suggestion as to the function of the bloodstream. Clearly, its main purpose was to transport things around the body. The question was, what things? Air, most likely (oxygen as such was not then known). Perhaps heat as well, helping regulate body temperature. Maybe it moved around all kinds of things. Possibly it transported nutrition from organ to organ, too, just like Galen had taught? If so, then perhaps it would be feasible to dispense with solid food altogether, and start injecting nourishing liquid substances like broth and watery pudding intravenously instead? This was the theory of Richard Lower (1631–91), a Cornish physician and correspondent of Robert Boyle. Inspired by Wren's activities, Lower tried to feed a starving dog by pumping milk and soup direct into its bloodstream, but admitted that the procedure failed – most of Lower's experimental soup-dogs died, their milky, broth-infused blood curdling within their very veins. He did not know that, while a substance like alcohol can pass unchanged straight into a living creature's bloodstream, soup and broth have to be exposed to various complicated chemical processes within the stomach, in which the nutrients they contain are broken down into much smaller molecular components. As such, you could no more fill a dog's blood full of soup to feed it than you could inject its veins full of Pedigree Chum.[5]

Horse-Hobby

Dogs were not the only animals whose circulatory systems were investigated during the Early Modern period. In 1733, Stephen Hales (1677–1761), a priest and Fellow of the Royal Society, published his book *Haemastatics*, an investigation into the nature of blood pressure in which, without any shame whatsoever, he revealed to the world his unusual hobby of tying live horses down to gates on their backs, cutting them open and then sticking brass pipes and glass tubes constituting pressure gauges into their veins and arteries, before draining them of all their blood. From these endeavours Hales managed to extract numerous interesting conclusions, including that an average horse might contain around 18 litres of blood, that its pulse rate increased as it suffered more pain, and that the more blood was extracted, the lower the animal's blood pressure fell. He also claimed that whenever the horse being drained sighed, as I'm sure it did quite often, its blood pressure rose momentarily, leading to his odd recommendation that all sad and depressed persons could cheer themselves up immediately simply by groaning, thus increasing the force of the blood in their veins and so filling them up with vim and vigour.[6]

Given such cruel and unusual methods, perhaps we should be glad that the scientists of the period stuck purely to experimenting upon animals rather than moving on to messing about with human beings instead – except, eventually, they did. Richard Lower, while he may have failed to transfuse soup, was the first man in history to have successfully transfused blood between living beings, as he described in a series of letters written to Robert Boyle in 1666. Lower had taken three dogs, two to act as forced blood donors and one to be the recipient. By bleeding this latter animal from the jugular vein and then transfusing blood from the cervical artery of each of the two donor dogs in turn, repeatedly, until these latter two animals died, Lower believed he had succeeded in completely replacing the living animal's blood with fluid from its unfortunate friends. The recipient dog itself happily survived, having the wound in its vein sewn up and then jumping down from the operating table and pawing its master affectionately before rolling around on the floor to clean the claret off as if nothing had happened.[7] This, thought Lower, was a great success – and the next step was to try out a similar experiment upon a human being. Getting hold of a convenient lunatic named Arthur Coga – a divinity student from Cambridge who suffered from 'a harmless form of insanity', apparently – he decided to see what would happen if he transfused some blood from a lamb into his veins. Coga's own opinion was that his somewhat tempestuous nature would be alleviated by the procedure, as 'sheep's blood has some symbolic power, like the blood of Christ, for Christ is the Lamb of God'. The hope was that the vital fluids of the proverbially meek and mild creature would produce a noticeable improvement in Mr Coga's agitated condition. Regrettably, they did not; but at least he didn't die.[8]

Christopher Wren, too, would have been quite happy to experiment on humans, if only he could have found a willing subject. In 1657, the French ambassador in London airily offered him 'an inferior domestic' of his, whom he thought 'deserved to have been hanged' in any case, to have his blood transfused. Wren seized upon the proposal, though the trial was thankfully abandoned after the poor servant, seeing the hideous equipment which was to be used on him, dropped down and fainted on the spot.[9] At least things had moved on slightly since 1492, when the hapless physicians of the dying Pope Innocent VIII (1432–92) had attempted to perform a much more rudimentary form of blood transfusion upon the pontiff by drawing large amounts of blood from three young boys and then pouring it down his mouth, resulting in the unsurprising deaths of all four parties.[10] It was once a traditional belief that every person had a natural allotted lifespan, and that if this lifespan was cut short by accidental or violent death – having your veins cut open by the Pope's quack doctors, for instance – then your blood would still contain reserves of vital life energy which could be consumed to add to an older person's dwindling supply. People used to attend public executions, mop up the blood spilled on handkerchiefs, and then wipe it across wounds in the hope of making them heal better.[11] Such beliefs, we now know, are simply superstitions – but the winnowing out of fable from fact is often only possible by subjecting such ideas to practical trials, much as some of these operations may now seem highly unethical to us.

Even the mystery of how digestion worked was only finally solved through the use of live experimental subjects. Lower had conclusively established that soupy substances could not be digested via the bloodstream, but it took a chance medical mishap involving a patient of the US military surgeon William Beaumont (1785–1853) for science to work out more fully how the chemistry of the stomach actually functioned. One day in 1822, a Canadian fur trapper named Alexis St Martin (1802–80) was accidentally shot in the stomach, and treated by Beaumont for his injuries. St Martin was expected to die, but made a surprising recovery. However, while his wound largely healed over, he also developed a permanent fistula, or abscess-derived opening, from the inside of his belly to the outside of his flesh. This left a smallish hole just below his chest wide open, thus giving interested parties a direct window into St Martin's most recently eaten dinners. Rather than sewing this convenient viewing panel up, Beaumont took advantage of the splendid opportunity to advance medical knowledge and signed St Martin up to be his handyman. However, his dubious contract also obliged St Martin to allow his employer to perform a series of bizarre medical experiments upon him. From now on, Beaumont was free to feed his employee meals and shove instruments inside the hole to extract food-samples at every stage of the digestion process, before then subjecting them to close analysis. At other times, he simply tied bits of food to pieces of string and dropped them in through St Martin's belly-hole like post through a letterbox, retrieving them at intervals to see how the digestion process was going. Naturally, this all caused St Martin considerable pain and distress. He ended up running away to Canada in 1825, but Beaumont had him recaptured and brought back home to be subjected to further abuse. Nowadays, you suspect that Beaumont may have found himself on the receiving end of a lawsuit for his actions, but when his book on the matter, *Experiments and Observations on the Gastric Juice and the Physiology of Digestion*, was published in 1833, few people saw fit to object to any of it.[12]

Animal Wrongs

What are we to conclude from all this strangeness? That people in the past were stupid and cruel? Not necessarily. Several contemporary colleagues of Wren, Boyle and Lower did feel somewhat queasy about the course they were pursuing. For example, Robert Hooke (1635–1703), the Royal Society's chief experimentalist, participated in various hideous-sounding operations, including a pair of public vivisections (one carried out with Lower's help) in which a live dog's windpipe or lungs were slit open and bellows inserted into the openings, in order to prove that neither the movement of the ribcage, nor even of the lungs themselves, was actually necessary for breathing to function. In spite of the general praise these findings raised, Hooke was not terribly keen on the pain and suffering he had been obliged to inflict – during one of the trials, he had actually had to remove the live dog's ribcage from its still-beating chest. As such, Hooke was always on the lookout for alternative ways of doing things. In his famous 1665 book *Micrographia*, for instance, in which the wonders of the world beneath the microscope were first revealed to an amazed public, one of the reasons Hooke gave for praising the new instrument was that it allowed scientists to pry into Nature's wonders 'undisturbed' rather than 'dissecting and mangling creatures whilst there is yet life within them', which he saw as being a great moral advance.[13]

Hooke knew full well that some people at the time viewed him and his friends as being both monsters and madmen. In 1676, the great dissector had attended a performance of *The Virtuoso*, a new play by Thomas Shadwell (*c.* 1642–92), the future Poet Laureate. As proceedings began, Hooke, to his horror, found that several members of the Royal Society – including himself – were being humiliatingly mocked live on stage in Shadwell's absurd character of Sir Nicholas Gimcrack. Gimcrack was an overtly comic figure, a gentleman scientist given to various mad pseudoscientific schemes such as developing a method to swim on dry land and bottling air from various regions before storing it all away in his cellar instead of wine. Worse, Gimcrack also directly imitated the experiments of Richard Lower by transfusing sheep's blood into a human being – albeit in the hope of creating an entire flock of human sheep from whom to harvest wool, rather than in an attempt to cure mental illness. The experiment goes awry, however, and Sir Nicholas' victim ends up growing what Shadwell calls 'a Northampton sheep's tail ... from his anus, or human fundament'. This was a joke, but a joke based upon the exaggeration of genuine contemporary fears; Robert Boyle himself had wondered whether putting one person's blood into another person's body might alter the recipient's hair colour, so a sheep's blood giving a man a sheep's tail was only taking matters to their logical conclusion.[14]

Stephen Hales' experiments on horses – and his other habit of skinning frogs alive to see how their muscles worked[15] – were also the subject of contemporary disquiet. Admittedly, he only ever chose to tie down and forcibly exsanguinate horses that were on their way to the knacker's yard anyway,[16] but this simple nod towards experimental ethics didn't satisfy everyone. The poet Thomas Twining's (1735–1804) verse 'The Boat', for example, described the Middlesex parish of

Teddington, where Hales had once lived and worked, as being nothing less than the lair of a mad scientist:

> Green Teddington's serene retreat
> For philosophic studies meet, [well-suited]
> Where the good Pastor Stephen Hales
> Weighed moisture in a pair of scales,
> To lingering death put Mares and Dogs,
> And stripped the skin from living frogs.
> Nature he loved, her Works intent
> To search, or sometimes to torment.[17]

Hales may well have tormented his equine victims, but his work did, nonetheless, search out certain truths of Nature, truths which, we can now see, have all kinds of practical implications for modern surgery and medicine. The fruits of this programme of vivisection, though, would take many, many years to ripen. As the medical historian David Wootton has put it, for people at the time, the work of men like Wren and Hales – and even of the great William Harvey – provided 'no therapeutic benefits' at all; 'not a single life was saved, not a single illness abbreviated'. In fact, Wootton concludes, all the information gained was at the time 'entirely useless', other than in the sense of providing knowledge for its own sake. The main contemporary practical application of Hales' work on blood pressure, for example, was to allow surgeons to perform bloodletting more effectively – a treatment we now know to have been wholly worthless.[18] So, was there any real point in these men doing what they did? Yes, but not for a long time.

The Naked and the Dead

Even the experiments such men carried out upon humans eventually provided some final benefit for mankind. Leonardo da Vinci (1452–1519), for example, was a great anatomist as well as a great artist, but restrictions upon corpses available for him to dissect meant that he sometimes had to extrapolate from work he had performed with dead cows instead – with the bizarre result that some of his illustrations of human anatomy actually contain bovine features![19] Galen, too, stymied by the dissection of corpses having been illegal in ancient Rome, had based many details in his own books of anatomy upon his investigations of dead macaques, leading physicians down a blind alley for a thousand years.[20]

Restrictions upon dissecting human corpses continued across Europe for much of the next two millennia, with one main exception. Beginning in the 1300s, the practice became seen as not only acceptable in Italian universities but, from the 1500s, as actively entertaining in nature, with audiences of hundreds turning up to see cadavers split open in town squares, or even inside churches. Soon, Italians became the acknowledged anatomical experts, and those studying medicine elsewhere were at a real disadvantage. Living in the town of Louvain during his youth, for instance, the great Belgian anatomist Andreas Vesalius (1514–64) was so starved of corpses that, seeing a decomposing criminal hanging from a gibbet one day in 1536, he was unable to restrain himself and immediately pulled off the arms and legs to take back home. Later that night, Vesalius returned, stole the rest of the body, boiled the bones free of flesh, and then reassembled them all as a full skeleton. So inexperienced were Vesalius's former university teachers in Paris that, when they had finally managed to get hold of a carcass one day, their grave-robbing student actually had to leave his seat and show them what to do. Once safely installed in Italy's Padua University from 1537 onwards, Vesalius showed little apparent shame at his activities, openly admitting to corpse theft in print and writing about how he let his bone-hungry students get skeleton keys made so they could enter cemeteries unhindered after dark. He even boasted about stealing the body of a particularly attractive prostitute, thus getting an even more intimate view of her insides than her clients had done. Visitors to his home, meanwhile, could often find his kitchen filled with the dead, their bones being boiled clean in big pots. Various small body parts, like ears and nose tips, he would simply remove with a knife and string together on necklaces as handy teaching aids.[21]

Vesalius wasn't the only such offender. So eager for fresh specimens was the anatomist Gabriele Fallopio (1523–62) that he once managed to convince the Tuscan authorities to give him a criminal to execute personally himself so he could pop him straight onto his dissection table while the corpse was still warm! As always, though, Vesalius had to go one better. Snatching someone who had supposedly just 'died' in an accident, he didn't let the fact that the poor man's heart was still beating stop him from reaching into his chest and removing it for a closer look.[22] Ethically dubious though he was, Vesalius is generally acknowledged as the greatest anatomist of all time. His 1543 *De Humani Corporis Fabrica* is a masterpiece, filled with the first genuinely accurate labelled diagrams of the human body. Vesalius was quick to suspect Galen had dissected only apes, not humans, demonstrating some 300 of the ancient Roman's anatomical errors within the *Fabrica*'s gloriously illustrated pages. Previously, dissections had largely functioned as three-dimensional illustrations to Galen's inaccurate writings; now, they stood as contradictions of them. Vesalius may even have cut open monkeys in public himself, too, comparing them to human corpses and thereby showing precisely how Galen's mistakes had originated.[23]

Would the world really be a better place today if doctors still thought that human beings had the partial physiology of cows and monkeys? Would it really be a good thing if we had no practical knowledge of the workings of blood pressure, or were unaware that the heart is a kind of pump? Blood transfusions, intravenous injections and other such beneficial procedures all have their origins in the cruel work of people like Wren, Boyle and Lower throughout the 1600s, and accurate knowledge of the human body truly begins only with Vesalius the skeleton thief. Their actual experiments themselves, though, were undeniably brutal and disturbing, and the sort of thing which could obviously never have been allowed to take place during more modern times ... could they?

A Dog's Life

In fact, the probable high point (or low point) of bizarre and distressing experiments being performed upon animals and human beings, both living and dead, was actually the twentieth century. Surely the most famous modern tests made upon animals were those of the Russian scientist Ivan Pavlov (1849–1936), winner of the Nobel Prize in 1904. Pavlov's name is common currency these days, and everyone knows of his basic ideas about so-called 'Pavlovian conditioning', which tells of the existence of conditioned reflexes. Today, Pavlov's dogs, the animals upon which he chose to perform the majority of his trials, are just as well known as he is. Rather fewer people, though, are perhaps aware of precisely what it was that Pavlov did to them.

In order to work out what made dogs tick, Pavlov first had to slice them open. In the words of one recent historian of animal-related experiments, Rom Harré, this entailed nothing less than 'the surgical reshaping of the bodies of his dogs under anaesthetic into pieces of scientific apparatus'.[24] Like Robert Hooke before him, Pavlov did have some qualms about the work he was doing. Disliking the sight of blood, he insisted an assistant clean it all up throughout surgery, and tried his best to restore his dogs back to their original state once the experiments were over – or, at least, to turn them back to normal 'in so far as the nature of the procedure permits'.[25] With many of his procedures, though, this was a bit like saying that someone would soon be back up on their feet again following a double leg amputation. Ultimately, Pavlov erected a monument to his many dead dogs upon which he had engraved the following message:

> The dog, man's helper and friend from prehistoric times, may justly be offered as a sacrifice to science, but this should always be done without unnecessary suffering.[26]

No 'unnecessary' suffering, perhaps, but I doubt Pavlov's animals actively enjoyed their fate. Pavlov's specific interest was the role nerves played in digestion, and one of the main inspirations for his work, curiously enough, was William Beaumont's research into the big hole that had once opened up in his butler's stomach. Imitating Beaumont, Pavlov created artificial fistulas in various parts of his dogs' digestive systems, thus allowing him to make detailed observations proving that stimulation of certain nerves would help prompt the secretion of gastric juices. The problem was that these fistulas kept healing over, thus removing the windows into the dogs' guts. Pavlov's solution was to imitate the way that natural orifices like nostrils keep themselves permanently open, sewing a small mucous membrane around the wounds.

The results of these procedures could be very peculiar indeed. For instance, in order to collect gastric juice from one set of dogs without contaminating it with samples of food and thus ruining it for chemical analysis, each animal had its oesophagus detached from its stomach. Then, the animals were fed mince. The act of eating and swallowing it stimulated certain nerves in each dog, thus causing gastric juices to be secreted, which were then collected. No contaminating food, though, ever reached the stomach; it simply fell out from the end of the severed oesophagus through one fistula, and plopped down onto a waiting dish. The dogs, naturally, were delighted and ate the same mince a second time, not understanding what was going on. This happened over and over again, with each act of swallowing stimulating nerves and so producing yet more gastric juice samples for Pavlov's assistants to collect from a second fistula leading into the belly. No matter how much meat the poor dogs ate, however, their hunger could never be satisfied and, unless Pavlov later provided surgical intervention, they would have no choice but eventually to starve to death – despite spending their every waking minute eating.[27]

Two Heads Are Better Than One

When the Bolsheviks came to power following the Russian Revolution of 1917, Pavlov was arguably the most famous scientist at work in the new People's Paradise. Whilst supporting the revolutionary regime's commitment towards science, Pavlov was nonetheless a firm critic of Lenin's government in other ways, writing letters to commissars including phrases like, 'You believe in vain in the all-world revolution ... You are terror and violence ... Have pity on the Motherland,' which, had he lived to see Stalin achieve high office, would doubtless have seen him shot.[28] Nonetheless, following his death in 1936, Pavlov was reclaimed by Soviet propagandists as having been one of them, and the influence of his idea of reshaping animals into living pieces of lab apparatus lived on among some of the regime's more heartless scientists and doctors.

Take the grotesque work of the Russian surgeon Vladimir Demikhov (1916–98), who in 1954 revealed to an appalled public that he had created a two-headed dog – just one short of the full Cerberus. Previously, Demikhov had succeeded in performing the world's first successful liver, lung and heart transplants, but didn't see why he should stop there. By taking an adult mastiff and grafting the head, neck and front legs of a puppy onto it, then wiring them up so they shared a blood supply, Demikhov sculpted a new artificial breed of animal, albeit one with the approximate life span of a mayfly. Due to tissue rejection and infection, no two-headed dogs managed to last more than a few days; indeed, they could not possibly do so seeing as, in a curious echo of Pavlov, the puppy's head was not connected up to the adult mutt's stomach, meaning that, whilst it did chew and swallow food, it had no means of actually digesting it. Perhaps this sense of unquenchable hunger was why the smaller animal took to constantly biting its adult host's ears. The ultimate aim of Demikhov's researches, apparently, was to create a supply of brain-dead 'human vegetables', upon whose prone bodies he hoped one day to be able to graft and grow a storehouse of spare limbs and organs for any Soviet citizens needing transplants.[29]

Grafting an extra head on to a dog was one thing; attempting to cut a dog's head off from its body and keep it alive was quite another. This, however, was the claimed ability of Sergei Bryukhonenko (1890–1960), of Moscow's Institute of Experimental Physiology and Therapy, who in a twenty-minute 1940 cine-film demonstrated the supposed powers of his amazing new invention, the *autojektor*, a device designed to infuse oxygenated blood through veins and arteries with the aim of resurrecting the dead. Bryukhonenko claimed this device really worked, and so produced an undeniably dead item to place on his operating table before the camera – the severed head of a dog. Attaching this up to the *autojektor* and switching it on, Bryukhonenko showed viewers he had the authority to reach beyond the grave. The dog's head, or so it appeared, began to breathe. To prove it could feel pain, the doctor's assistants promptly poked things into its newly opened eyes, producing the expected reaction, and, to demonstrate that this was not simply some kind of automatic nerve response, daubed citric acid on its nose. The head, irritated by its sourness, made the apparently conscious decision to lick it off. For an encore, Bryukhonenko then had a second dog placed on the table, drained of all its blood and left to lie dead for ten minutes. Then, he flicked the switch on his *autojektor*, waited a few moments, and proudly demonstrated to viewers that the dog was now breathing.

Then, following an inter-title reading 'A FEW DAYS LATER ...' we cut to a shot of the resurrected dog (and another Christ-like canine) wagging their tails in full recovery.[30]

That, at least, is what the film *seemed* to show; seeing as its first Western screening was to an assembly of US military scientists in New York in 1943, it was probably largely an exaggerated Soviet propaganda exercise. The *autojektor* did exist – it was a kind of primitive heart-lung machine of the kind surgeons now use to keep patients alive during certain operations – and it really could keep a recently severed head looking animate, or a surgically removed heart beating, for a certain short period of time. However, it could not truly resurrect the dead. After all, as has recently been pointed out by the science writer Robert Swain in his own discussion of Bryukhonenko's work, seeing as brain damage begins to kick in on any corpse after only ninety seconds of death, any genuine returnee from the Other Side would almost certainly be severely physically and mentally handicapped, not bright-eyed and bushy-tailed.[31] Sergei Bryukhonenko, though, was not the first scientist to have tried to bring animals back from the pet cemetery ...

The Cat Came Back

Today, cat videos seem to make up around nine-tenths of the Internet. Had any footage existed of the experiments performed by a once notorious Prussian physician named Karl August Weinhold (1782–1829), however, it is a fair bet that they may not have received the accustomed level of popularity among the general public. In an 1817 book, Weinhold detailed his weird experiments with something he called 'bimetallic electricity', a phenomenon which relied upon the fact that placing different metals in contact with one another can, under certain circumstances, produce a galvanic or electrical charge. Knowing that such charges were capable of stimulating nerves and muscles even in dead animals, Weinhold decided to test this knowledge to destruction.

First of all he got hold of a three-week-old kitten and chopped its head off. As expected, the animal died. Equally predictably, when Weinhold then applied galvanic batteries to the dead kitten's muscles, these muscles began to respond automatically. Then, however, he went one step further and cored out the animal's spinal cord, before filling up the hollow with an amalgam of silver and zinc. This liquid, he said, spread out from the spinal cord to the kitten's nerves, with the galvanic charge present acting to stimulate the headless corpse's heart, which began to beat again before the animal stood up and jumped around wildly of its own accord. Eventually, the headless cat died a second time, this time for good. Weinhold's next move was to get another kitten and, instead of decapitating it, simply to remove its brain, scooping it out with a small spoon. Then, as well as filling up the hollow spinal cord with liquid metal, Weinhold filled the cavities left by the brain with his galvanic amalgam. According to him, this electrically charged cat then 'got into such a life-tension that it raised its head, opened its eyes, stared for a time, tried to get into a crawling position, sank down again several times, yet finally got up with obvious effort, hopped around, and then sank down exhausted', a performance allegedly lasting some twenty minutes. Weinhold was convinced that this undead animal could both see and hear, so took his cutlery to yet another kitten's grey matter. Through further experiment, the peculiar Prussian proved to his own satisfaction that the cat's eyes were still sensitive to light (he thrust burning lamps near its face), and its ears still sensitive to sound (when he loudly threw a bunch of keys down on to a table it winced).

Weinhold was adamant that these zinc-brained cats were genuinely, though only temporarily, alive. This, surely, was the proof you could indeed bring back the dead, albeit as brain-damaged automatons – wasn't it? Sadly not. Nobody has ever managed to successfully replicate Weinhold's wonders (has anyone ever really tried?) and he is generally written off as being either a deliberate fraud or, at best, guilty of wishful thinking. The only other person I can think of who has tried to resurrect dead cats was the British occultist Aleister Crowley (1875–1947), who famously claimed to have tested out the idea of a cat having nine lives by subjecting one to nine different modes of murder (arsenic, chloroform, hanging, burning, stabbing, drowning, defenestration, smashing its skull and slitting its throat) 'in the interest of pure science'.[32] Weinhold was only marginally less of a scientific charlatan than Crowley, however. A strange and unhappy man with abnormally long limbs, a female voice, no facial hair and deformed genitalia which suggested he may actually have been a hermaphrodite, he is described as having been a rather uncompromising man/woman whose previous claim to fame had been making the proposal that all poor males in Prussia be forcibly fitted with a tight metal ring around their testicles to prevent them siring any children. He wasn't seriously listened to, then or now, and probably doesn't deserve to have been.[33]

Goodbye, Lenin!

In Soviet Russia, though, things were different. The influence of a truly strange form of scientific mysticism nowadays known as 'Russian Cosmism' – which we shall examine properly in this book's conclusion – was initially very strong in the years following 1917's revolution, and the idea of science conquering death considered a plausible possibility. The Bolshevik leader Vladimir Ilyich Lenin's

(1870–1924) death in 1924, for example, was considered by some to be a purely temporary state of affairs, particularly by one Leonid Krasin (1870–1926), the Soviet Commissar of Trade and head of the so-called 'Immortalisation Commission', the organisation given responsibility for the creation of Lenin's tomb. Krasin's big idea was to freeze Lenin's cadaver with the aim one day of restoring him to life. This was no metaphor. Three years earlier, at the funeral of another prominent Communist, Krasin had given a comforting speech in which he claimed:

> I am certain that the time will come when science will become all-powerful, that it will be able to recreate a deceased organism ... And I am certain that ... that time will come, when the liberation of mankind, using all the might of science and technology, the strength and capacity of which we cannot now imagine, will be able to resurrect great historical figures.[34]

Great historical figures, of course, just like Lenin. Commissioning the architect Alexey Shchusev (1873–1949) to design the great man's mausoleum, Krasin did so in the knowledge that he was a disciple of the Russian-Polish artist Kazimir Malevich (1878–1935). Malevich, inspired by occult teachings as well as by Cosmism itself, also declared that Lenin was immortal by taking it upon himself to create a new shape in his memory – the cube. You may have thought that cubes already existed, but Malevich's new one was a very special cube indeed. Rather than existing in only three dimensions, like decadent Western capitalist cubes, Malevich's new cube – or 'cube', for the sake of clarity – apparently existed in four, or at least he said it did. In this new occult fourth dimension, said Malevich, death did not exist. Thus, by getting Shchusev to make Lenin's tomb in the shape of Malevich's 'cube', his body would be stored within a deathless realm until Soviet science had advanced far enough to wake him back up again.[35] Sadly, Nature had different ideas. The primitive German fridge-freezer Krasin had Lenin placed in within his crypt actually started to destroy his body tissue, and his hands began to turn green. Krasin tried fitting the freezer with double-glazing for further protection but this didn't work either, and ultimately the Champion of the Proletariat had to be chemically embalmed instead – but not before his brain had been extracted, sliced into 30,000 pieces and preserved in alcohol by the Politburo in the presumable hope of being eventually able to artificially recreate it.[36] To this very day, however, Lenin still remains dead.

This, then, was the partial intellectual atmosphere in which Sergei Bryukhonenko had found himself working. So popular was Malevich's 'cube' among the Soviet hierarchy that it was ordered that thousands of miniature 'cubes' be manufactured and distributed among the workers, who were supposed to place them in a corner of their home, and reflect happily upon Lenin's timeless immortality.[37] Clearly not everyone fell for this insanity, but several early Soviet high-ups really were to some extent Cosmists, and as such directly encouraged research into raising the dead. 'Lenin Lived, Lenin Lives, Lenin Will Live!' went one slogan, but it was meant some day to apply to all Russian people, too. Some useful idiots apparently even made excuses for the Soviet dictator Josef Stalin's (1878–1953) exploits in mass murder by claiming that his paranoid programme of executing millions of dissidents, both real and imagined, was merely a temporary measure and that, once the Worker's Paradise had been fully achieved, they could simply be resurrected and rehabilitated – Stalin wasn't slaughtering them permanently, you see, merely sending them off to spend time in a sort of celestial sin bin for their own good.[38]

Cosmism predated Communism, though, and there were several earlier Russian attempts to restore and prolong life prior to Bryukhonenko. Like him, the people behind them were not necessarily Cosmists themselves, but were picking up, whether consciously or not, on a messianic strain in Russian society of which both Cosmism and socialism were concrete expressions. For example, there was the work of one Aleksei Kuliabko (1866–1930), who in 1907 had developed a means of artificial blood circulation which allowed him to bring recently severed fish heads back to life. By 1929 he had decided to move on to experimenting with humans, apparently succeeding in making the heart of a man who had died during surgery the previous day begin beating again for some twenty minutes. Kuliabko's assistants were so disturbed when this happened that, after the corpse let out a kind of wet rattle from the throat, they fled the room in terror. They need not have bothered. The man was still dead, and the beating of his heart a mere mechanical response, although the event was still a significant achievement.[39] Hearing of this success, in 1934 Bryukhonenko had a try at reviving humans himself, attaching the corpse of a man who had committed suicide three hours earlier to his *autojektor*. Allegedly, the dead man woke up. His heart began beating, he opened his eyes and, supposedly, stared at his 'saviours' for two whole minutes before Bryukhonenko, profoundly disturbed, pulled the plug and let him die a second time, thus raising an interesting legal question: if he really had been brought back to life and then killed once more, then was the cause of death really suicide or was it murder?[40]

Electric Dreams

The answer, of course, is still 'suicide', because Bryukhonenko hadn't really brought the man back to life at all. People had already been making dead bodies move around for the best part of two centuries. In nineteenth-century Germany, for instance, so many doctors were electrocuting severed heads in the hope of reanimating them that the pastime actually had to be made illegal![41] At no point, however, had a dead man or animal ever actually been resurrected, only caused to dance around wildly like a fleshly marionette.

It had been widely known that applying an electric current to the dead could cause muscular spasms, ever since the Italian anatomist Luigi Galvani (1737–98) had published his *De Viribus Electricitate in Motu Muscularis* in 1791. In 1780, Galvani had been dissecting frogs at Bologna University, during which the crural nerves in their hind legs had been exposed. Also in the dissecting room was a friction generator, used to create an electric current. At one point, quite by chance, one of Galvani's assistants happened to touch the exposed crural nerves in one of the frogs' legs with the metal tip of his scalpel, thus channelling current from the generator into them. As the amphibian's legs were suddenly cast into a violent jolt, a spark was simultaneously emitted from the friction machine, causing Galvani to conclude that atmospheric electricity was being transmitted direct into the frog through the metal conductor of the scalpel, something which led him to set up a strange experiment in his garden. Getting a number of conveniently available dead frogs – supposedly, his wife liked to eat them – he shoved brass hooks through their spinal cords, hung them from iron railings during a storm, and waited to see if electricity from the lightning clouds would make them twitch. It did indeed, but to Galvani's surprise the legs still jumped around even when no storm clouds were present. Deducing that the rubbing of the brass hooks against the damp iron railings was generating the electricity which made the legs dance, Galvani went public with his conclusions. Ultimately, Galvani's theorising contained one major error – he thought most of the electricity was already stored within the frogs' muscles as 'nervo-electric fluid', and was only released by the metal hooks, rails and scalpels allowing for its jerky discharge through completing a circuit – but his accidental discovery of bimetallic electricity led directly to Galvani's great rival Alessandro Volta (1745–1827) developing the first batteries, or 'galvanic piles', which made use of just such a phenomenon.[42]

With Galvani and Volta's findings in mind, experiments in applying electricity to body parts quickly began to proliferate all across Europe; researchers in Turin, for example, discovered that if you placed a human eyeball on an electrified plate, its pupil would contract.[43] Following the Revolution in France there were plenty of severed heads to go around, meanwhile, which led to numerous bizarre speculations emerging about whether or not the head could live on for a few moments following separation from the body on a guillotine. Supposedly, executioners had witnessed severed heads trying to talk or pulling strange faces, and one rumour had it that, after the head of the revolutionary assassin Charlotte Corday (1768–93) had been pulled from its basket and slapped by a detractor, its cheeks flushed red in outrage. Doctors – including Joseph-Ignace Guillotin (1738–1814), original champion of the dreaded device – investigated, and concluded that such stories, if true, represented only mechanical nerve actions in the manner of Galvani's frog legs, not the momentary continuation of life itself.[44]

The most famous such experiments were conducted by Galvani's nephew, Giovanni Aldini (1762–1834). Setting up the Galvanic Society after his uncle's death, Aldini began experimenting upon larger animals than frogs, getting hold of a severed ox-head, exposing its brain and then applying electric current to different sections, making the ox pull a range of funny faces. Aldini was also the first to electrocute human corpses, beginning in 1802. Acquiring three recently beheaded criminals, Aldini put on a public show near Bologna's law courts, working with an early neurologist, Carlo Mondini (1729–1803), to demonstrate how attaching a galvanic pile to specific parts of the human brain would automatically produce certain facial grimaces. He thought this procedure might provide the basis of a cure for epilepsy – although presumably without having beheaded the patient first. While Aldini stated he was aiming to 'produce, reanimate and, so to speak, control the vital forces' with these trials, however, he was careful never to claim he was aiming to resurrect the dead.[45] His main purpose with such trials was to advance the field of medicine. After electrocuting his own head and finding that it didn't actually kill him, just induced insomnia, Aldini began a programme of shocking depressed people, hoping thereby to jolt them back into happiness – the first instances of the now discredited psychiatric practice of Electro-Convulsive Therapy.[46] In particular, he thought that a short, sharp electric shock might be able to resuscitate people who had recently drowned (the previous treatment for such cases had been to blow smoke up the patients' anuses).[47]

The general public, though, could easily get the wrong impression, as was proved by Aldini's most famous demonstration, performed on the freshly hanged corpse of a murderer named George Forster before London's Royal College of Surgeons on 17 January 1803. The results of applying electrical stimulation to Forster's face were described in a report in *The Times* for 22 January: 'The jaw of the deceased criminal began to quiver, and the adjoining muscles were horribly contorted, and one eye actually opened.' Applying the galvanic pile elsewhere, 'the right hand was raised and clenched, and the legs and thighs were set in motion'. When Aldini shoved an electrified rod up Forster's bum, the reaction was even stronger; as Aldini himself put it, stimulating the corpse's rectum 'almost gave an appearance of reanimation'. Reportedly, several members of the audience genuinely believed the murderer was about to rise again.[48] If this had happened, then it would have proved a challenge to the law. After all, if a man had been sentenced to death, and then came back, would he have to be hanged again, or set free? The press at the time were in no doubt; Forster had been condemned to 'hang until he be dead', even if it took two goes to make the state permanent. Apparently, there was a legal precedent; in 1752, a convicted murderer named Ewen Macdonald had been ineptly dispatched, and woke up on the dissecting slab. Rather than running away screaming, the surgeon present calmly picked up a mallet and beat him to death instead. Cause of death – justice.[49]

Resurrection Men

One man who came to believe it really might be possible to reanimate dead criminals was Dr Andrew Ure (1778–1857), a Glasgow chemist, physician and academic. Scotland during his day was a centre of unusual medical beliefs, with the quack Edinburgh doctor John Brown (1735–88) gaining numerous disciples for his unusual creed that all animals, humans and plants were nothing more than giant candles in disguise. Just as candles slowly burned down once lit, said Brown, so all living beings were possessed of their own metaphorical wick which dissipated over time via a process of slow combustion. This wick was an invisible chemical life force Brown termed 'excitability', hidden away in the body's nervous system. When people got ill, this made the life force burn away either too quickly or too slowly, he said, resulting in ultimate death. Health was thus redefined simply as the wick burning normally, and Brown hoped to develop a scale on which the level of deviation from this point of equilibrium would be designated purely by number. So, instead of being diagnosed with 'flu, a sick person would just have had a special thermometer inserted, and would be told that their candle was twelve degrees away from a point of normality. That, at least, was Brown's dream; the thermometer in question was never actually developed. He did, though, manage to develop a reliable cure for nearly all ailments – drugs and drink. People whose candles were burning too slowly he gave alcohol, to speed up their nerves, and people whose flames were too quick were given opium, to slow them down through stupor. Brown himself sought total chemical balance, dissolving opium in whiskey and drinking it to ensure his candle remained in tip-top order, even while giving public lectures; naturally, some suspected he was just an alcoholic drug addict who sought an excuse to keep on abusing his favoured substances.[50]

Like Brown, Andrew Ure also pursued a path somewhere between medicine and chemistry, publishing a textbook, *A Dictionary of Chemistry*, in 1821, and helping found the Pharmaceutical Society of Great Britain in 1841. Among Ure's rather stranger personal beliefs, though, was his idea that Victorian factory hands were, to a man, evil layabouts, while their slave-driving employers were saints in human form. Any illnesses from which this subhuman scum suffered were not caused by overwork or insanitary conditions, said Ure, but arose from their always eating too much bacon, an over-rich foodstuff they were only able to afford in the first place because their wages were much too high. Indeed, he even went so far as to claim that, far from being bad for you, working in a cotton-mill could actually make you immune to cholera![51] Undoubtedly Ure's most peculiar belief, however, was that he had the power to raise the dead.

On 4 November 1818, Ure, working in concert with Dr James Jeffray (1759–1848) of Glasgow University, got hold of the corpse of a freshly hanged murderer named Matthew Clydesdale, and began subjecting it to an evening of incredible weirdness. Jeffray was best known as the inventor of the chainsaw – originally a surgical instrument![52] – and so was hardly averse to chopping things up, though it seems on this occasion he deputised his assistant, James Marshall, to expose the corpse's nerves for him. Clydesdale was brought into a public anatomy theatre filled with skeletons dangling from the ceiling, plonked down in a chair, and then Ure, armed with a galvanic battery, began to stimulate him, with dramatic results. In Ure's own words:

> On moving the second [electrified] rod from the hip to heel, the knee being previously bent, the leg was thrown out with such violence as nearly to overturn one of the assistants, who in vain attempted to prevent its extension.[53]

So far Ure was merely imitating Aldini, albeit in slapstick fashion, but then he tried something new; by applying a rod through an incision into Clydesdale's diaphragm, Ure appeared to restart the murderer's lungs. The audience, seeing a dead man breathe (sort of), began to get nervous, a condition exacerbated by Ure's attempts to shock the corpse's face into a number of different expressions like a stage actor during rehearsals. Then, through stimulation of the ulnar nerve, Ure made the corpse's fingers move about dexterously, as if playing an invisible violin. One of Ure's assistants grabbed the moving hand and tried to force it closed, but such was its strength he could not. Finally, by applying electricity to the tip of one of Clydesdale's fingers, Ure made it extend and his arm shoot out in front of him so it looked as if the dead man was pointing at the audience in accusation.[54] Some of those present genuinely thought he was alive; one man fainted, and others fled or were sick. Evidently, once the spectacle had concluded (they still had the delight of seeing Ure shove tubes up Clydesdale's nose and then electro-shock them to see what would happen) the audience started to spread exaggerated rumours about the event. Indeed, the most famous account of the demonstration, written some fifty years later, makes the astonishing and completely untrue claim that after being galvanised Clydesdale opened his eyes, pulled his tongue, stood up and had to be killed all over again by Dr Jeffray (who was dressed like a bishop for some reason) stabbing a scalpel into his throat.[55]

The strange thing was that Andrew Ure seemed to believe that he really could have resurrected Clydesdale and made his heart beat again if only his body hadn't been largely drained of blood beforehand, a standard surgical procedure always made before dissection; in his view, the only thing that could prevent such a thing being done some day were legal problems centring around saving criminals' lives. 'This event, however little desirable with a murderer, and perhaps contrary to law, would yet have been pardonable in one instance, as it would have been highly honourable and useful to science,' he wrote. Indeed, Ure actually described a hypothetical device which, he said, might be able to restart a recently stopped heart through electrical stimulation, a device which, had he bothered to make one, would actually have worked – in theory, Ure had devised the world's first defibrillator, over a century too early. Had he just had a little more determination, Dr Ure really could have raised the (very recently) dead after all![56]

Shocking Stories

Getting the human heart to beat again after death was the main problem facing experimenters on fresh corpses from Aldini onwards, and for a long time it seemed as if galvanism was the only answer.[57] This was because electricity during this period was considered to be a kind of wonder-cure for all kinds of things. For example, in 1779 a quack named Dr James Graham (1745–94) opened an infamous 'Temple of Health and Hymen' in London, where a number of semi-naked young women given florid names like 'Hebe Vesteria, the Rosy Goddess of Youth and Health' delivered lectures from a wondrous 'Electrical Throne' and tried to entice impotent and frigid customers to spend good money to lay on a so-called 'Magneto-Electric Bed'. For £50, Graham would allow the sexually dissatisfied to lie in it with their frustrated sweetheart and then run a massive charge of electricity through the item of furniture, forcing their drooping organs to spring into life again, or so he said. A further £50 would gain you access to the even better 'Celestial Bed', which was so highly charged it could apparently get a woman pregnant. For those in search of a cheaper thrill, meanwhile, Dr Graham would happily have one of his glamorous assistants wire your genitalia up to a battery for only half a crown. Alternatively, you could just attend Dr Graham's advertised 'Feast of Very Fat Things', though I wouldn't like to speculate as to what precisely *that* was ...[58]

It seemed as if electricity could restart a great many things, then, but not the human heart. Rumours about such feats being possible, though, began to proliferate across nineteenth-century Europe. Most famously, Mary Shelley's (1797–1851) *Frankenstein*, arguably the first-ever science-fiction novel, was directly influenced by whispers of unholy experiments in necromancy taking place within mad scientists' laboratories. Writing about the work's genesis, Mary and her husband, the Romantic poet Percy Bysshe Shelley (1792–1822), told of how it had been partially inspired by conversations about 'some of the physiological writers of Germany' (possibly a reference to Dr Weinhold and his metal-brained cats), and some sadly fictional trials supposedly performed by Erasmus Darwin (1731–1802), the more famous Charles' grandfather, who was meant to have caused some vermicelli to come to life and 'move with voluntary motion' by enclosing it within a special glass case.[59] 'Perhaps a corpse would be reanimated; galvanism had given token of such things; perhaps the component parts of a creature might be manufactured, brought together, and endued with vital warmth,' speculated Mary in the introduction to her masterpiece.[60]

Even Percy Shelley had been moved to carry out electrical experiments himself, after hearing about the activities of Antoine Nollet (1700–70), a French monk and galvanism expert who had once lined up 180 members of the Royal Guard in front of Louis XV in his palace at Versailles, got them to hold hands, and then hooked them up to an electric current, making the soldiers all leap uncontrollably into the air. Reportedly, Louis was most amused.[61] Shelley's own youthful trials were just as irresponsible. Once, he set fire to the family butler. Another time, he ran a current through his two sisters, leaving them blackened and charred. Away at Eton, he created his own friction generator, whose charge hurled his teacher across a corridor when he placed his hand upon Shelley's door handle. Studying at Oxford, he used electricity to ignite dishes of alcohol and gunpowder, and killed a stray cat by flooding it with current. According to one Oxford contemporary, Shelley was busily consuming all available literature about both galvanism and witchcraft, thinking these fields to be compatible, and playing about with generators, standing on a protective glass-footed stool then deliberately electrocuting himself for fun. Shelley also talked about his desire to create a giant array of metal kites and fly them during a thunderstorm, hoping thereby to draw down lightning to some predetermined point on the ground and blast it. Thinking men to be merely lumps of 'electrified clay', no doubt he was psychologically open to believing all he read about such matters.[62]

In subsequent years many early galvanists have been labelled as having been 'the real Dr Frankenstein', often quite inaccurately; Andrew Ure, for instance, is sometimes said to have been the inspiration for Mary's character, though seeing as his experiments with Matthew Clydesdale's corpse took place several months *after* the novel had been published in January 1818, this is impossible. Instead, as the researcher Andy Dougan argued in his own recent examination of Ure's experiments, it is much more likely that, by publishing her book, Mary 'might have created a climate in which Ure felt he was able to pursue a subject that had intrigued him for some years, should a legitimate opportunity arise'.[63] Either way, the fictional Dr Frankenstein undoubtedly gave critics a new lens through which to view bizarre scientific experiments of all kinds from now on; rather than saying that Frankenstein had acted like Karl Weinhold, for instance, it now became possible to argue that Karl Weinhold had been acting like Frankenstein.

Science of the Lambs

Outlandish rumours about what other people may or may not have been up to in their laboratories, then, have often been a spur to creation, both scientific and literary. In a similar way, wild whispers about Sergei Bryukhonenko and Aleksei Kuliabko supposedly having succeeded in raising the dead and reanimating severed heads began to spread like wildfire throughout American universities in the 1920s, and led directly to an even stranger series of trials along those same lines being pursued by a man named Dr Robert Cornish (1903–63), of the University of California.

Cornish's big idea was that by tying dead animals and people to a special see-saw and then tipping them up and down on it really quickly over a period of time, you could get the blood pumping back around their body again, the device thus acting like a giant artificial heart. His first attempt occurred in 1933, when he spent a full ninety minutes see-sawing a dead heart-attack victim to no particular effect. The next time, he tried putting an electric blanket over a drowned man to help reheat his body during the trial, but it just started to cook his flesh instead. Chastened, Cornish began experimenting upon dead sheep, researching the ancient Japanese judo technique of *katsu*, in which a skilled practitioner delivered a series of special blows to precise points on the body to revive someone incapacitated during combat. Finding another doctor who had learned the method himself, Cornish gassed a lamb, tied it to his see-saw and, after a period of vigorous tipping, told his friend to karate-chop it back into consciousness. This did not work either, and in 1934 Cornish started suffocating dogs instead, before pumping them full of adrenalin while see-sawing them around and performing mouth-to-mouth resuscitation. Eventually, Cornish succeeded in bringing a freshly killed fox-terrier back from the dead ... and putting it straight into a coma. After thirteen days the dog recovered consciousness, however, and was christened 'Lazarus' (a name Dr Cornish gave to all his lab-dogs – even those which died permanently). There was a problem, though; the animal, brain damaged from its ordeal, was now seriously handicapped. 'That dog will be an idiot the rest of his days,' moaned his unsympathetic owner. Most of the other dogs he resurrected went blind. The University of California, highly disturbed by these results, threw Cornish out.

Refusing to despair, Cornish speedily agreed a deal with Universal Pictures for a movie to be made about his toils, 1935's *Life Returns*. The bizarre plot centres on a depressed scientist whose son loses all respect for him after he fails to raise the dead. When the little boy's dog dies the scientist starts to go insane until in steps Dr Robert Cornish, playing himself. Helpfully, our hero is able to use his see-saw to resurrect the dog, the film ending with genuine footage of Cornish at work on such an animal in his new home surgery. The boy's pet is revived, the father's sanity restored, and all live happily ever after (except for the dog itself, which would actually have been a drooling moron). Inexplicably, the film flopped, and Cornish slipped out of the public eye for a while.

In 1947, however, he was back. Inspired by Bryukhonenko's work, the doctor had built his own *autojektor* from a series of components seemingly sourced from Albert Steptoe; it is hard to imagine a working heart-lung machine being assembled from an old vacuum cleaner, radiator tubing, a discarded iron wheel and some 60,000 metal shoelace-eyes, but that's exactly what Cornish claimed to have achieved. Hearing that he had spent quite some time unsuccessfully trying to arrange to have fresh corpses delivered to him for experimentation, a child-murderer named Thomas McMonigle wrote to Cornish from San Quentin's death row, asking politely if he could bring him back to life with his new *autojektor* after a pressing appointment he had to keep in the gas chamber. Sadly, the prison governor refused to let the experiment occur, and McMonigle stayed dead. The problem was that Cornish needed to operate on the man immediately, but it took an hour for all the deadly gas to dissipate from the chamber; had he gone in there to help McMonigle straight away, explained the governor, then Cornish would have needed reviving too. Nonetheless, Cornish was back in the headlines, and some fifty or so members of the general public contacted him, volunteering to be killed and revitalised in

McMonigle's place. Some asked for large sums of money to be given to them prior to being murdered, but it is unclear exactly how they would have been able to spend their fee.[64]

They Tried to Make a Monkey Out of You

The way that Cornish and Bryukhonenko interchanged quite comfortably between working on animals and experimenting upon human beings was very telling, both about the doctors themselves and the societies they inhabited. Not only did such trials somewhat parallel the path trodden earlier by figures like Christopher Wren and Richard Lower, they also spoke of the way in which, in Soviet Russia in particular, people increasingly found themselves being treated as human guinea pigs to be cruelly experimented upon in the name of the forthcoming socialist-scientific utopia. Many individuals in America objected to Dr Cornish's ghoulish trials, but few people in Russia ever got the chance to complain about their own native Frankensteins – at least not openly. Had they done so, they may quickly have found they became the next batch of human guinea pigs themselves.

Take the disturbing history of Soviet research into toxins. Various secret laboratories-cum-prisons were established in Russia, where convenient inmates were taken to be poisoned in the 'noble' cause of advancing the state's ability to carry out assassinations more efficiently. One of the main men to pursue such research was Grigory Mairanovsky (1899–1964), a biochemist who ran his own lab in Moscow to which were attached several cells whose doors were fitted with peepholes for observational purposes. Each day, new consignments of prisoners were shipped in and fed poisons in the guise of medicine, or hidden within food. Then, Mairanovsky's men would time how long it took them to die. Sometimes it took minutes, sometimes weeks. To test out all variables, Mairanovsky got hold of any and all physical types to murder: tall people, short people, thin people, fat people, those who were already ill, and those in perfect health. To block out the screams of agony, the lab assistants bought a radio and played it full-blast. These trials were considered to be justified and 'not illegal' by those behind them on the grounds that they were 'being performed on people sentenced to execution as enemies of the Soviet government'.[65] In other words, they were expendable lab rats – literal 'non-persons', as the old saying went.

Such thinking had existed for some time prior to Mairanovsky setting up his lab, though; in 1933, the Russian writer Maxim Gorky (1868–1936), celebrating the creation of Stalin's new Institute of Experimental Medicine, commented that 'hundreds of human units' would be required to be experimented upon within.[66] By describing them as 'units', Gorky reduced such people down to a status even lower than animals. This was akin to the attitude of Japanese military scientists during the Second World War who reportedly sometimes 'cried tears of regret' when their 'valuable experimental materials were wasted' – that is, when their human captives died before they had gleaned all the data they sought to extract from their torture.[67]

Incredibly, so prevalent did the idea of using human beings as experimental animals become in the USSR that at one point attempts were made by scientists to actually cross people with apes, supposedly in order to create a race of man-monkey super-soldiers for enlistment in the Red Army, a story partly true and partly exaggerated. Its origins lie with the work of a genuine Russian scientist, Ilya Ivanovich Ivanov (1870–1932), who gained great success artificially inseminating horses around the year 1900, then moved on to cross-breeding various closely related species, such as zebras and donkeys. In 1910, he speculated it might even be possible to cross-breed humans with apes, but in 1917 came the Revolution, and Ivanov fell from favour. In the 1920s, though, the atheistic Soviets finally gave Ivanov the go-ahead to try out his proposed scheme – reputedly in order to prove Darwin right and the Church wrong. By 1926 Ivanov was in Guinea, where he made three separate attempts to inseminate native female chimpanzees with human sperm. This failed, so Ivanov tried a different tactic; rather than men impregnating female monkeys, maybe women could be impregnated by male apes instead? Making the astonishing suggestion that he just try and squirt some monkey sperm inside female patients at a nearby hospital without their knowledge or consent, Ivanov found little local favour for his plan. Returning to the Soviet Union, Ivanov set up a primate lab in the Black Sea port of Sukhumi, where most of his specimens died from cold. Only one ape survived, an orang-utan named 'Tarzan'. Ivanov found a woman willing to have Tarzan's baby, but before fertilisation could take place Tarzan died too. After he asked a Cuban ape breeder to sell him some monkey sperm, the American press got hold of the story and Ivanov again lost support from the Soviet authorities, being exiled to Kazakhstan, where he died in 1932. Ivanov's primate lab survived him, though, later becoming the prime source of monkeys for Russia's space programme.

This much is definitely true. However, in 2005 *The Scotsman*, citing certain unnamed 'secret documents' supposedly just found in Moscow, sensationally reported that Josef Stalin himself had personally ordered Ivanov to make a race of monkey-men to serve as slave-labour shock troops in his army. According to the article, Stalin desired 'a new invincible human being, insensitive to pain, resistant and indifferent about the quality of food they eat'. It's an incredible story ... but it appears untrue. No Russian-language newspapers reported the tale prior to *The Scotsman*'s revelation, and nobody seems to know what these alleged 'secret documents' actually were, or where they came from. There is no evidence that Stalin ever met or spoke to Ivanov, nor that he even knew specifically what he was up to. In 1980 *The Times* had reported that the Chinese Communist Party had allegedly tried and failed to inseminate chimps with human sperm in order to 'found a race of helots for economic and technical purposes', and it would seem that the story of Stalin's ape soldiers is really just an updated variant of that old yarn.[68]

Devil-Doctors

The twentieth century's other great political monsters, the Nazis, also treated human beings like experimental animals. In the death-camps, Jews and others – subhumans to the Reich – were subjected to a variety of bizarre and cruel experiments in the name of science. Prisoners were given burns, exposed to toxins and disease, forced to drink sea water, shot with poisoned bullets, frozen alive, had body parts needlessly removed and transplanted, were vivisected, sterilised and injected with petroleum. Trainee doctors, needing experience, were allowed to practise on healthy Jews just to develop their skills. Bodies were stripped of fat and their skeletons used as teaching aids. In need of lumps of flesh in which to grow bacteria, chunks cut straight from executed prisoners were used as a free source of meat, with more 'valuable' animals being needed as food during wartime. Likewise, when troops required blood transfusions, it was very simple to slit prisoners' carotid arteries, bleeding them to death and taking whatever was needed. Seeing as the inmates were going to die anyway, many Nazi doctors reasoned that they were now little more than 'raw materials'; or, as one camp inmate put it, 'the cheapest experimental animal ... cheaper than a rat'.[69]

More abnormal even than these crimes were some of the trials carried out by Nazi doctors on their captives purely in order to test out their own bizarre pet theories. Before the war, Dr Johann Kremer (1883–1965) had been a Professor of Anatomy at Münster University, where he had been formally rebuked for publicising two quite mad ideas. Firstly, he felt that children could develop traumatically inherited deformities in the womb – an idea more familiar from folklore than science. Joseph Merrick (1862–90), for example, the famous 'Elephant Man', was rumoured to have been the result of his mum being petrified by a pachyderm during pregnancy, while even today well-endowed persons are told, 'Your mother must have been frightened by a horse!' More originally (but equally wrongly), Kremer also proposed that white blood cells were really decayed tissue cells from other parts of the body. Upon arriving at Auschwitz in 1942, Herr Doktor had plenty of opportunity to investigate these areas, along with the issue of starvation, in which he had an especial interest. Selecting suitable prisoners to be killed in his lab, Kremer had them tied to a table, given a lethal injection straight to the heart, and then cut out various organs for preservation in glass jars. After the war, Kremer became notorious for his macabre diaries, in which descriptions of horrible medical procedures were immediately juxtaposed with itemised accounts of his meals that day:

> The most horrible of horrors ... I took and preserved ... material from quite fresh corpses ... liver, spleen, and pancreas ... Today, Sunday, an excellent dinner: tomato soup, half a chicken with potatoes and red cabbage (20kg of fat), sweet pudding and magnificent vanilla ice-cream.[70]

Unlike his victims, Kremer himself was never going to starve. One of his assistants concluded that Kremer 'looked upon the prisoners as so many rabbits', and it is hard to disagree.[71]

Even more striking is the case of the most notorious Nazi medic of all, Dr Josef Mengele (1911–79). Besides all his infamous experiments on twins, dwarfs and other such physiologically curious persons, Mengele also found time in Auschwitz to make his own, wholly unique, contribution to the annals of pseudoscience; namely, his strange investigations into eye colour. Seeking out gypsies who had been born with heterochromia of the iris (one eye blue, one brown), Mengele gave orders for their eyes to be cut out after death, preserved and then sent to him to add to his own collection. It appears that one of the aims of this sick exercise was a fantasy Mengele had about trying to make brown-eyed people into blue-eyed Aryan types. Some Germans were born with the 'correct' blonde hair, but also had 'incorrect' brown eyes. Therefore, Mengele got hold of a number of gypsy child prisoners of this same type and began forcibly injecting doses of blue methylene into their eyes to see what happened. What happened was that they died, went blind, or suffered severe ocular infections and horrible pain. Their eyes, though, remained resolutely brown.[72]

It seems that Mengele was able to rationalise these acts by thinking of his prisoners as animals. One favoured Auschwitz inmate, an artist named only as 'Eva C', specifically claimed that 'I was a pet!', and described her owner as being like an inspector in charge of a dog pound:

> He would point out maybe a pile of dirt ... in one of the cages ... and admonish the keeper [to] wash up that excrement there ... to keep it clean, to keep it healthy, to keep the dogs healthy, to keep them well-fed. Look, this one doesn't have water, you'd better give them some food ...[73]

Most of the 'dogs' who ended up in Dr Mengele's pound, of course, eventually ended up being put down once they had outlived their usefulness to their 'inspector'.

The Nazi doctors were also useful to the higher authorities they served – even when the Nazis themselves had gone. Dr Ferdinand Sauerbruch (1875–1951) was once a genuine surgeon of high repute who had carried out surgery on famous figures and habitually waived his fees when operating upon the poor. However, after war broke out he began training military field surgeons, for which he received a decoration from Hitler himself, and was thus charged with collaboration in 1949. Sauerbruch was exonerated, though, and was put to good use by the new Communist regime in East Berlin, where his hospital was located. Unfortunately, in old age he rapidly lost his faculties and began pulling out patients' brain tumours with his bare hands, killing them. 'The finger is still the surgeon's best instrument!' he once declared triumphantly, waving a bloody tumour around in his fist. Seeing as he had a habit of lashing out at people with a scalpel, few dared challenge Sauerbruch openly, even though his mind had by now become so clouded he was unable to recall which of his relatives were dead and which were alive, struggled to hold coherent conversations and forgot to reattach organs inside his patients during surgery.

Finally forced into retirement, Sauerbruch stubbornly refused to stop practising, performing operations on his dining table which were free of both charge and anaesthetic. He carved up bodies like mutton, chopping off breasts and removing tumours with unsterilized equipment before sealing wounds with his wife's sewing kit. Unsurprisingly, not all operations proved successful. Sauerbruch carried on like this for five years, with neighbours getting so disturbed by the constant flow of screaming and dead bodies that they complained to the Board of Health – who initially delayed prosecution, issuing a mere warning. After all, the famous surgeon lent valuable prestige to the Communist regime. To quote Dr Josef Haas, Director of the East German Academy of Sciences:

> In the coming struggle of the proletariat [the Cold War] ... millions will lose their lives ... It is trivial whether Sauerbruch kills a few dozen people on his operating-table. We need the name of Sauerbruch![74]

Moral Freefall

Most Nazi doctors made no real contributions to science. Many knowingly pursued overhyped dead-ends in the death camps simply to avoid being sent to the front.[75] Rather more problematic, however, was the work of Dr Sigmund Rascher (1909–45), whose horrific experiments produced a body of data which was actually highly useful – and not only to fascists. Rascher was a physician in the Luftwaffe and SS who used his influence as an acquaintance of SS head Heinrich Himmler (1900–45) to gain permission to carry out a series of experiments simulating the effects of high altitudes within Dachau concentration camp. Tests performed on monkeys proved unsatisfactory, so in 1942 Rascher was given a pressure chamber and began locking prisoners inside it to simulate the effects of high altitude upon Nazi airmen and parachutists. By altering pressure changes quickly or slowly, Rascher could mimic both gradual ascents and total freefall, and see what such states did to the human body. It turned out that rapid changes in air pressure proved deadly. One prisoner's lungs exploded, while others began to rip their own heads apart with their bare hands due to the unbearable stress they felt inside their skulls. Heart attacks and brain embolisms killed around eighty victims; one man, thought to be dead, woke up on the dissecting table whilst his chest was being cut open. When he heard of this event, Himmler suggested that anyone who survived the trials should be shown 'mercy' and given life imprisonment rather than death. Rascher, deeming them to be 'only' Poles and Russians, ignored him.

Even weirder were Rascher's experiments in freezing. Wishing to know the best way to treat pilots who had bailed out over the cold North Sea, Rascher forced prisoners to stand outside naked in winter for as long as fourteen hours, or else immersed them in tanks of icy water. Then, he tried to revive them from unconsciousness by placing them in hot baths. Himmler, though, had a different suggestion. The wives of North Sea fishermen, he declared, resuscitated their half-drowned husbands by taking them to bed and using their 'animal warmth' upon them. To test the theory, Rascher brought in female gypsies from Ravensbrück, stripped them, and then forced hypothermia victims to lie in a kind of human sandwich between them. It was determined that the revival process was speeded up if the gypsies had sex with the man, but hot-water baths were still the most effective option. Aldini had to use a hanged corpse as substitute for a drowned man to revive; Rascher didn't. He just froze people, and then had them raped back into consciousness.

Further experiments, meanwhile, resulted in a blood-clotting agent called Polygal. To see how well it worked, Rascher shot people through the neck or chopped their limbs off without anaesthetic, after first administering the drug. He also helped develop the cyanide capsules which his master Himmler would later swallow to evade justice when captured by British troops. Rascher's downfall finally came not due to his incredible cruelty, his alleged collection of stylish handicrafts made from human skin, nor even on account of his apparent murder of his own lab assistant, but because of his embarrassing habit of kidnapping babies. Claiming to have found a new way to extend the female childbearing age, Rascher started publicising the 'fact' that his wife had given birth to three children despite pushing fifty. Impressed, Himmler used photos of the Rascher brood for propaganda purposes, and was shocked to later discover that the babies had, in fact, been either stolen or bought to help Rascher get ahead in his career. Rascher was ultimately imprisoned in Dachau, where he himself had previously caused so much misery, and summarily shot in April 1945. You know you must be a bad person when even Heinrich Himmler thinks you've gone too far.

Rascher's research, however, while horribly immoral, produced valid results which proved useful not only to the Luftwaffe but, after the war had ended, to many others as well. Data from his work allowed people to establish safe procedures for parachutists, and to help revive hypothermia victims. Several Nazi scientists who had worked with Rascher ended up at NASA. The case still poses serious questions for medical ethicists – and, indeed, for humanity in general.[76]

The Long and Winding Road

So far, we have covered a multitude of different topics, from electrocuting corpses to reanimating dogs. How, precisely, are these tales all related, however, other than through their extreme oddness? Well, this rambling story so far has been deliberately rambling. In science's quest for immortality and extended life, we have undoubtedly already met plenty of dead-ends and quacks, but have also read about some genuine discoveries. Sometimes, as we have seen, these were ignored, sometimes made use of immediately, and sometimes only put to any use years afterwards. We have seen experiments both good and bad, in terms of procedure, ethics and application. We should also have observed that apparently separate fields like medicine, chemistry, electricity – even politics and literature – don't develop in isolation from one another. Sometimes one needs to leap ahead before the other can advance; sometimes one even needs to go backwards. This, I would suggest, is the true path of science, technology and medicine – a profoundly crooked one. The writer Arthur Koestler (1905–83) once summed up the situation as follows:

> The progress of science is generally regarded as a kind of clean, rational advance along a straight ascending line; in fact it has followed a zigzag course, at times almost more bewildering than the evolution of political thought.[77]

How right Koestler was – and it seems the course is still zigzagging. The latest bizarre experiment in the spirit of men like Bryukhonenko is about to take place in China. Those behind it are an Italian doctor, Sergio Canavero, and a Chinese hand-transplant surgeon, Ren Xiaoping. Their ambitious plan is to perform the world's first-ever human head transplant. The Chinese Communists have stumped up £1 million to fund the project, and the doctors have found their first willing subject, a Russian named Valery Spiridonov, who suffers from spinal muscular atrophy and wants his head chopped off and stuck onto a healthy man's body. According to Dr Canavero, the procedure would 'change the course of human history' by 'curing incurable medical conditions'. If you have cancer, then for Canavero this won't matter; so long as there are no tumours in your head, you can just have it cut off and transplanted onto a donor. Of course, the problem of who exactly is going to donate their healthy body to this scheme appears insoluble – until you consider that in China the corpses of executed criminals are routinely harvested for organs.[78]

If this works, Canavero and Xiaoping will become famous. If it doesn't – and the general opinion is that it won't – then they will end up bracketed by posterity with loons like Robert Cornish. After all, previous attempts at head-transplants, by the neurosurgeon Dr Robert White (1926–2010) in America, did not go well. In the 1960s, White cracked a rhesus monkey's skull open, hooked its brain to an artificial blood supply after severing its arteries, and found it still registered neural activity. In 1970, 'Humble Bob', as he liked to call himself, went one further. He beheaded two monkeys, and sewed one lucky animal's head onto the other's body. The head could eat, see and respond to stimuli, but couldn't control the body it was grafted onto, leaving it quadriplegic. Also, it quickly died. During the 1990s, White planned to perform similar procedures upon the wheelchair-bound physicist Stephen Hawking (b. 1942) and paralysed *Superman* actor Christopher Reeve (1952–2004), but never got much further than practising for the operations by swapping around the heads of corpses in a local mortuary. Amazingly, in 1981 the devout Catholic White established the Vatican's Commission on Biomedical Ethics at the behest of Pope John Paul II (1920–2005).[79]

In 1987, in an attempt to prevent any further such trials taking place, an American attorney adopting the pseudonym of Chet Fleming took out a patent on a machine designed to keep severed heads alive. He never actually built the device, though, doubting that such a thing would really work. However, Fleming was sure that his design was the first logical step towards achieving this undesirable goal and that, if any mad scientist ever tried to build something similar, he would be infringing Fleming's patent and so could be sued and his research stopped in its tracks.[80] He didn't anticipate that, nearly thirty years later, Professor Rene Anand, of Ohio State University, would for the first time succeed in growing artificial human brain tissue from stem cells and keeping it alive in his lab entirely without the aid of Mr Fleming's home-made *autojektor*. The tiny brain is a near-exact replica of that of a five-week-old human foetus, and Anand intends to expand the process for wholly benign purposes. He wants ultimately to grow artificial brains showing symptoms of Alzheimer's and Parkinson's, then expose them to various drugs to see which are most effective or harmful. The worry is what other, less benevolent, persons might try and do with such technology; no sooner had Anand announced his breakthrough than people were speculating whether or not disembodied brains had any legal rights, and whether it might be possible to graft them on to computer chips to create leaps in processing power.[81]

The year 1997 saw another such bout of media-fuelled medical panic, when a team of American scientists led by the brothers Joseph and Charles Vacanti revealed they had successfully managed to grow a human ear on the back of a lab mouse. This was presented at the time as 'genetic engineering gone mad', even though it involved no manipulation of genes whatsoever. The 'ear' was not really an ear at all, but a piece of cartilage manipulated to grow in the external shape of one. The ear had no functional qualities, and was based upon pre-existing techniques in which virtually two-dimensional strips of tissue were grown from cells in a laboratory and then later grafted onto burn victims as a kind of substitute skin. The Vacantis' idea was to develop this technique further by creating a sort of temporary, biodegradable three-dimensional scaffolding which would allow these sheets of cells to grow into more complex structures, like the exterior of the human ear. The role of the mouse was thus not to grow this ear from out of its back at all, but to provide it with the necessary nourishment to allow the cells to grow around the scaffolding. In fact, the scaffolding and cells were surgically transplanted onto the mouse, rather than sprouting up from it through manipulation of its genes, as many reports implied. Basically, then, the 'Vacanti Mouse' was a kind of living Petri dish in which the ear cells were cultured.[82]

The Vacantis had all kinds of ideas in mind for the future application of their technology. They cannot, though, have anticipated that in 2015 an Australian artist called 'Stelarc' would announce to the world that he had found a team of surgeons willing to repeat the procedure on himself, growing an artificial third 'ear' on his arm for weird artistic purposes. Next, Stelarc plans to have a small Internet-enabled microphone implanted within his new lughole and to broadcast the sounds of his life to anyone who wants to listen, all day, every day (Wi-Fi coverage permitting). 'Imagine if I could hear with the ears of someone in New York, while seeing with the eyes of someone in

London', said Stelarc, predicting such activities could really catch on soon.[83] Conclusive proof that you can never entirely predict what will be done with any scientific discovery at some point in the future. Koestler's path of unpredictable progress just keeps zigzagging on ...

Vulgar Notions

The patron saint of the scientific zigzag – and of this book – is one of the most eccentric experimenters of all time, the rather brilliant Sir Thomas Browne (1605–82). A Norfolk physician, philosopher, antiquarian and amateur proto-scientist acclaimed by contemporaries as 'a person of great knowledge',[84] Browne wrote several books, their topics ranging from religion to gardening, but the work with which we shall be most concerned is his 1646 masterpiece *Pseudodoxia Epidemica*, a massive catalogue of widely believed fallacies, or 'vulgar errors', together with details of the often incredibly odd experiments Browne devised in order to test or disprove them. Imagine if someone had got hold of a book of old wives' tales and tried pulling an ugly face when the wind changed to see if it really would stick, or had deliberately walked beneath a ladder to see if it brought him bad luck. That, basically, is what Browne did.

Browne certainly had a wide range of potential beliefs to choose from. People have always believed weird things, so you could very easily make a long, long list of vulgar errors. Here are some of my own personal favourites, ancient and modern:

- According to *Ruralia Commoda*, a gardening manual written in the early 1300s by Petrus de Crescentiis (*c.* 1233–*c.* 1321), an Italian lawyer, cucumbers are scared of lightning and 'shake from fear' during thunderstorms. Petrus also claimed that cucumbers were equally terrified of olive oil, so placing this substance near one would make it 'bend away just like a bow', turning it nice and curvy.[85]
- During the seventeenth century, physicians believed that opium put people into a stupor because its particles were round, not jagged, and thus 'soothed the nerves' as they passed through an addict's bloodstream, making them go all dreamy.[86]
- During the eighteenth century, some authorities believed birds did not wee; instead, any liquid they ingested was taken straight from their bladders and transformed into feathers. Furthermore, seeing as they were always flying about, any excess moisture in them would be dried out by the wind anyway.[87]
- Galen thought hair was made of soot; pores in the skin got blocked, leading to a build-up of soot in the bloodstream beneath. Then, once enough pressure had built, the soot was expelled all at once in a long string, as hair. The blacker your hair, the higher

the temperature in your body, he said, making the soot smoky. Blondes thus had icy hearts.[88]

- It was once believed the breast milk of black women tasted better than that of white maidens as 'brown women are hotter than others', and so their tasty fluid had already been heated to the optimum temperature, like we do now with babies' bottles.[89]
- Female bears were long thought to give birth to shapeless lumps of eyeless, quivering flesh which they then had to sculpt into the correct shape of a baby bear with their tongues. This belief is the origin of our modern term 'lick into shape'. In actual fact bears do lick their young, but this is just to clean them.[90]

No doubt the reader could easily add several other cherished fallacies of their very own.

Freedom to Experiment

Many of the vulgar errors investigated by Sir Thomas Browne were drawn from respected books by classical authors. Others, though, were just things the local country folk believed, such as the peculiar notion that hanging a dead kingfisher from a string would transform it into an accurate weathervane. Getting hold of such a dead bird, Browne suspended it from the ceiling with a silken cord in 'an open room … where the air is free'; he found it just dangled around at random. Obtaining another, he strung it up next to the first. The animals twisted in different directions. Dead kingfishers, Browne had conclusively demonstrated, had no special ability to illustrate wind direction.[91] What about storks, though? Was it true that they could only prosper within republics, as an old saying had it? By searching out proof that they had been seen living quite happily within monarchies such as France, Browne showed that, like all birds, storks had no perceptible desire for regicide.[92] Sometimes, his experiments were inconclusive. Knowing of the local belief that bitterns made their distinctive calls by blowing on hollow reeds in the Norfolk fens and using them as makeshift bassoons, Browne captured one and imprisoned it within his yard, where he denied it any reeds at all. The bird never made a sound; but that did not prove that they used reeds as pipes, Browne said. Maybe it was just too depressed to sing?[93] And that was just his investigations into birds …

However, no inherent hard-line sceptic, Browne himself possessed a faith in many vulgar errors we ourselves might now find bizarre. Believing in palm-reading, he saw no reason why it should not be possible to tell the future of monkeys and moles by such means as well, seeing as they too had naked palms.[94] He also thought it plausible that elephants could not only speak, but had even sometimes 'written whole sentences' with their trunks. As evidence, he cited an account of someone who said he once heard an elephant shout, 'Hoo, Hoo!' at him.[95] A problem Browne was particularly obsessed with was whether or not ostriches could eat metal. General opinion said they could, but no ostriches lived in Norfolk so it was hard for Browne to find out for sure. His son Edward, however, *did* have access to an ostrich! He was a physician in the royal court, and managed to acquire one from a flock King Charles II had been given as a gift from abroad. Excited, Browne wrote off to Edward with a suggestion; would he mind wrapping a lump of iron up in some pastry like a metallic sausage roll and seeing if the bird would guzzle it? Sadly, nobody knows if this experiment ever actually took place, but if it had done, the answer would have been ... yes, ostriches *can* eat iron! In order to digest food properly, the birds have to swallow stones to sit inside their stomach, and there is no particular reason why swallowing one small piece of iron should not do the job just as well.[96] Browne even made an early attempt to invent the telegraph. Familiar with new knowledge about magnetism, Browne was intrigued by the notion that, if you rubbed two pins down with the same magnet, they would become twinned. By making two Ouija board-like alphabet circles and placing a pin at the centre of each, he hoped that by turning one around to face a particular letter he would make the other turn too, in an exactly corresponding fashion. Then he could dispatch one board off to London and send it messages by manipulating the magnetised twin-pin on his own device. It was a brilliant plan, flawed only by the fact that it did not work.[97]

From today's perspective, Browne can come across almost as a lunatic, but this was not the case. Browne inhabited an age filled with boundless curiosity – a time when, as we have seen, it was considered perfectly acceptable to try and fill a dog full of soup or transfuse blood from a lamb into a madman. Such were Browne's contemporaries; he was not abnormal. Far from being a nutcase, he made several notable contributions to science. He appears to have been the first person to perform chemical experiments upon eggs, for instance, seeing what would happen to the embryos within when he added substances like vinegar or saltpetre, or exposed them to varying temperatures.[98] In addition, Browne invented large numbers of scientific words: 'electricity', 'botanist', 'medical' and 'hallucination' are all his. He sits in twenty-fifth place for coining new words in English. One of his best creations was the word 'anomalous', and he was certainly always on the lookout for anything that might have fitted that description.[99] His home 'elaboratory', as he called it, was not merely a secure cell built to contain a madman. As Browne's recent biographer Hugh Aldersey-Williams put it, 'Browne takes us back to a period in science when unknowns were all around and yet could be probed with relative ease.

We too easily forget this today, and allow ourselves to think these enquiries silly.'[100]

But they were *not* silly. If things still had to be discovered, they still had to be discovered; we should thank God there were men as inquisitive as Browne around to discover them.

An IgNobel Cause?

Possibly the closest thing to Browne's escapades we have today are the much-loved IgNobel Prizes, awarded each year to those people who have performed the weirdest experiments – those that, in the IgNobel awarding body's own words, 'cannot or should not be reproduced'. To take a random selection, in 2005 Dr Victor Benno Meyer-Rochow and Dr Jozsef Gal won a prize for their paper 'Pressures Produced When Penguins Poo – Calculations on Avian Defecation', while another pair of brave scientists, Edward Cussler and Brian Gettelfinger, won the chemistry prize for proving conclusively that humans can swim just as quickly in a pool of sugar syrup as in a pool of water. The most recent awards at time of writing, in 2015, included prizes given to Chile's Dr Rodrigo Vasquez for demonstrating that if you stick an artificial tail to a chicken's bum it will walk like a dinosaur, and Michael L. Smith of Cornell University, who volunteered to be stung by bees on various parts of his body, from his nostrils to his penis, then produced a descriptive pain scale to classify how much each jab hurt.[101]

Are these discoveries worthwhile? Yes! Of course, everyone would rather have a cure for cancer than detailed knowledge about penguins going to the toilet, but all knowledge is worthwhile, surely, and you never know when even the most obscure piece of information will come in useful, especially to zookeepers. And yet, despite this fact, today the greatest (honorary) IgNobel laureate of all, Sir Thomas Browne, is almost unknown, neglected by mainstream historians of science and treated with scorn by figures like the arch-rationalist Richard Dawkins (b. 1941).[102] Why is this? Possibly it is because, as I said earlier, Browne embodies within himself the spirit of the scientific zigzag – and science today is really supposed to be more of a straight line. Even Browne's writing style is somewhat non-linear. Often, he will begin by talking about one thing, go off on a tangent and then bring the subject back around to where he began again; the attentive reader might notice I have tried to do a similar thing in this present book, as a kind of tribute.

Is there really a perfectly straight line of scientific progress? Not exactly. Scientists today know far more than Thomas Browne knew, in terms of sheer volume of data and information. Browne knew nothing of evolution, nuclear fusion, or germ-theory. But he also knew nothing of future scientific dead-ends like Ilya IIvanov's attempts to sire monkey men, nor Dr Josef Mengele's tries at altering gypsies' eye colour. Both these men, living later in history, knew more overall than Sir Thomas Browne did; and yet each made some incredible new mistakes, in both their science and their ethics. Browne also knew nothing of caloric, or phlogiston, or the luminiferous ether, all substances once believed in by the majority of respectable scientists who came after him – and now all exposed as wholly imaginary. This may sound highly critical, but this book is certainly not anti-science. It just attempts to show that science sometimes functions as a kind of myth, which is capable of supporting innumerable different projections of imagined content out from people's heads and on to the body of neutral facts and knowledge of which it genuinely consists.

As we shall soon see, the current alliance of interrelated myths it supports are those of universal progress, liberal humanism and Dawkins-style hard-line materialism and atheism. It need not necessarily be this way, though, and throughout the years there have been many persons who have made a deliberate effort to rebel against such notions. Some of these people have been genuine scientists, and have made real discoveries on to which they have nonetheless then projected their own, often quite aberrant, personal philosophies and belief systems. Others have simply been lone loons, whose pseudoscientific theories have never stood up to the scrutiny of anyone but themselves. Some of these concepts have been accepted by very few; others by millions. History is littered with their wreckage, and this book tells the story of some of the very strangest. If the whole idea of scientific progress is actually structured after a Darwinian model, as I shall ultimately be arguing, then many of these ideas have proved to be history's dodos – but even dodos deserve their own memorial, I think.

Science Fictions: Technological Utopia, Objectivity, and Other Scientific Myths

A new scientific truth does not triumph by convincing its opponents and making them see the light, but rather because its opponents eventually die, and a new generation grows up that is familiar with it.
Max Planck[1]

Nobody ever really likes to visit any invalid's sickroom – but the sickroom of a young Regency lady named Lydia Baines must have been even more unpleasant to walk into than most. Quite apart from anything else, the smell was atrocious; but this was not because Miss Baines' flesh was rotting, or because her bladder or bowels had failed, or because her clothing was caked in blood and vomit. It was because her sickroom was full of cows.

Lydia Baines was the patient of an unusual medical man called Dr Thomas Beddoes (1760–1808), who had also made a name for himself as a poet, writer, chemist and general scientific investigator. Ever eager to unravel the secrets of creation, there was nothing about the world around him which escaped Beddoes' notice, no observation too trivial to be noted down for future reference. For example, while performing charitable work among the poor of Bristol during the 1790s, there was something very curious indeed he had begun to notice about one of the greatest killers of his day – consumption. Consumption (or tuberculosis) is a truly horrible condition, a disease of the respiratory system which once took thousands of lives across Britain every year. Happily, it can now be cured fairly easily, through antibiotics. Unhappily, during Dr Beddoes' day, it could not. Recovery seemed essentially random in nature, and there was no reliable cure. Appalled by the terrible toll the affliction took, Beddoes called consumption 'the perpetual pestilence of our island', and vowed to do his best to devise some means of dependable treatment.[2]

Poor Cow

Any clue, no matter how small, seemed worth pursuing. Therefore, when Beddoes made the curious observation that none of his patients appeared to be employed as butchers, he became intrigued. Rather than being thin, cadaverous and hollow-cheeked, like the average sufferer from tuberculosis, butchers were stereotypically thought of as plump, jolly, red-faced types with levels of physical robustness comparable to those of their proverbially fit pet dogs. Possibly this was attributable purely to their superior diet, with butchers having easy access to fresh meat. But what if there was something else to this phenomenon, something everyone else had completely missed? Approaching his local slaughtermen and asking for their health tips, Beddoes was particularly struck by the testimony of one well-fed fellow, who said that whenever he cut open a sheep, he found it 'very wholesome to swallow the steam' which poured forth from its still-warm flesh and blood. It was 'the smell of meat' which 'kept [butchers] from disorders', he reckoned, a theory which Dr Beddoes was all too willing to take on board.[3]

Possibly 'the smell of meat' could be bottled and administered direct to patients, Beddoes speculated, but an even better solution might be to introduce the source of such odours directly into the invalid's presence. Cows, he thought, were the ideal candidates. Dr Beddoes felt that the breath of cows was likely to be high in some unknown and unspecified substance which had amazing health-giving properties. It was well known that both meat and milk were good for you, he reasoned, and when the cow breathed in the surrounding air and then breathed it out again, maybe some invisible tinge of these substances was exhaled out into the atmosphere together with its breath. A patient living in a room surrounded by cows would thus, with every gasp they made, be inhaling air seeded with health-giving cow essence, perhaps helping heal the tubercular ulcerations in their lungs – something Beddoes called 'cow-house therapy'. And that was how poor Miss Lydia Baines came to be confined to her sickbed amid an all-day symphony of mooing.

Sadly, the prescribed programme of treatment did her little good, neither mentally nor physically. For one thing, the constant close-up presence of the cows was deeply disturbing to her, while the nature of some of their gaseous emissions proved rather less wholesome than had been anticipated. The unavoidable appearance of large quantities of cow-dung splattered all over the floor was yet another hitch, the whole episode leaving Baines with the distinct impression that she was being treated purely as an experimental guinea pig, not a distressed human being in need of succour. This was an unfair assessment, however, seeing as Beddoes genuinely did want to help; recognising the flaws in his plan, he happily took some of Baines' complaints on board. Accordingly, in 1799 he applied to take out a loan of £500 to add a specialist 'cow-therapy ward' to the hospital-cum-laboratory which he ran, a place rather marvellously christened 'The Medical Institution for the Benefit of the Sick and Drooping Poor'.

Dr Beddoes' basic idea was to help such sick and drooping persons by building a stable adjoining a series of small treatment rooms on the hospital's ground floor. Here, a hole would be cut into the connecting wall in each treatment berth, through which the cows in their corresponding stalls would be encouraged to poke their heads and burp on the dying. Should the sight of the cows' large, unblinking, saucer-like eyes become too much for the invalid, meanwhile, a convenient curtain would be provided which could simply be drawn across the animal's face to prevent any further invasion of privacy – useful indeed if you were shy getting dressed. Naturally, these curtained stalls would also be constructed in such a way that only the cows' breath would be able to drop down into the patients' faces this time around. As a further added benefit, said Dr Beddoes, the cows' great body warmth (and that of their dung) meant that the nearby herd would also act as a free form of central heating for the invalids, who, being consumptives, could ill afford to be exposed to any chills. He himself had made personal experiments in using cows as portable radiators, he said, and pronounced the experience 'delicious'. As far as can be told, this loan was never granted.[4]

Mad Cow's Disease

Undoubtedly, this all sounds rather insane now. However, this would be to utterly ignore vital matters of context; it seems that during this period the idea of cows having diverse health-giving properties was simply 'in the air', as it were, and continued to be so for quite some time. I have no idea whether a certain aristocratic eccentric who liked to be known as Helena, Comtesse de Noailles (1826–1908) – despite the fact she was English, not French – ever knew of Dr Beddoes' experiments in cow-house therapy, for instance, but reading about her bizarre lifestyle today, it certainly sounds as if she did. Living well into her eighties, Helena always attributed her surprising longevity and general good health to the beneficial influence of bovines, with the only beverage she deigned to drink being milk, a liquid she deemed to be a sure-fire means of preventing the development of alcoholism (even though this same milk was often topped up by her with cognac and champagne). Whenever she went on a picnic, the comtesse insisted on taking a few handy cattle along with her to be milked directly into a glass. She was also, just like Dr Beddoes, extremely keen upon the idea of breathing in cows' various gaseous discharges, tying up an entire herd directly beneath her open window each night so the bracing smell of methane could waft in, rejuvenating her while she slept.

So keen was Helena upon this unusual health regime that she even saw fit to *buy a child* and inflict it upon her, too. The adoptee in question was a little Italian girl named Maria Pasqua Abruzzesi, who, according to one version of the story, was happily swapped by her previous owners (or 'loving parents', if you prefer) in return for the price of a small vineyard. Getting delivery of the imported purchase back home in England, Helena sent her new human pet off to a religious boarding school where the nuns were specifically instructed to keep a healthy cow to hand to give fresh milk to Maria and break wind upon her. In order that the methane – and fresh air in general – would be able to circulate properly across her skin and body, the nuns were ordered to waive the usual uniform rules and force Maria instead to wear a loose, ancient Greek-style tunic and sandals, so that these curative gases would even be able to waft about between her exposed toes. Worse, if poor Maria wished to receive her fair share of her new mother's inheritance, then she would be obliged to continue to follow certain aspects of this cruel system during adulthood as well. Even when the comtesse finally keeled over in 1908, her will directly specified that, if Maria wanted the £100,000 she was due, then she had to agree to a series of incredibly weird conditions which included always dressing in white during summer, and never tying her shoelaces.[5]

Clearly, Comtesse Helena was insane – other incredibly paranoid health advice she gave her adopted daughter included the tip that she kill all nearby trees before they killed her – but she was not the only person to have breathed in cow methane for her health down the years. The New Zealand-born novelist Katherine Mansfield (1888–1923), for example, was yet another tuberculosis sufferer who was advised to look for an unlikely cure-all inside a cow's back passage. Having fallen under the spell of a Russian mystic named G. I. Gurdjieff (1866–1949), in 1922 Mansfield moved to Gurdjieff's rather dubious 'Institute for the Harmonious Development of Man' in Fontainebleau, France, where the esteemed guru ordered a course of treatment for his trusting disciple which could have come straight from the mouth of Dr Thomas Beddoes. On 1 January 1923, an impressed Mansfield wrote a letter back home to her father, telling him all about it:

> Did I tell you in my last letter that the people here have had built a little gallery in the cowshed with a very comfortable divan and cushions? And that I lie there for several hours each day to inhale the smell of the cows? It is supposed to be a sovereign remedy for the lungs ... the air is wonderfully light and sweet to breathe, and I enjoy the experience. I feel inclined to write a book called *The Cowiness of the Cow* as a result of observing them at such close quarters.[6]

The book in question never made it into print, sadly, because a week later Mansfield was dead. However, just because cow-house therapy didn't work, it did not mean that cows had no health benefits at all to them. Indeed, there was a far more pressing and notable controversy surrounding the alleged medicinal properties of cattle which was playing itself out during Dr Beddoes' day too – one whose influence is still being felt today.

Poxy Ideas

Nowadays, everyone knows the name of Edward Jenner (1749–1823), and reveres him as being the foremost father of the vital science of vaccination. During the 1790s, however, he was equally well known as being an irresponsible quack-like nutter who went around the countryside forcibly injecting little boys with some disturbing substance which had previously been living somewhere inside a cow. The story of Jenner, a Gloucestershire physician, is well known. Basically, Jenner had noticed that dairy maids tended not to catch the deadly disease of smallpox, something he attributed to them often contracting a much milder form of the infection, cowpox, from the cattle they milked. If this observation was true, then perhaps deliberately giving a person a small dose of cowpox would mean they would then be unable to later develop full-blown smallpox? In 1796, Jenner had tested this theory out by approaching a milkmaid infected with cowpox, jabbing one of her sores with a needle, and then shoving the infected barb into the arms of a small boy named James Phipps. Phipps quickly developed a mild dose of cowpox, but soon recovered. Then, Jenner took a dose of actual smallpox and stabbed that into his bloodstream with a needle, too. The boy thankfully proved immune to infection, and Jenner was doubtless much relieved to have escaped going down in history as a child-murderer.

Today, we know that such methods work because tiny doses of a weakened or mild form of a particular disease allow the body's immune system to accustom itself to identifying and then fighting off the malicious germs involved – getting in a bit of handy practice which allows it to easily see off any larger infection of the same sort in the future. Jenner's principle proved so effective that, in the case of smallpox itself, the world was made 100 per cent free from the disease in 1980, as the result of a global vaccination programme carried out on behalf of the World Health Organisation. However, at the time of Jenner's original experiments the underlying principles of how the process actually worked were but dimly understood, with the end result that many credible persons simply did not believe that it *did* work. For example, Jenner was a Fellow of the Royal Society, but when he wrote up his experiments and sent the account off for publication in the Society's journal, the organisation's president at the time, Sir Joseph Banks (1743–1820), flatly refused to print it. The evidence for Jenner's claims, he said, was far too flimsy for inclusion in such a prestigious organ. A leading naturalist, Banks was no idiot but, to him, the very idea of vaccination seemed ridiculous, and many others were of the same opinion.[7]

One of the most prominent sceptics as regarded Jenner's theory was none other than Thomas Beddoes. Beddoes pointed out that he had seen plenty of country people using something similar to Jenner's methods as a folk remedy for many years beforehand, a folk remedy which quite frequently did not work. This was probably because these traditional folk methods centred around infecting people with old and dried-out cowpox-riddled tissue rather than freshly gathered material as Jenner had done (Jenner's method was specifically termed 'vaccination' to distinguish it from these older, more unreliable methods of 'inoculation'), but Jenner himself speculated that Beddoes' real issue with his idea was one of professional jealousy. After all, the similarities between their two self-invented therapies were obvious; both had their origins in the observation that certain classes of workers – butchers and milkmaids – seemed immune to certain types of disease, and both sought the hidden cause of this immunity in cattle. The fact that Dr Jenner's intuition turned out to be correct and Dr Beddoes' to be wrong must have been somewhat galling to the latter. Eventually, though, as further evidence continued to accumulate, Beddoes finally felt obliged to admit that he was mistaken.[8]

A Method in His Madness

The idea of cow-house therapy may seem wholly comical to us now, but this is largely a result of that marvellous quality, hindsight. Thomas Beddoes was a noteworthy figure, and far from a crank. He was the father of preventative medicine in this country, being one of the first doctors to argue that it was better to try and stop disease breaking out in the first place by improving methods of hygiene and nutrition, rather than simply waiting to treat people when they got sick – received wisdom now, but not at the time. A 2009 biography of Beddoes, *The Atmosphere of Heaven* by Mike Jay, has conclusively proved the man's genius, but even that commendable effort seems unlikely ever to rescue the man from his current status in the history of science as a mere piece of comic relief. Put yourself in the shoes of a reasonably educated observer of the time, though, and you may possibly have seen fit to put your faith initially in Dr Beddoes' cow therapies rather than Dr Jenner's cow vaccines. The historical record clearly shows that it wasn't immediately obvious to contemporary observers which man's idea was the better, or even if either idea had any worth at all. With the existence of germs still utterly unknown, the actual theory underlying both sets of experiments was really just guesswork. As trials progressed and more and more evidence began to accumulate for the truth of Jenner's ideas, though, and very little evidence indeed was produced for Beddoes' notions, it gradually became obvious which side to place your bet on – even to Dr Beddoes himself.

By accepting the conclusions towards which the evidence eventually led him, no matter how professionally painful this may have been, Beddoes would have felt he was merely following the precepts of Francis Bacon (1561–1626), the English essayist, statesman, lawyer and early scientific theorist, some of whose most celebrated statements Beddoes appended to the front of his many books. Bacon is generally seen as being the architect of what later became known as 'the scientific method', something which held that dispassionate experiment was the key to gaining knowledge – a method to which Bacon himself was so committed that he supposedly ended up sacrificing his life on the altar of discovery after catching a chill while filling a dead chicken up with snow during a first-hand investigation into techniques of refrigeration. As Beddoes understood it, the Baconian method meant coming up with a hypothesis and then setting out systematically to test it in a series of trials which would prove objectively whether it was actually true or not. (Although actually Bacon felt that a hypothesis should not be decided upon beforehand, but simply emerge naturally from the act of experimentation itself – one of the main differences between Bacon's system and scientific methodology as it is often practised today.)[9] According to this way of doing things, no matter how much an experimenter may *want* his theory to be true, if the results of his experiments show it not to be, then he would simply have to accept that he was wrong and rethink his ideas accordingly. If you reflect back to your school science lessons, then you will instantly realise that this is still how society teaches its citizens that scientific investigations should be – and by implication invariably are – conducted even today.

At school, though, the experiments we are told to perform always have an extremely obvious outcome to them which will be in accordance with various theories our teachers have already explained to us beforehand. Nobody, when testing the effect of hanging weights down from a metal spring, comes up with the hypothesis that, the more weight you add to it, the less it will extend. With Beddoes and Jenner, the situation was clearly different. They made educated guesses as to what might happen, but they didn't know anything like for certain that they would be right. For all they knew, their treatments might actually have killed people. Both men would undoubtedly have been prosecuted had they pursued such cavalier methods today. And yet, for Francis Bacon, writing in his essay *Of Innovations*, such experiments, though of uncertain outcome, had nonetheless always to be performed, for otherwise how would knowledge ever manage to advance? In Bacon's view:

> As the births of living creatures at first are ill-shapen, so are all innovations, which are the births of time ... Surely medicine is ever an innovation, and he that will not apply new remedies must expect new evils.[10]

In that same essay, Bacon also made his famous statement that 'time is the greatest innovator',[11] the implications of which have since been taken to be self-evident. The more detached scientific experiments that are made, the more hypotheses tested, the closer mankind will get to knowing the full truth about the world – which must, by definition, therefore be considered a form of wholly *scientific* truth. The Baconian method is painted as being purely objective, and will reveal, in the end, all 'ill-shapen' hypotheses like those of Thomas Beddoes to be merely abortive stillbirths, while the genuine 'new remedies' of persons like Edward Jenner flourish and grow into healthy, bouncing, perfectly formed babies, recognised and loved by all. That, at least, is the myth as we are currently being peddled it.

It's Getting Better All the Time

When I talk about science as we are usually encouraged to think about it being somehow like a myth, I do not mean this in a literal sense, exactly. Yes, the Earth orbits the sun; yes, hydrogen is the lightest element in the Periodic Table; yes, the air we breathe is made up of 78 per cent nitrogen. I make no claim that generally accepted scientific facts are myths, merely that some of these facts, and in particular the parent philosophy which discovered, propagated and sustains them, can also be seen at times to *function* as myths.

This is not an original idea. Perhaps its most significant contemporary proponent is John Gray (b. 1948), the author of best-selling books like *Straw Dogs*. These are usually described as being searing critiques of the currently prevailing Western worldview of liberal humanism; broadly put, the idea that human beings are inherently rational actors, and that, in spite of certain unfortunate incidents which may occasionally occur, the world is getting better all the time, largely due to the unstoppable rise of secular values and their habitual handmaiden, science. When the American academic Francis Fukuyama (b. 1952) famously started writing books proclaiming *The End of History* just after the end of the Cold War, predicting that every country in the world would inevitably end up adopting the governmental model of peaceful, non-authoritarian free trade-based democracies like the US and the UK, then that's liberal humanism. Another word for this particular viewpoint may perhaps be 'naivety' – or at least so John Gray would say. Mr Gray used to be the Professor of European Thought at the LSE, and at some point during his tenure appears to have come to the conclusion that most current examples of European thinking are based upon false premises. This argument sounds like it is about politics, but it also about science. Viewing modern-day Western democratic values as inherently rational in their nature, some recent thinkers have tried to paint democracy and science as being somehow twin expressions of the same thing – namely, the logical mind of contemporary Western (that is to say, scientific and secular) man. For example, in his suggestively titled 1996 book *The Demon-Haunted World: Science as a Candle in the Dark*, the American cosmologist Carl Sagan (1934–96) had this to say about the matter:

> The values of science and the values of democracy are concordant, in many cases indistinguishable. Science and democracy began ... in the same time and place, Greece in the seventh and sixth centuries BC ... Both science and democracy encourage unconventional opinions and vigorous debate. Both demand adequate reason, coherent argument, rigorous standards of evidence and honesty.[12]

That sounds like the mainstream political process of no present-day Western democracy that I know of. Nonetheless, Sagan tries to tie the very existence of American democracy itself to the fact that several of its most significant Founding Fathers were scientists. Benjamin Franklin (1706–90) was one of the leading lights of the then new field of electrical physics, and Thomas Jefferson (1743–1826), too, America's third President, was a man of science, with wide-ranging interests from astronomy to palaeontology. Just as significant to Sagan, though, was the way that such men actively designed the structure of the American constitution by reference to scientific concepts like the self-regulating circulation of blood in the human body, and the way that mechanical balances worked. Sagan does have a point. The Founding Fathers were, as he says, 'creatures of the European Enlightenment', which clearly influenced the way they tried to construct their new American utopia.[13] He quotes the historian Clinton Rossiter to the effect that 'free inquiry, free exchange of information, optimism, self-criticism, pragmatism, objectivity – all these ingredients of the coming Republic were already active in the Republic of Science that flourished in the eighteenth century'.[14]

But were they? Sagan approvingly quotes the second US President John Adams (1735–1826) as saying that 'all mankind are chemists from their cradles to their graves ... The Material [and thus human and political] Universe is a chemical experiment', and yet elsewhere criticises his passing of the 1798 Alien and Sedition Acts, which banned all 'false or malicious' criticism of government policy, and thus in effect ended the right to that free exchange of views which was apparently so beloved of both science and democracy.[15] Sagan also quotes from a letter of Jefferson's, written a few days before his death in 1826, in which he praises the 'light of science' for having conclusively demonstrated that 'the mass of mankind has not been born with saddles on their backs'.[16] Strange, then, how slavery continued in the country until 1865, and was not finally abolished until a disastrous – and hardly entirely rational – civil war had played itself out, killing 620,000.

Nevertheless, Sagan is surely correct to praise such men as being giants, certainly in comparison to the pygmy politicians the West has today; but then, if the standard of political discourse has veered down a cliff in recent decades, then that hardly says much for the idea of universal progress, does it? The fact that technological advances have helped precipitate this decline – the rise of the news media having given birth to the contemporary disease of policy frequently being governed more by the twenty-four-hour news cycle than by rational long-term planning, for instance – also seems to contradict the idea that scientific developments inevitably bring positive social developments in their wake. Perhaps what Sagan really means is that the values he hymns are those to be found in an *ideal* democratic state, not an actual one – but, naturally, such an ideal state does not, and never will, exist. As Sagan admits, 'democracy can be subverted more thoroughly through the products of science than any pre-industrial demagogue ever dreamed'.[17] Worse, recent academic research seems to have demonstrated that science can flourish in totalitarian states every bit as much as it can in the West.[18] North Korea is poor and autocratic, but has atom bombs. Scandinavia is rich and democratic, but has none – it prefers to divert its resources to other ends. Who put the first man in space? The Soviets. Who built the first ballistic missiles? The Nazis. Democracy has no monopoly upon scientific discovery or industrial achievement.

Sharia Science

Nonetheless, the current myth of liberal democratic values and the values of science being joined together at the hip continues to be perpetuated. After all, if human beings are inherently rational actors, then they need to have access to an inherently rational mode of thought, do they not? And, as people like Francis Bacon have shown us, science *is* that inherently rational mode of thought, yes? Not always. Even Bacon himself, in his 1620 text *Novum Organum*, admitted that scientists do not always proceed in a wholly dispassionate and objective manner:

> The human understanding is no dry light, but receives infusion from the will and affections ... For what a man had rather were true he more readily believes ... Numberless in short are the ways, and sometimes imperceptible, in which the affections colour and infect the understanding.[19]

This quote is actually cited in Sagan's book in relation to flaky pseudoscientific things, like Atlantis and crop-circles, which Sagan (rightly) did not believe in. But could it not also apply to some things that he *did* believe in, too? Apparently, there was once something called 'the Enlightenment', after which science gradually but inevitably pulled us all up from out of the stinking primeval swamp of superstition, ignorance and religion, and improved our lives forever – an irreversible process, so they say. Sitting here in a rich, safe, Western democracy with easy access to food, fuel, shelter, advanced technology and free healthcare, this sounds a plausible enough analysis. But try telling all that to someone who lives in Syria or Iraq, whose own scientific 'Golden Age' was now nearly a thousand years ago, and where ISIS fanatics have recently restructured university curricula so that only something called 'Islamic Science' can be taught – a catalogue of highly irrational beliefs such as that Jews are biologically descended from pigs.

Can there be such a thing as 'Islamic Science'? No; or, at least, not in any *objective* sense. If a fact is a fact, then it's a fact for Christian, Muslim, Hindu and Jew. And yet, in 1993, Sheik Abdel-Aziz Ibn Baaz (1910–99), then the Grand Mufti of Saudi Arabia, issued a *fatwa* officially decreeing that the world was flat, and that anyone who said otherwise was a sinner.[20] However, let us take a more subtle example than this. Dr Ibrahim B. Syed (b. 1939), an Indian-American Muslim holding a professorship in medicine at the University of Louisville, is a demonstrably knowledgeable man. He also believes in genies. As a good Muslim, he is required to; genies (or *djinn*) are mentioned repeatedly in the Koran, where they are deemed to have been made from 'smokeless fire' by Allah Himself. For Dr Syed, though, they are better thought of in purely scientific terms. In a recent paper, 'The Jinn – A Scientific Analysis', the professor patiently explained that genies are probably made from plasma and are born inside the sun. Plasma is the so-called 'fourth state of matter', being neither liquid, solid nor gas. It consists of atoms stripped of their electrons. When an atom is stripped of its electrons, it becomes a positively charged ion, and these ions then mix with the free-floating negatively charged electrons in a kind of unstable and super-heated ionised gas ... which isn't precisely a gas. Plasma is rare on Earth, but can be found in space, for example inside stars like our sun. Seeing as plasma is super-hot and luminous but emits no smoke, Dr Syed says that this must be the smokeless fire from which Allah made the *djinn*. When 'patterns of magnetic force' interfere with this plasma inside the sun, Syed claims it imposes a special internal structure on the plasma which turns it into genies.[21] He has no conclusive proof of this theory, however; you basically just have to take it on trust – like all religion.

No non-Muslim could write a paper like that ... could they? Actually, it turns out they could. The American physicist David Bohm (1917–92) was the father of plasma physics. During some of his early experiments, he found that the shed electrons from plasma behaved less like individuals than as a kind of collective mind. Studying them, he frequently had the impression they were in some sense alive (although not necessarily literally). More recent experiments in Romania have backed this unlikely idea up, with the electrons inside lab-created argon-based plasmas seemingly grouping themselves into self-replicating 'cell-like forms' which appeared to 'grow' and deliberately interact with one another. According to the speculations of one of the scientists involved, Dr Mircea Sanduloviciu, while such cells were more *analogous* to living things than actually alive as such, it could still be the case that 'the emergence of such [cells] seems likely to be a prerequisite for biochemical evolution'.[22] However, as observed by the researcher Paul Devereux (b. 1945), who has collected numerous fascinating accounts of people's alleged encounters with plasma-like forms behaving in a seemingly intelligent manner, such ideas are 'resisted by the bulk of mainstream physicists' not necessarily because they are based upon faulty observation, but because they 'harbour a range of awkward implications' – that is to say, they undermine the currently prevailing scientific doctrine of materialism.[23]

Dr Syed has projected a Muslim myth of living genies on to plasma forms; most Western scientists have projected a secular myth of utter deadness on to them instead. Might it not be possible that the actual reality lies somewhere in-between?

Passing the Acid-Test

This may sound too obvious a point to even need making, but sometimes the most obvious points are the most ignored: what is thought to constitute science in any given society is inevitably in some sense a reflection *of* that same society, especially given that the word 'science' is often now thought to be synonymous with the word 'truth'. 'We hold these truths to be self-evident' go the famous opening words of the US Declaration of Independence; and yet it is equally self-evident that the particular truths which are held to be self-evident by any group of people change over time and from society to society. The first 'self-evident' truth enumerated by the Founding Fathers of 1776, for instance, holds that 'all men are created equal' because 'they are endowed by their Creator with certain unalienable rights'.[24] But what's all that about mankind having a 'Creator'? That's one truth, I think, which no longer seems quite so self-evident at all – certainly not to most liberal humanists!

For an excellent example of another scientific 'truth' which was a product purely of its times, let's pick on poor old Dr Thomas Beddoes again. As a good Baconian, Beddoes thought that you should make use of the process of objective experimentation in order to solve any potential conundrum, even the question of – as he put in the title of a now long-forgotten 1792 essay – 'The Complexion of Natives of Hot Countries'. Progressive inhabitants of a liberal humanist society as we all doubtless are, this paper now makes for uncomfortable reading. Essentially, Beddoes' essay asks the question of why it might be that black people should be black at all. In case you're wondering, the real answer is because black people have a higher level of melanin content in their skin than white people do, melanin being a naturally occurring bodily pigment which is of some aid in protecting against the sun. Dr Beddoes, however, thought different. His theory was that all those dark-skinned races inhabiting tropical regions of the globe were really just white men who had gone a bit rusty.

Beddoes' whole idea of rust was very much mistaken, but entirely in line with the prevailing science of his day. For Beddoes, rust was essentially a form of slow 'burning' of the metal upon which it appeared. During this process of burning, the metal's oxygen content became released from it, he thought, leaving the previously shiny surface covered in a layer of blackish-brown residue. Light was implicated as an agent that could release oxygen from various substances under certain conditions, too. Beddoes knew that some plants discharged more oxygen when placed in strong sunlight, and saw equal significance in the fact that a solution of nitric acid and silver turned black when exposed to illumination, something he interpreted as due to the oxygen being drawn out of it by the light. Presumably, then, when white people got a tan, this was because of the sun gradually causing their skin to lose some of its oxygen content. Given this, if oxygen could be reintroduced into a black man's skin, then might it not turn him back white for a little time? The chemistry of the day (incorrectly) assumed that all acids were veritable reservoirs of oxygen. What better way to make a black man turn pale, then, than to cover his skin in acid? Visiting Oxford, Dr Beddoes, in his own words, 'prevailed upon a negro' to help him test this painful idea out, somehow managing to persuade the poor fellow to shove his hand into an acid solution. Pulling it out again, the 'negro' found that his fingers were indeed now white, a colour they stayed for several days. Today, we would instantly be able to perceive that this was because they had been burned and bleached by the acid, but Beddoes took it as proof he was correct in his conjectures.[25]

Dr Beddoes' paper obviously could not be published today for two different reasons. Firstly, there is the inarguable fact that the basic theory it contains is entirely incorrect. The storehouse of scientific knowledge has moved on, and the very idea of black people rusting now seems every bit as nutty as cows curing TB. The second reason, meanwhile ... well, do I need to point it out? Even at the time there were those who considered Beddoes' experiment to be distasteful, not least the unnamed 'negro' himself, who in later years took to stationing himself outside Beddoes' hospital and begging for charity from passers-by, telling them the disturbing tale of how he had been callously duped into being used as a guinea pig by the local mad scientist. It would have been a hard-hearted individual indeed who refused him a spare shilling.[26]

Social Sciences

By the standards of his day, Dr Beddoes was hardly a racist; he was a prominent anti-slavery campaigner who flirted with early forms of Fair Trade, suggesting that people stop buying slave-produced West Indian sugar and try importing more ethically sourced maple-syrup from North America instead. You might almost say he was an early liberal humanist – apart from his belief in God, of course.[27] It is clear he was a good and charitable man, who, being a creature of his times, simply felt there was nothing much wrong with his treatment of the 'negro'. Most of his peers would have agreed. He was never prosecuted for his actions, nor hounded from his chosen career. Compare this with the recent sad case of Sir Tim Hunt (b. 1943), a Nobel Prize-winning biologist who helped develop treatments for cancer, before being chased out of his honorary professorship at University College London by a gaggle of self-righteous harridans, puritanical Twitter mobs and other such ranks of the professionally offended, for the way lesser 'crime' of making a weak joke about girls crying in laboratories during a June 2015 speech – a gag which was specifically labelled as such, and which was actually supposed to encourage women not to be put off from pursuing their own scientific careers. The latest news about Sir Tim is that he and his wife (another eminent scientist, incidentally) are leaving the country to go and live in Japan, partly in order to escape the needless PC witch-hunt.[28]

What it is considered legitimate to talk, theorise or even think about shifts quite radically from age to age. Science and scientists do not exist in a vacuum, and certain theories and experiments are considered acceptable only within the context of certain times and places. To return to the topic of skin colour, for instance, during the twentieth century the fact that black people have more melanin within their bodies has been reinterpreted by a number of non-white pseudoscientists in order to further the modern concept of 'black pride'. For example, Leonard Jeffries Jr (b. 1937), a professor of Black Studies at New York's City College, became infamous during the 1990s for claiming that blacks should be rechristened 'Sun People' and whites 'Ice People'. In Jeffries' view, the Sun People's higher level of melanin content meant that they were able to 'negotiate the vibrations of the universe' more successfully than Ice People were, making them more peaceable and co-operative. White Ice People, however, having less melanin, were less in tune with Mother Nature, and so inevitably ended up being more competitive and individualist and thus prone to warfare and violence, he said, things which did not come naturally to black Sun People. Did anyone tell Idi Amin?[29]

Even weirder are the theories of an American psychiatrist named Francis Cress Welsing (b. 1935), inventor of the 'Cress Theory of Color Confrontation', which argues that all white people are albino mutants genetically programmed to commit acts of genocide against other races, due to their fear of being contaminated or wiped out through interracial breeding. Claiming that the entire white male establishment is absolutely terrified of black men's genitalia, Welsing began an obsessive programme of reinterpreting prominent Western cultural symbols as being coded representations of this phobia. The Washington Monument, for example, that big white stone needle which towers 169 metres above America's capital, is really a giant white penis, dominating a city predominantly inhabited by blacks and helping keep them in line. The cross of Christianity, meanwhile, is in fact a representation of a black man's penis, chopped off from his body; or, as Welsing puts it, 'a brain-computer distillate of the white collective's fear-induced obsession with the genitals of all non-white men ... which have the power to genetically annihilate the white race'. Furthermore, the 1933 film *King Kong* is apparently nothing more than a fable about the dangers of interracial sex, and the way that certain white female primatologists like Dian Fossey (1932–85), of *Gorillas in the Mist* fame, spent all their time in the company of apes simply reflected their repressed desire to mate with black men – or 'gain possession of the Black Phallus', as Welsing preferred to phrase it. White people's genocidal impulses, said Welsing, were all directly attributable to their lack of melanin, this deficiency being 'the neurochemical basis of evil'. All other races in the world have more melanin than whites, though, and so are more sensitive and religious, having more 'soul' and 'rhythm', as well as a 'sixth sense' which enables them to communicate with Nature. As 'evidence' for these claims, Welsing cites the work of the black American scientist and inventor George Washington Carver (*c.* 1860–1943), who made certain obscure discoveries about peanuts. According to her, Carver's insights were derived not simply from careful lab-work but also from his melanin-suffused skin enabling him to psychically 'communicate with the energy-frequencies' which were emanating outwards from his peanut plants, and which very kindly told him all their secrets.[30]

Welsing is certainly very committed towards her theory – some might say overly so. This, for example, is her explanation of why both chocolate and professional ball games are really metaphors for white men's profound and enduring envy of black men's testicles:

> On ... St Valentine's Day ... the white male gives gifts of chocolate candy with nuts ... If his sweetheart ingests 'chocolate with nuts', the white male can fantasise he is genetically equal to the black male ... Is it not also curious that when white men are young and vigorous, they attempt to master the large brown balls [of American football], but as they become older and wiser, they psychologically resign themselves to their inability to master the large brown balls? Their focus then shifts masochistically to hitting tiny white golf balls in disgust and resignation.[31]

An entire cottage industry seems to have sprung up around these bizarre ideas, with titles like *Melanin: The Chemical Key to Black Greatness* pouring onto American bookshelves from the 1960s onwards, corresponding with the rise of the black civil rights movement. Probably the strangest was the psychology professor T. Owens Moore's

1995 tract *The Science of Melanin: Dispelling the Myths*, a book which, ironically, ended up spreading quite a few myths of its own. Moore's big 'insight' was that the mysterious dark matter and dark energy which are said to make up around 95 per cent of our universe have so far escaped the comprehension of white scientists because … they are really melanin, hence the name! Apparently, melanin's 'special bioelectric properties' help maintain our universe, and only black people can tune into it properly because (in Moore's inaccurate view, anyway) all black people are born with 'kinky or wire-like hair'. This wiry hair, he says, 'is an evolutionary advance' over the 'matted or animal-like hair' of white men, which acts 'like an antenna to absorb more readily those naturally occurring electromagnetic waves' which emanate from the dark matter-infused universe. Thus, black people, whose hair acts as a kind of psychic radio, are more in tune with the cosmos, whose secrets white astronomers will never be able to uncover. Furthermore, seeing as 'melanin in the skin and nervous system helps to provide metaphysical experiences', it would appear that black people are, once again, pre-programmed to be 'more in tune with Nature by living in peace and harmony with the world and its inhabitants', unlike those dastardly white folk who go around causing wars and environmental disasters.[32]

Such mad ideas, you have to say, are actually quite racist in nature – switch the words 'black' and 'white' around in these books, and they could not have been published without fear of prosecution. Go back a hundred years, however, and the situation would have been reversed. Then, a number of equally pseudoscientific books, 'proving' that black people were violent savages who were inferior to civilised whites, were considered respectable enough to be printed and sold in high-street stores. Every age and society, it would seem, develops its most appropriate flavour of pseudoscience.

Fascist Fission

The Nazis are another case in point. When considering the nature of the subatomic world, they talked of both 'Jewish Science' and 'Nordic Physics', the latter supposedly being somehow better than the former. You may have thought that the structure of the atom was a simple matter of objective fact, and exactly the same in every respect for both Jews and Germans (and, indeed, for Jewish Germans) – but apparently not. In the opinion of Adolf Hitler (1889–1945) himself, 'There is very likely a Nordic science, and a National Socialist science, which are bound to be opposed to the Liberal-Jewish science.'[33]

Following the Nazis' rise to power, there was an astonishing turnover in the staffing of German universities. As many as one in four physics professors were dismissed or resigned, accused of being Marxists or Jews, or of preaching the 'wrong' kind of science.[34] One Nazi official, giving a speech before university-lecturers in Munich, gave them the following advice: 'From now on the question for you is not to determine whether something is true, but to determine whether it is in the spirit of the National Socialist revolution.'[35]

If it wasn't – then you were out. A cheeringly ironic consequence was that several exiled Jewish scientists helped America develop the first atom bomb ahead of Nazi Germany and its fêted 'Nordic physicists'.[36] What precisely *was* Nordic Science, though?

The key figures in its development were a pair of Nobel laureates, Philipp Lenard (1862–1947) and Johannes Stark (1874–1957), both of whom felt distaste for the new direction physics had taken since the incredible discoveries of Albert Einstein (1879–1955) – who was, of course, a Jew. Lenard understood neither Einstein's relativity theory nor the world of quantum physics it ushered in, and decided its increasing global acceptance was the result of a secret Jewish conspiracy. According to him, Jews did science in a different, unhealthy way compared to Aryans. 'Jewish' quantum theory made no sense, and was a kind of abstract mysticism, whose conclusions could not be proven experimentally; Einstein was thus an advocate of relativity in more ways than one. Quantum theory, which studied a subatomic world so tiny it could not even be seen, was a sort of voluntary mental illness, a web of theoretical fantasy. Racially pure Germans had more sense. Realising that Nature was holy, they set out to observe it practically with their own eyes, worshipping it by discovering its secrets. Jews and Marxists, however, wished to abuse their knowledge of Nature by turning men into slaves to technology. Johannes Stark agreed, co-authoring an article in 1924 praising 'The Hitler-Spirit and Science' before Adolf had even become Führer.[37] Seeing as Hitler had no scientific training whatsoever, though, and went around saying mad things like 'A new era of the magical explanation of the world is rising, an explanation based on will rather than knowledge' and 'There is no truth, either in the moral or the scientific sense', this may not have been a terribly good idea.[38] Indeed, such statements sound very like a subjective form of mysticism themselves.

In a 1934 article, 'National Socialism and Science', Stark claimed that only Aryans could make genuine scientific discoveries, obsessively interpreting portraits of famous scientists from the past as demonstrating their 'Nordic-Germanic' features, even if they were patently foreigners. Here, he also introduced his concept of *Deutsche Physik*, or 'German Physics', one which emphasised hale and hearty practical experimentation over weak and degenerate hypothetical theorising. Apparently, Aryans' superior physiques allowed them to perform such experiments more easily than weakling Jews and Communists; the Nazi scientist's well-toned body 'does not shrink from the effort which the investigation of Nature demands of him', Stark declared.[39]

Stark's idea of *Deutsche Physik* won the direct approval of Hitler – but was this madness really the reason the Nazis failed to get the atom bomb? No. In the view of the science historian Philip Ball, Hitler's 'grandiose statements' on the value of Nordic Science 'had as little real influence on the way affairs were conducted at the daily, prosaic level as do the proclamations of the Pope on the dealings of a local Catholic church'.[40] In his own study of the subject, Ball shows conclusively how, while scientists were required to pay verbal obeisance to such ideas, in practice what the Nazis wanted was results, and, seeing as it was only 'Jewish' quantum theory which could get them an atom bomb, scientists were completely free to make use of all its findings. Had a Jew or a Marxist invented the aeroplane, there would still have been a Luftwaffe. For example, Germany's leading physicist, Werner Heisenberg (1901–76), clearly made use of Einstein in his own war-time lectures and researches – he just tactfully didn't cite him by name.[41] Another Nazi physicist, Ernst Pascual Jordan (1902–80), even reformulated quantum theory into a new, fascist form, saying that its Jewish 'subjectivity' was actually a cheering harbinger of Nazism's inevitable triumph over despised Enlightenment values like rationality and reason.[42]

Man vs Machine

Jordan had a point; contrary to Stark and Lenard's beliefs, early quantum theory was actually highly compatible with Nazi ideology if viewed in a certain twisted way. Ideas about the world being a kind of illusion created by its observer, and of hidden secrets lying beneath the visible realm, were deliberately promoted by some quantum physicists to make science seem more romantic in the face of an increasingly industrialised world. Early twentieth-century writers like Oswald Spengler (1880–1936), author of 1918's *The Decline of the West*, were busily promoting the idea that Western culture, once a vital and noble thing, had recently decayed into a mere 'civilisation', an empty shell in which technology and commerce had replaced spirit and Nature, draining all of life's romance away. This, for instance, was Spengler's view about what scientists had done by inventing their blasted industrial machines:

> ... man has become *the slave of his creation*. The machine has forcibly increased his numbers and changed his habits in a direction from which there is no return ... It forces the entrepreneur not

> less than the workman to obedience. *Both* become slaves, and not masters, of the machine, which now for the first time develops its devilish and occult power.[43]

Mankind's only hope, Spengler says, is that the engineers who build, design and maintain the devil-machines might one day rebel in a spirit of Romantic mysticism:

> Suppose that, in future generations, the most gifted minds were to find their soul's health more important than all the powers of this world; suppose that, under the influence of the metaphysic and mysticism that is taking the place of rationalism today, the very elite of intellect that is now concerned with the machine comes to be overpowered by a growing sense of its *Satanism* ... then nothing can hinder the end of this great drama.[44]

The machines, it seems, will be smashed! In such an intellectual climate, science came to be denigrated by many leading cultural figures, and even by some scientists themselves, who wished to infuse a sense of holiness back into it – an opinion shared by Lenard and Stark. If they hadn't been so distracted by their rabid anti-Semitism, perhaps the two men might actually have found in quantum theory something to their liking. Nils Bohr (1885–1962), for instance, one of the early giants of the quantum realm, promoted the idea of 'complementarity', which held that in the subatomic world two different and apparently contradictory observations could be considered equally valid; it just depended how you looked at them. Light, for instance, could behave either as a wave or as a particle in the laboratory, and so was said to have 'wave-particle duality'. In public statements, Bohr implied that this same idea could be applied to spheres like ethics, law and religion, too. While the Dane Bohr was certainly no Nazi – he hated Hitler, and fled to Britain to aid the Allied cause in 1943 – such claims, in the words of Philip Ball, nonetheless 'made quantum physics a political matter'.[45]

By implying that, at some fundamental level, the world was really what the observer made of it, physicists like Bohr were not only describing genuine lab findings, but also trying to fit in with the anti-industrial, neo-Romantic *zeitgeist* of their era. The trouble was that Nazism to some extent grew from out of the Romantic movement, too; the Third Reich wasn't only hi-tech U-boats and V-2 rockets, but also nocturnal torch-lit parades in which people dressed up as Parsifal and the Knights of the Round Table. Consequently, some right-wingers began claiming that quantum theory was in itself inherently fascist. Just as Nazis like Hitler embodied the Nietzschean 'will to power', remaking the fallen, decayed Germany they saw around them into a new and different world, so the quantum physicist, remaking the reality of light by observing it in different ways in his laboratory, became another kind of neo-Romantic hero.[46] Clearly, then, the idea of what exactly constituted both Nordic and Jewish Physics lay entirely in the eye of the beholder. In the final analysis, Jews and Marxists were dismissed from university posts not because they were mystics, but because they were Jews and Marxists. *Deutsche Physik* was essentially for show.

Period Dramas

Even in the modern West, science can find itself becoming politicised. During America's anti-Communist witch-hunts of the 1940s/50s, for instance, some quantum physicists were hauled up before McCarthyite committees on the spurious grounds that, seeing as their theories were 'revolutionary', so too might their politics be.[47] Even dafter was the case of the inadvertently amusing Belgian philosopher Luce Irigaray (b. 1930), who has written much about the philosophy of science. Perhaps I shouldn't criticise Irigaray unduly, as her basic idea is not dissimilar to my own:

> Every piece of knowledge is produced by ... [scientists] in a given historical context. Even if that knowledge aims to be objective, even if its techniques are designed to ensure objectivity, science always displays certain choices [and] exclusions.[48]

However, Irigaray then goes on to say that these exclusions are 'particularly determined by the sex of the scholars involved'. Or, in other words, science is sexist! One recent story about sexism in science concerned comments made in 2015 by Professor Averil Macdonald, of the University of Reading, who claimed that only 31.5 per cent of women think fracking for shale gas should be allowed in the UK, compared to 58 per cent of men, because females simply 'don't understand' the science involved. Because they were less prone than men to continue with science-based subjects post-GCSE, Macdonald said women's opinions about issues like fracking were more likely to be based upon 'feel' and 'gut reaction', rather than actual facts. Furthermore, as 'women are always concerned about threats to their family more than men', being 'naturally protective of our children', they would be more primed to see dangers in such techniques where none really existed.[49]

Examining such patronising statements, maybe we can admit Luce Irigaray does have a certain point. However, she goes rather further than this, and claims that the *actual laws of physics* are in themselves inherently sexist! Most famously, Irigaray denounced

Einstein's $E=mc^2$ formula as 'a sexed equation'. Here, energy equals mass multiplied by the speed of light, squared. But the speed of light, says Irigaray, is a masculine concept, and by including it in his equation, Einstein 'privileges the speed of light over other speeds that are vitally necessary to us'. Not only is the equation used when men make nuclear weapons (missiles being phallic), it also 'privilege[s] what goes the fastest' (men being faster on average than women). Even weirder, she also claims that fluid mechanics has historically been a neglected field because it subliminally reflects the way women have periods, while solid mechanics has been privileged because men have genitalia that can become hard. Worse, maybe mathematics itself is prejudiced against the fairer sex? According to Irigaray's disciple Suzanne Damarin, maths privileges 'linear' concepts like distance and acceleration over 'the dominant [female] experiential cyclical time of the menstrual body'. Speaking of a 'female mind-body', Damarin appears to suggest that schoolgirls might have problems understanding geometry properly because they have periods.[50] In a bizarre echo of Averil Macdonald's claims, in 1993 Irigaray also wrote that, seeing as having periods apparently links women to 'cosmic rhythms', it makes them particularly sensitive to the figurative rape of Mother Nature by man and his science:

> If women have felt so terribly threatened by the [nuclear] accident at Chernobyl, that is because of the irreducible relation of their bodies to the universe.[51]

That really is some very silly thinking indeed. Still, what do you expect from a woman?

Trials and Errors

Maybe, though, you would prefer to dismiss such cases as being mere *perversions* of science, not science as it should truly be pursued by its knights in shining white lab coats. Up to a point, yes – but the distinction between an outright fraud and certain other, more reputable scientists, is not quite as clear-cut as we might like to imagine. Take the issue of recording experimental results during laboratory research. I remember an occasion, long ago, during a practical experiment at school, when I recorded some results which I really wasn't supposed to have done. Precisely what the experiment involved I now forget, but the results I got clearly contradicted the hypothesis (in reality a well-established physical law) I was supposed to be testing. Obviously, I had done something wrong. The solution was obvious – doctor the results. Even though my later write-up of the trial was thus a complete deception, it was judged to be correct by the teacher in a way which my more honest (and incompetently derived) findings would not have been. The 'correct' result was already known beforehand; I may as well have just produced the write-up, and left the Bunsen burner and test tubes sitting in the cupboard where they belonged.

So, where some scientists who doctor their results are merely crooked, I was both crooked and incompetent. Or was I? The case of R. A. Millikan (1868–1953), winner of the Nobel Prize for Physics in 1923, would suggest otherwise. Millikan performed an experiment into the nature of electrons which involved suspending tiny electrically charged drops of oil, one by one, within an electric field before subjecting them to x-ray bombardment. Through this means he managed to work out the precise nature of the charge on any individual electron. Happily, the results as Millikan published them entirely supported the hypothesis which he – and the majority of his peers – had formulated about the topic beforehand. Unhappily, however, the results as Millikan published them were not the full ones. Following his death, researchers obtained his papers and looked through them. The original records of Millikan's oil-drop trials showed that 60 per cent of his actual results did not confirm his theory at all. Instead he had simply selected the minority 40 per cent of results which did back him up, published these, and written the others off as 'mistakes'. The experiment itself was rather fiddly to perform, and Millikan guessed that any oil drops which did not behave as he expected were purely the result of his own cackhandedness.[52]

Imagine I have a bag of sweets, and predict that they are all red. I then tip them out and find that the majority are actually white. Unable to cope with this finding, I theorise that these white sweets are *really* red after all, but that their colouring agent has faded away inside the bag. I then eat these white sweets to hide the awkward data, and present only the red sweets to you as evidence of the truth of my original theory. This is exactly what Millikan did with his own results. As it is, we now know that Millikan most definitely cheated – and yet also that he was right to do so. Subsequent trials performed by others proved that Millikan was entirely correct in his original theory after all. He was also, though, fundamentally dishonest. After all, if he had interpreted his findings objectively, rather than through the subjective lens of his own desire, then he would have reported that his hypothesis was wrong – even though, as we now know, it actually wasn't! In light of all this, I now see that my own childhood efforts at blatant scientific fraud were in fact highly commendable.

Against the Prevailing Climate

Perhaps I should apply for a job in the ever-expanding field of climate-change science. In August 2015, a paper was published in the *ICES*

Journal of Marine Science by two Australian researchers named Christopher Cornwall and Catriona Hurd, which reviewed some 465 studies into the likely effects of acidification upon the world's oceans. For many years now, scientists have been warning that increasing levels of carbon dioxide in the atmosphere are being absorbed by our oceans, with potentially dire consequences. At the moment our oceans are slightly alkaline in nature, but it is predicted that, by the end of the century, they might become mildly acidic due to CO_2 absorption, killing off coral reefs. The 465 studies which Cornwall and Hurd assessed all backed this hypothesis up – but, according to their own ensuing meta-analysis of these studies, only twenty-seven used what they described as being 'an appropriate experimental design'. That's less than *6 per cent*! In the opinion of Cornwall and Hurd, the vast majority of research they studied 'used either an inappropriate experimental design and/or data analysis, or did not report these details effectively'.[53]

However, this does not mean that the idea of climate-change potentially having a devastating effect upon our oceans is in itself necessarily a false one, any more than R. A. Millikan's obfuscations negated the truth of his own pre-existing theories about electrons. Cornwall and Hurd were at pains to report that the 'overwhelming evidence' that seas are indeed becoming acidified still stands; it is just that many of the *individual* studies which supposedly acted to further 'prove' this were intrinsically flawed. One of the most common failings was that the scientists involved had failed to eliminate the risk of observer bias; namely, many seem to have actively *wanted* their research to turn out in a certain way, doing their best to produce a body of results that fitted in with the prevailing consensus on the matter. Presumably, if they had reported results suggesting the opposite, there may have been consequences for their careers or credibility. They may, like Millikan, have thus chosen, consciously or not, to have arrived at the correct results by incorrect means.

Climate-change is a hugely complicated issue, whose precise twists and turns, and the vast mountains of data pointing in one direction or another, are simply beyond the capacity (and will) of laymen like myself to process. Most people just have to take the word of scientists upon such matters essentially on trust – but, as Cornwall and Hurd proved, you can't always do so. However, the reality or otherwise of climate-change is not the issue I wish to discuss here – instead, what really interests me is the question of whether the belief many scientists, politicians and campaigners have in it is really genuinely objective in its nature. Very frequently, I think it is not. Purely for the sake of argument, imagine that the idea of climate apocalypse turns out to be wildly exaggerated, like the sceptics claim. In a hundred years' time, someone writing a similar book to this one will be able to have some real fun with some of the more bizarre-sounding news stories about the topic. For example, there was the infamous 2006 report from the UN suggesting that, if we really wanted to save our dying planet, then not only did we now have to find a way of reducing our reliance upon fossil fuels, but also a means of safely plugging up cows' overactive anuses. As Thomas Beddoes once found, cows like to break wind. It turns out that methane, though, is even worse for the planet than it is for an invalid's lungs – at least according to the UN's report, which claimed that cattle's bum-holes did far more damage to ailing Mother Nature than all the planes, trains and automobiles in the world put together. Methane, it seems, is even worse for the atmosphere per unit than CO_2 is, leading to some innovative solutions to the problem being proposed; some farmers, for example, have experimented with having their cows live inside a giant plastic bubble which funnels all their burps and farts into a generator to be transformed into electricity, an idea which was ultimately deemed too inefficient and 'inhumane' to be applied worldwide. Far more effective was feeding cows garlic, which has the effect of making their emissions somewhat milder, sweeter-smelling and environmentally friendly.[54]

The weirdest global warming story I ever saw, however, involved the topless glamour model Linsey Dawn-McKenzie, who in 1998 announced to an astonished world that her breasts had grown from an already implausible 37FF bra size to even larger 37H proportions. *The Sport* newspaper, after providing extensive photographic evidence of the fact for their more environmentally-aware readers to study for themselves in private, contacted one Dr Stig Samuellson from the Danish Institute for Ecological Studies and asked him if it could all be down to climate change. His alleged response was that 'there's nothing to say that this girl's breasts are not sensitive to changes in climate temperature'. Even better for readers of *The Sport*, added Dr Samuellson, 'If the Earth carries on getting hotter, perhaps [all] breasts will get bigger.'[55] Sensible people will recognise this as a simple tabloid joke; but is it really that much more laughable than the idea that the world will one day be destroyed by cow farts? Any amused future chronicler of such scare stories would be forgiven for thinking that the current outbreak of climate-change hysteria is essentially a form of secular religion. If this is indeed the case, though, then heretics to the cause would obviously have to be punished – and, unsurprisingly, they often are.

Environmental Economics

One excellent example of a climate-change heretic being burned at the stake (presumably using some kind of low-emission, smokeless fuel) is that of Matt Ridley (b. 1958), a former Science Editor of *The Economist* and currently an opinion-columnist in *The Times*. Like his namesake Bishop Ridley (*c.* 1500–55), the sixteenth-century English Protestant divine, Mr Ridley has also been persecuted for spreading a message of unacceptable heresy to nonconformists. Initially, he took the idea of climate apocalypse at face value, but gradually began to change his mind about the certainty of the science involved. In a 2015 article, he explained his current position thus:

> I am a climate lukewarmer ... I think recent global warming is real, mostly man-made and will continue, but I no longer think it is likely to be dangerous and I think its slow and erratic progress so far is what we should expect in the future. That last year [2014] was the warmest yet, in some data-sets, but only by a smidgen more than 2005, is precisely in line with such lukewarm thinking.[56]

In Ridley's view, 'No prediction, let alone in a multi-causal, chaotic and poorly understood system like the global climate, should ever be treated as gospel' – and yet it very often is. Ridley's honest scepticism has led not only to abuse from those who disagree with him, but also the apparent sabotage of his place in public life. Ridley is a Member of the House of Lords and, given his scientific background, has occasionally applied to fill certain unpaid public-sector roles in which his training would prove useful. However, he seems sometimes to have been excluded from making the shortlist for such appointments – which, it is important to note, *were not even directly climate-related in their nature* – purely because of his lukewarm attitudes towards the topic. As Ridley has said:

> In the climate debate, paying obeisance to scaremongering is about as mandatory for [obtaining] a public appointment, or public funding, as being a Protestant was in eighteenth-century England.

And scaremongering, he says, there certainly is. The Intergovernmental Panel on Climate-Change (IPCC) actually gives a range of possible consequences of global warming, running from the virtually harmless to, as Ridley puts it, 'the terrifying'. As he explains, however, most Western politicians ignore the less sensationalist options on the IPCC's menu and go straight for the 'terrifying' option. He quotes the US Foreign Secretary John Kerry, for instance, as calling climate-change 'perhaps the world's most fearsome weapon of mass destruction' – and we all know how good American politicians are at recognising *those* when they see them. The most interesting accusation made against Ridley, though, is that, as he says, 'I stand accused of "wanting" climate-change to be mild because I support free markets', as exemplified by the fact that he owns land in Northumberland upon which two coal mines currently operate; in October 2015, eco-protestors calling themselves 'Matt Ridley's Conscience' invaded one, saying the mine only 'lines the pockets of millionaires like the climate-sceptic Lord Ridley'.[57] (Ridley receives a fee for allowing coal mining on his estate, but, so as to avoid charges of hypocrisy, has also consistently refused to allow wind-farm developers and solar-panel operators to build their devices there, believing the government subsidies which he would receive for doing so to be a waste of public funds – which doesn't sound like the attitude of a shameless money-grabber to me.)

When that arch-free-marketeer Ronald Reagan (1911–2004) came to power in 1981, he initially considered doing without any official White House science advisers because, in the words of one of his aides, 'We know what we want to do anyway, and they'll only give us contrary advice.'[58] Such cynical thinking didn't die with Reagan, and doubtless some right-wing climate sceptics do indeed actively desire global warming to be mild for reasons of both profit and ideology – but if Matt Ridley can stand accused of wanting climate-change to be a damp squib because he believes in free markets, then the flipside is that some left-wing environmentalists can equally be accused of wanting it to be catastrophic because they do *not* believe in free markets. A call for less consumption, of both energy and consumer goods, must inevitably call at the same time for a consequent restructuring of capitalism, and provides plenty of opportunity for incontinent public emoting about the supposedly wicked ways of Western industrialised nations. The Green Party in Britain, for instance, now contents itself not only with noble ecological causes such as saving the trees and the whales, but has also started pushing a series of absurd economic and social policies which veer so far to the Left they would make even Jeremy Corbyn blanch.

Both of the most extreme viewpoints upon this matter, while based ultimately upon a competing interpretation of genuine scientific data, function simultaneously also as myths. A position of utter confidence upon such an incredibly complex and uncertain issue at this point in time strikes me as being less truly detached in its nature than we are always being told science is. Matt Ridley's latest claim at time of writing, for instance, is that increasing levels of CO_2 might actually be good for us, as it increases crop yields; he cites several apparently reputable authorities (including one of the founders of Greenpeace!) who would back him up on the issue. This was actually once the mainstream opinion. The Swedish chemist Svante Arrhenius

(1859–1927), then the leading expert on such topics, said as long ago as 1908 that more 'carbonic acid in the atmosphere' would 'bring forth much more abundant crops ... for the benefit of rapidly propagating mankind'. How times change. A prominent French TV weatherman, Philippe Verdier (b. 1968), was taken off-air in October 2015 for daring to suggest that there might be just such 'positive consequences' to climate-change.[59] This is nothing less than censorship, even if Verdier should turn out to be utterly wrong in his opinion. My point is not that the idea of climate apocalypse is *necessarily* a fiction, but that the society we currently inhabit would not allow us to admit so, even if it was.

Change Is a Constant

The average person can't reasonably be expected to keep up with all this stuff. Fortunately, the notion of supposed 'scientific consensus' on climate-change – the kind of consensus so ceaselessly promoted to us by our governments – means that most of us don't feel we have to. The fact that there *is* no absolutely universal scientific consensus upon the matter is thus rendered entirely irrelevant; the current social meaning of the data is allowed to simply overrule its awkwardly uncertain scientific meaning. You could justifiably argue that the IPCC's worst-case scenario *might* be true, so it would only be sensible for us to prepare for the worst – but actual debate seems to be the last thing which is ever encouraged here. Such a gross oversimplification of the issue gives us a blissful freedom from ever having to think about it properly. For John Gray, that scourge of liberal humanist thought, this is one of the main human needs that science currently fulfils in the West. It gives an illusion of stability to our worldview which many find comforting. As Gray has it:

> ... science alone has the power to silence heretics. Today it is the only institution that can claim authority. Like the Church in the past, it has the power to destroy, or marginalise, independent thinkers ... In fact, science does not yield any fixed picture of things, but by censoring thinkers who stray too far from current orthodoxies it preserves the comforting illusion of a single established worldview ...[This] is undoubtedly the chief source of science's appeal. For us, science is a refuge from uncertainty, promising ... the miracle of freedom from thought; while churches have become sanctuaries of doubt.[60]

The idea that doomsday is around the corner might be frightening, but not half as frightening as the idea that we may not actually *know* whether climate apocalypse is coming or not. Matt Ridley's honestly held view that 'no prediction ... should ever be treated as gospel' thus actually becomes far more of a heresy to the modern mindset than a very firmly held opinion that climate-change is all just a bunch of pinko nonsense. At least the person who holds such a cartoonish view is still upholding the notion of scientific certainty; Ridley's mode of thought just doesn't fit in with our current social myth of the absolute certainty of science. We can see this in the way that so many people moan when the TV weather forecast proves incorrect. In actuality, this is merely an estimate of probability about what the weather will be, not an accurate prediction of absolute certainty. Some forecasters would like to make this clearer, saying things like 'there is a 62 per cent chance of rain tomorrow afternoon', but broadcasters consistently overrule them. Apparently, viewers don't want to be given probability estimates, they want to be given accurate, certain information – even when it does not exist. In short, they actively *want* to be lied to; lies are so much easier to grasp.[61]

Ignorance Is Bliss

A brilliant illustration of the general public's woeful level of scientific ignorance was provided by a 1997 Channel 4 comedy show called *Brass Eye*. Presented by the satirist Chris Morris (b. 1962), *Brass Eye* was a fake news programme in which scripted sketches and news reports alternated with genuine footage Morris had obtained of public figures whom he had fooled into saying bizarre things. Most relevant was the 'Science' episode, in which the TV racing pundit John McCririck was encouraged to give his opinion upon genetically engineered crabs giving birth to human babies, and the weather girl Tania Bryer fooled into warning viewers about a new plague of 'mutant-clouds' raining water upside-down into the sky, thus causing drought in Africa. The most elaborate scam focused upon something called 'heavy electricity' which was allegedly falling out of wires in Sri Lanka and squashing children, leaving them only eight inches tall. According to sitcom actor Richard Briers, being struck by heavy electricity was like 'being hit on the head by a tonne of invisible lead soup', which sounded most disagreeable. 'How can we sit around eating pies?' whilst this plague continued, asked Briers, outraged, demanding on camera that something be done about it immediately. Soon, something *was* being done about it; Morris tricked several of his dupes into sending protest letters off to the Sri Lankan Embassy in London, demanding that they act.

Worse, it appeared that the scourge was about to hit here, too; according to gameshow hostess Jenny Powell, the UK could soon be struck by an outburst of heavy electricity so massive that it would knock our entire island underwater. 'I should say the chances of this are negligible, about 20 to 1, but the point remains,' said

Powell, thereby proving an entirely different point about many people's absolutely terrible understanding of the science of statistics. Seemingly, celebs were happy to just take Morris at his word in his claims, no matter how preposterous. 'Non-scientific contributors were sure to check out all the facts,' said Morris in voiceover, across footage of daytime TV star Nick Owen reading a piece of A4 paper in a park and saying 'Heavy electricity? Is that what it's called?' Clearly Mr Owen did not check his facts that carefully, however, seeing as he then began emoting powerfully about the 'fact' that heavy electricity had recently dropped out of a wire onto some Sri Lankan cows, flattening them like pancakes as something called 'sodomised electrons' rushed up straight to the cow's 'head-end' (a technical anatomical term, apparently). Simply by whispering the magic word 'science' into these people's ears, Morris was able to trap them into saying and believing the most absurd things. It does tend to have that effect sometimes ...

In the same programme, Morris also managed to successfully persuade the Sunday newspaper editor Eve Pollard that side effects from poorly regulated Russian chemical weapons research had recently resulted in a Siberian woman giving birth to a two-foot-long testicle. According to Pollard's piece to camera, the chemicals released by the Russian military had been disrupting organ precursor cells in human embryos, chemically switching them so they all ended up making the same body part – in this case, a giant human ball. Over grainy footage of the gland in question being kept in a baby's cot and patted by a sympathetic medic, Pollard implied the thing was somehow sentient. While it had no mouth, and so could not be fed, a clearly distressed Pollard informed viewers that the testicle nonetheless had the capacity to feel constant pain. It would only live for two or three days, she said, but perhaps that was simply a blessing in disguise.[62]

Eve Pollard was an actual newspaper editor, not a humble gameshow hostess. And yet, she still fell for it. Label something as 'science' and, no matter how absurd, you can bet that someone, somewhere, will be willing to believe in it. When asked in 2007 whether people like Pollard should not have been able to spot straight away that they were being asked to talk rot, Morris replied, 'They should really, because ... there's no fun in so perfectly mimicking reality that you pass off something that's made up that might as well be real. Whereas if all the labels on it are saying "This is fucking nonsense," then you've got a gap to play with.'[63]

If there was indeed a gap between reality and fantasy in the mad things Morris got his victims to believe, then they singularly failed to mind it. In another episode of *Brass Eye*, radio DJ Neil Fox was memorably persuaded to knock a nail into the shell of a dead crab to hammer home to worried viewers the idea that such creatures supposedly shared large amounts of genetic material with paedophiles. Fox then helpfully backed this insane assertion up by intoning the phrase, 'Of course, there's no real evidence for this – but it is SCIENTIFIC FACT.' It's hard to add anything to that, really.

Gene Genies

Even genuine scientists can sometimes fall victim to believing unlikely things which happen to fit in neatly with the mood of the times. Often, such missteps can best be categorised under the heading of 'hubris'. One recent example centred upon the Human Genome Project, a kind of biological map of all our species' genes, the completion of which was undoubtedly an incredible technical achievement. The problem around which all the subsequent hubris centred was not the genome map itself, but the exaggerated claims made about what mankind would soon be able to do with it. When the map was first published in June 2000, for instance, the then US President Bill Clinton (b. 1946) held a White House press conference where he implied that it would enable us to treat and cure 'most, if not all, human diseases', and even prevent some of them breaking out in the first place.

Sadly, however, this medical utopia has as yet singularly failed to occur. Nobody ever really thought that within two short decades mankind would be declared entirely disease-free, but progress towards this point really hasn't got as far as many commentators at the time might have hoped. Disappointingly, the presumed underlying genetic basis for many diseases has not yet been found, perhaps because data from the genome map is so incredibly complex or, as some more pessimistic souls have suggested, perhaps because many of those diseases might not actually have any underlying genetic basis in the first place.[64] A layman cannot be expected to know the true answer to this conundrum – but what we can be expected to recognise is an instance of hyperbole when we see it. At the time, the prestigious scientific journal *Nature* was so enthused by the work of the Human Genome Project that one of its editors predicted that, by the end of this century, the new science of genomics would 'allow us to fashion the human form into any conceivable shape. We will have extra limbs, if we want them, and maybe even wings to fly.'[65]

Maybe we will be able to make a race of giant sentient testicles, too? Biologists themselves differ in their views about what the genome map will allow them to pull off. The South African-born biologist Lewis Wolpert (b. 1929), for instance, believes firmly that the human genome explains, in principle, the entire subsequent organism, so that by analysing a fertilised egg we will one day be able to entirely predict what the ensuing baby will turn out like. Then, doctors could easily step in to alter the egg, ironing out any genetic flaws so that the future infant will be perfectly healthy – or, theoretically, so that it also has wings or extra legs, presumably.[66] While this is not an uncommon viewpoint, other experts would disagree. In 2009, Wolpert met up with a dissenting biologist, Rupert Sheldrake (b. 1942), to debate the matter as part of the Cambridge Science Festival. According to Sheldrake, the whole idea of being able to entirely predict the baby from the egg was a virtual impossibility because, among other things, any future doctor would need to be able to 'predict the structures of cells on the basis of the interactions of hundreds of millions of proteins and other molecules, unleashing a vast combinatorial explosion, with more possible arrangements than all the atoms in the universe'.[67]

If Sheldrake is right, then such a thing seems about as impossible as knowing 100 per cent for sure the extreme long-term future of the Earth's climate; but similar fantasies of human omniscience have popped up repeatedly throughout the history of science. Sheldrake himself provided one particularly absurd example in his recent book *The Science Delusion*, quoting the eminent Victorian scientist and promoter of Darwin T. H. Huxley (1825–95) to the effect that:

> If the fundamental proposition of evolution is true, that the entire world, living and not living, is the result of the mutual interaction, according to definite laws, of the forces possessed by the molecules of which the primitive nebulosity of the universe was composed, it is no less certain the existing world lay, potentially, in the cosmic vapour, and that a sufficient intellect could, from a knowledge of the properties of the molecules of that vapour, have predicted, say, the state of the fauna of Great Britain in 1869.[68]

If this was true, however, then how much information would you need in order to be able to do so? Even presuming that Huxley was correct in principle, in practice the mathematics of such a feat would be so complex you would need a supercomputer as large as the known universe to process it all. The field of science which is most honest about this fact is quantum physics – an area so notoriously complex that, as the old saying goes, 'If you think you've understood it, then you haven't understood it.' The most famous quantum principle which seems to refute the notion of hard-line, Huxley-style scientific determinism is known as 'Heisenberg's Uncertainty Principle', after the German physicist who formulated it. The idea is well known, and states that, when it comes to the subatomic world, no observer can hope to know the precise position and speed of a particle at any one time. The more precisely we calculate the one, the less precisely we can calculate the other; so, we can know its speed but not its location, and vice versa. There is then, as it has often been put, a certain 'fuzziness in Nature' – or at least in Nature as we are currently able to observe it.[69] Heisenberg, after formulating this idea, told one colleague, 'I have refuted the law of causality!' 'As a matter of principle,' he later wrote, 'we cannot know all the determining elements of the present.'[70] In fact, Heisenberg's claims to have 'refuted the law of causality' were slightly overstated; they just rendered it impossible to work out fully in practice. If you were hubristic enough to try and predict what a certain specific particle would be doing in a hundred years' time, then you would obviously have to know such things as where it was now, what its velocity was, in which direction it was travelling and so forth. You *can't* know all these things, though, said Heisenberg – you could guess, but, like the weather forecasters forced to lie to us on TV every evening, it would simply be a matter of probability.

A Burning Question

Such ideas can seem disturbing to us, as they go against the currently prevailing social myth that science will one day be able to solve all our problems, and formulate a picture of the world that is both eternal and unchanging. Only religion can ever hope to fulfil the latter part of that bargain but, as has often been observed, science can plausibly be viewed as being a kind of modern, secular form of religion with scientists themselves filling the role of high priests – high priests whose function is to condemn those who put forward new and contentious theories in such a way that, as John Gray said earlier, is similar to that once pursued by the Catholic Church against heretics.

The clearest example of this kind of behaviour at work in recent decades occurred in 1981 when Rupert Sheldrake's book *A New Science of Life* was first published. To simplify, Sheldrake proposed that Nature had a kind of memory, with its supposedly eternal laws really being more like learned habits; so a rose grows in the shape it does due to the earlier way that all other previous roses have grown in the past. A rose, it might be said, has an inherited habit of 'rose-ness'. Whether the theory was true or not, it certainly captured the public's imagination, and in a now infamous issue of *Nature*, the journal's editor, Sir John Maddox (1925–2009), declared that Sheldrake's book ('an infuriating tract') was 'the best candidate for burning there has been for many years' – a fate which, he was careful to point out, he didn't think even Hitler's *Mein Kampf* deserved. Sheldrake's book, creditably, contained a section outlining several experiments which could be performed to test his hypothesis out – tests Maddox condemned as 'impractical', not in the sense that they could not be performed, but 'impractical in the sense that no self-respecting grant-making agency will take the proposal seriously'. Or, in other words, why bother to test out a strange new idea when you can just automatically dismiss it instead? As Brian Josephson (b. 1940), recipient of the Nobel Prize for Physics in 1973, later wrote in to complain, in such phrasing *Nature*'s editor showed 'a concern not for scientific validity but for respectability'.[71] In a 1994 BBC interview, Maddox rather backed this impression up, saying, 'Sheldrake is putting forward magic instead of science, and that can be condemned in exactly the same language that the Pope used to condemn Galileo, and for the same reason. It is heresy.'[72]

Then why not test it out and disprove it? Perhaps part of the answer lies in the fact that Sheldrake did later get to see a number of his proposed trials be performed – doubtless funded by some grant-making organisation with extremely low levels of self-respect, but rather higher levels of intellectual curiosity – and achieve some results which he interpreted as being positive in nature.[73] Many of these experiments were very entertaining; inconclusive attempts were made to see whether crosswords were easier to complete the day after they had been published, for instance, when the 'habit' of other people solving them had already been developed.[74] Sheldrake's son Merlin and his schoolfriends even tackled their GCSE papers backwards on the grounds that the earlier questions, having been attempted previously by other candidates, would thus have been easier to answer![75] Most famously, Sheldrake also interpreted various examples of animal behaviour – like the way that various sheep across Britain independently learned how to roll across cattle grids and away to freedom – as being evidence for his novel idea.[76] Alternative explanations are perhaps available for such phenomena, but there is no doubt that Sheldrake's researches have helped highlight some apparently genuine natural anomalies. No wonder Maddox wanted his book burned.

The Power of Positive Thinking

Proposing the burning of heretical texts is only one side of the religion of science. A much odder aspect can be found in the thinking of Auguste Comte (1798–1857), founder of a school of thought known as 'Positivism'. In essence, this philosophy proposed that there were three distinct stages to mankind's intellectual development. First was the Theological Stage, when man knew no better than to attribute all events to the actions of gods and spirits. Then came the Metaphysical Stage, when invisible but inanimate forces like gravity were understood to govern everything, with the universe recast as a gigantic machine. Finally, there was the Positive Stage, which during Comte's lifetime was really just beginning; by this point, man knew enough about science to take control of the world he lived in and make life better for all. For Comte, this three-stage development of civilisation was an actual scientific law; he drew upon Darwin to declare the 'Ascending Evolution of Humanity' was taking place. As his new motto for the Positive age, Comte adopted the following insight:

> The whole succession of man, during the long series of ages, should be considered as One Man, who continues to live and who continually learns.[77]

This 'Positive Stage', of course, was simply an early version of the unacknowledged religion of liberal humanism, given a fancy name. The main difference is that for Comte the religious aspect of humanism did *not* go unacknowledged. Instead, recognising that the absence of God from his worldview left it a little arid and unsatisfying emotionally, he decided to set up what he called 'the Religion of Humanity' as a remedy. This was no metaphor. Comte made serious plans to convert Paris' Notre Dame Cathedral into a Temple of Reason, which he planned to call the Central Church of the Occidental Republic. Inside, he intended to hold bizarre rituals in which a legion of specially selected humanist saints like the Scottish economic theorist Adam Smith (1723–90) would be worshipped instead of the usual gang of God-botherers like St Peter and St Michael. Most tellingly, Francis Bacon would become a kind of new secular Messiah, presiding over mankind's scientific regeneration.

Furthermore, traditional saints' days were to be replaced with a brand-new 'Positivist Calendar', a deeply confusing roll call of people from history whom Comte thought had advanced human evolution, in which, for example, the ninth month (of thirteen) was called 'Gutenberg', the first week of that month 'Columbus', and that week's first day 'Marco Polo'. So, instead of saying it was 1 September, you could say 'Today it is Gutenberg, Columbus, Marco Polo', a truly stupid idea. The final month, as the culmination of the year, had days named purely after famous scientists, representing science's status as the final end point of human evolution. Comte even devised a series of 'logical' sacraments to mark the key turning points in any person's life. In his view, a person could only really be initiated into humanity properly aged fourteen, as small children were very silly, and even grown men only achieved full humanistic maturity at the age of forty-two, an event which also had to be ceremonially marked. At death, meanwhile, instead of a funeral there was to be a secular 'Ceremony of Transformation' at which, rather than commemorating a person's ascent into Heaven, his or her family members were supposed to recognise that their corpse would soon be rotting down into mere atoms. Officially, a dead person would be transformed 'from objective presence to affectionate memory' and, seven years later, would get their name inscribed in a 'Public Roll of the Worthy'. So confident was Comte that this new scientific religion was better than the old godly ones that he actually wrote to the head of the Jesuit Order asking if he would like to sign up to it, but it seems he preferred worshipping Christ, not Comte.

Although an atheist since fourteen, Comte approved of the basic unifying spirit of New Testament Christianity, thinking Christ was an early proto-Communist who preached 'equality and fraternity'. While the dogma of Catholicism was wrong, he said, and involved praying to mere plaster images, if allied to science, technology and politics, its old monastic spirit could become a new kind of 'social science' – an infallible way to run the world properly, for the benefit of all. Instead of preaching about Heaven, why not have Heaven on Earth instead? Everything in our world and our minds was reducible down to number and mechanics, Comte thought, so surely if we just made the right calculations, then all would be perfect. Scientific thinkers, he said, must become both spiritual leaders – telling people to contemplate the beauty of the stars through astronomy – and politicians, developing government into an 'exact science'. Proposing the existence of a kind of 'Social Physics', Comte ultimately became the founder of sociology, to the eternal gratitude of *Guardian* readers everywhere. Feeling that he had 'evolved' since quitting smoking and coffee, Comte decided to take up a life of monk-like asceticism. Hoping to live to a hundred and witness 'the public inauguration of the Religion of Man' occur in Paris, Comte had only two meals a day, one of which was basically just milk, and the other of which concluded with the ritual consumption of bread, in solidarity with the starving millions who had none. Weirdly, he began a daily ritual of reciting sections of the medieval mystic Thomas à Kempis' (*c.* 1380–1471) *Imitation of Christ*, but with the words 'Jesus' and 'God' crossed out and changed to 'Humanity'. Before bed, meanwhile, he would read Dante's *Divine Comedy*, which he saw not as a journey from Hell to Heaven via Purgatory, but through the three stages of mankind's evolution.

Comte's commitment to the quality of rationality was so extreme that, ironically, he eventually became insane. After marrying a friendly teenage prostitute named Caroline Massin, Comte began to suffer the delusion he was an ancient Greek hero and started throwing knives at her in their Paris apartment. Following a spell in an asylum, Comte's mother suggested her son's madness was a punishment from God for the couple undergoing a civil marriage, not a Church one, and Caroline agreed to remarry the madman in a ceremony conducted by a Catholic priest. Outraged, Comte signed his name in the marriage register as 'Auguste Comte, Brutus Bonaparte' and jumped into the River Seine. Soon, he turned against his spouse, separated from her, and began filling his numerous books and pamphlets with totally irrelevant descriptions of her former exploits in whoredom. Styling himself 'the Grand Priest of Humanity' – how he signed all his letters – he eventually took up with one Clothilde de Vaux, a much younger woman who, perhaps sensibly, consistently refused to have sex with him. Seeing in this woman's virtue a real-life counterpart to the (in his view) fictional Virgin Mary, after Clothilde's early death in 1846, Comte decided to make her another humanist saint in his Religion of Humanity, taking the extreme step of designating her favourite chair the official Altar of the Virginal Deity and worshipping it as the earthly bum-supporting relic of a real-life goddess. Prior to his death, Comte collapsed before the chair and refused to be moved for some time, declaring, 'I shall lie in my tomb' next to the sacred object. The major trouble with hymning Clothilde's virginity in this way, though, was that she was not actually a virgin at all, and had in fact already been married – she just refused to sleep with mad Monsieur Comte. No wonder he liked to worship her as an icon of rational thought.[78]

Social Engineering

In spite of his obvious insanity, Comte did have his disciples, including such modern-day idols of rationality as the English libertarian philosopher John Stuart Mill (1806–73), who liked Comte's (literally) progressive politics but eventually fell out with his mentor after refusing to address him as 'the Founder of the Universal Religion and Grand Priest of Humanity' in their correspondence.[79] Mill was the ultimate godfather of what later became liberal humanism in this country, and is now afforded by some a status nearing that of one of Comte's secular saints.

There were other people at the time who tried to make a religion out of science and engineering, too. Comte's early guru Henri de Saint-Simon (1760–1825), for example, had spoken of his followers as 'evangelists for the engineer' and 'apostles of the religion of industry' when setting up his weird quasi-socialist creed of New Christianity. According to the historian David F. Noble, Saint-Simon possessed 'the Baconian vision of redemption from labour through science', and it is hard to disagree.[80] Saint-Simon wrote pamphlets with titles like *The Workers' Political Catechism*, and spoke of a forthcoming world of happy workers who would be directed by an 'aristocracy of science and art', and felt that by allying science with Christianity, all war between the nations of Europe would end as they became partners in 'International Parliament' – an EU many years before its time.[81] The fact that Saint-Simon's scientific religion was basically a cult can be seen in what happened to it after his death, when several 'churches' were founded in his name, and one of his followers, one 'Father' Barthélemy Enfantin (1796–1864), began claiming to be the Messiah. Proclaiming he was as great as Moses and Mohammed, he started wearing a bizarre costume involving big white pantaloons and a special symbolic waistcoat which had to be fastened by an assistant from behind, thus demonstrating the need for all men to co-operate, and made calls to abolish marriage and private property. He also played a role in persuading the French diplomat Ferdinand de Lesseps (1805–94) to get on with the 'holy' task of cutting the Suez Canal through Egypt.[82]

Also around this time, the whole idea of a civil engineer as a distinct class of person emerged from Freemasonry, an esoteric movement partially devoted to the idea of creating a kind of 'New Man' and a new world fit for him to live in. Many of the most significant early civil engineers, like Thomas Telford (1757–1834), were also Freemasons.[83] Another evangelist for engineering was Amos Eaton (1776–1842), who wandered through America preaching a gospel of God, science and technology, until he was eventually appointed head of the country's first civil engineering school, the Rensselaer Polytechnic Institute, founded by Eaton's patron Stephen van Rensselaer (1764–1839), a prominent Freemason.[84] In religious iconography, God the Creator Himself had long been depicted as a kind of master craftsman holding compasses and square; now His proudest creation, mankind, would follow suit.[85]

Priests in White Lab Coats

An earlier example of this same kind of pseudo-religious scientific mentality at work can be found in an unfinished work of Francis Bacon, a strange text published posthumously in 1627 called *New Atlantis*. It is an attempt at utopian fiction, an allegory in which some shipwrecked sailors discover an unknown land, the titular 'New Atlantis'. Here, the perfect society had been constructed, ruled over by a scientific priesthood based in an ideal college termed Salomon's House, a gigantic laboratory furnished with all manner of special equipment where all branches of the arts and sciences were diligently researched for the benefit of mankind. The ultimate aims of the institution were hubristic indeed; to gain 'the knowledge of causes and secret motions of [all] things' and 'the enlarging of the human empire, to the effecting of all things possible'. Particularly desirable were 'the prolongation of life' by curing all diseases, and 'the transformation of bodies into other bodies', thereby enabling the artificial 'making of new species'. In New Atlantis, even dead scientists and inventors were treated like kings, being honoured with statues in memory of their contribution to human happiness. Even better, during life they got to wear long robes, like the holy men they were, and their leader was given the privilege of riding through the land in a sort of sun-chariot, symbol of scientific enlightenment, blessing the grateful proles as he passed them by.[86]

Bacon's fantasy may seem comical to us now, but it had real influence at the time. The Royal Society, founded in 1660 and the first such organisation devoted purely to the pursuit of scientific knowledge and research in England, was to a certain extent directly inspired by Bacon's fantasy. In its earliest years, a Baconian emphasis upon the importance of experiment and the consequent collection of data were the Royal Society's main concern, but some people, when they heard about its founding, slightly got the wrong idea. At the time, there were several people who automatically associated the Royal Society with a pan-European movement known as Rosicrucianism. This centred around a mythical figure named Christian Rosenkreutz who had reputedly devised a plan for a form of perfect universal religion in which science, politics, art and philosophy were all combined together to create a kind of Paradise on Earth. Supposedly, in 1407 Rosenkreutz had founded a secret order called 'the European Fraternity of the Rosy Cross' ('Rosenkreutz' means 'rosy cross', and a cross intersected by a rose was the society's symbol), whose members pledged to spend their lives hunting out the secret laws of Nature for the betterment of humanity, travelling across Europe in disguise, healing the sick and seeking new recruits. Because of their highly secretive nature, this benign Brotherhood later became known as the 'Invisible College'; a highly appropriate name, given that they probably didn't exist.[87] Interestingly, it was to a certain extent from Rosicrucianism that Freemasonry – and thus the art of engineering – ultimately sprang; of the Royal Society's first 250 Fellows, at least 89 were Freemasons, including our old friend, the dog-torturing Sir Christopher Wren.[88]

One other celebrated Englishman who apparently agreed with the Rosicrucians' basic aims was Francis Bacon. In 1972, the English historian Dame Frances Yates (1899–1981) published her classic book *The Rosicrucian Enlightenment*, in which she pointed out certain similarities between Bacon's Holy Order of priest-scientists in *New Atlantis* and Rosenkreutz's own supposed band of brothers. One of Bacon's Atlantean officials, for instance, wears a turban emblazoned with a small red cross – a Rosicrucian symbol – and the ruling caste themselves are described as going out into the world invisibly to collect knowledge, adopting the dress of the far-off lands through which they wander in order to blend in with the natives, just like the Rosicrucians were meant to do. Some people have used these facts to argue that Bacon was secretly a Rosicrucian himself, though actually there is no real evidence that he belonged to the brotherhood, merely that he knew about their legend and later allegorised it.[89]

The Invisible Men

It appears that other early scientists flirted with Rosicrucian ideals too – in particular, the founders of the Royal Society. In the years prior to the Society's foundation, many of the men who ended up as early recruits were meeting in private discussion-groups in any case, assemblies which were, significantly enough, described by the great chemist Robert Boyle, whom we met earlier, as gatherings of an 'Invisible College'.[90] In 1648, a number of these invisible people began congregating in the rooms of a Professor of Geometry named John Wilkins (1614–72) at Wadham College, Oxford. Wilkins was the author of a book, *Mathematicall Magic*, which mentioned Bacon frequently. The book also frequently mentioned such now naive-sounding topics as automata, or 'mechanical men', and talking statues, wonders which Wilkins thought men had once been able to build in the past, and which greater knowledge of science might one day allow men to recreate in the future. Wilkins' room itself contained a hollow statue within which the ingenious mathematician had managed to conceal some kind of long pipe, allowing an accomplice to make it appear as if it could speak. It wasn't an actual robot, but at least it was a start.[91]

Other like-minded souls met up in London's Gresham College, and it was during the aftermath of one of these conferences, in 1660, that the Royal Society itself was formally instituted, and the Invisible College became suddenly visible. In 1661, one of its members, a Freemason named Sir Robert Moray (1608–73), petitioned King Charles II for a royal charter for their new club, which was granted in 1662.[92] While this new organisation certainly counted both Bacon and Rosicrucianism among its direct intellectual forbears, in reality its aims were somewhat more limited than creating a genuine New Atlantis. After all, these were the years following the Civil War of 1642–51, when schemes for the restructuring of society were considered dangerous ones to have. Thus, Bacon became more fêted among its members for his focus upon the experimental method than for his alleged sympathy with Christian Rosenkreutz – at least openly.[93]

Not everyone quite understood these subtleties, however, and seemed to think of the society as being a real-life Rosicrucian Lodge in all but name. One such man was Johann Amos Komensky (1592–1670), known by his pen name of Comenius, a Bohemian (or Czech) mystic who had himself tried to create a system of universal knowledge he called 'pansophia', after Sophia, the ancient Greek goddess of wisdom.[94] Comenius' idea of pansophia grew directly out of his encounters with Rosicrucianism, and in 1668 he published a book in Amsterdam, *The Way of Light*, which he dedicated to the Royal Society, crediting its members with setting out to 'secure ... the empire of the human mind over matter'.[95] What? Over *all* matter? Perhaps this was just rhetoric upon a newly enthused Comenius' behalf. In 1623, he had written another text called *The Labyrinth of the World*, in which he painted a picture of a depressed and demoralised city, a confused labyrinth in which the promises laid out by the Rosicrucians had led only to failure and disappointment. This allegory is like a dystopian twin to Bacon's *New Atlantis*, in which the scientific priesthood have their robes ripped from them, and their false and hubristic promises laid bare. In it, Comenius tells of how he is walking through the marketplace one day when he hears a trumpet signalling the arrival of a man on horseback, who comes to tell the market-day crowd of the incredible achievements of 'some famous men', the Rosicrucians, who had 'remedied' all the world's 'insufficiencies' and 'raised the wisdom of man to that degree which it had in Paradise before the Fall'.[96]

Backwards Thinking

This idea may now seem odd, but we must remember that at the time history was envisioned as being essentially cyclical in nature. Instead of the endless progression of mankind along a straight line towards a better future, which is the firm and sure path we are taught science is leading us down today, to many Renaissance minds the idea of going forward in the spirit of discovery meant in fact only to go backwards towards the rediscovery of lost knowledge. For Francis Bacon, the point of science was not to arrive at an endpoint of absolute perfection and total scientific understanding of the universe, but instead to arrive back at the beginning of time, when humanity had that same knowledge anyway. Thus, for mankind to achieve mastery over Nature through science was only for him to return back to the original state of Adam in the Garden of Eden, a process Bacon called the 'Great Instauration'. Man's increasing knowledge and command over Nature was entirely justified for Bacon by an extract from the Book of Genesis in which God asks Adam to invent a name for 'every living creature'.[97] Through naming, taming and understanding Nature in this Adam-like way, argued Bacon, humanity would eventually be able to re-enter its original Edenic Paradise. You may have thought that the only reason Adam and Eve were thrown out of Eden in the first place was because they had tasted the forbidden Fruit of Knowledge at the behest of the Serpent, of course – but evidently Bacon disagreed. As he famously had it, 'Knowledge is Power'; power over Creation itself.[98]

Actually, this wasn't an entirely new idea. Johannes Scotus (*c.* AD 815–*c.* AD 877), an itinerant ninth-century Irish philosopher, had previously argued something very similar. To Scotus, no new discoveries of any kind were, strictly speaking, even possible; Adam had once had mastery over all technology, and so whenever we learn or discover any new skill, he wrote, 'we do nothing but recall to our present understanding the same arts which are stored deep in our memory' from the time of Eden.[99] When Christopher Columbus (1451–1506) first set foot in the New World in 1492, meanwhile, he genuinely thought he had rediscovered Eden and set about naming everything he saw, whether plant or animal, hoping thereby to restore human dominion over Paradise.[100] Modern-day taxonomists do the same, classifying plants and thereby aiding pharmacists hoping to transform them into medicine to expand our life spans.

A Biblical prophecy from the Book of Daniel,[101] meanwhile, suggesting 'many shall run to and fro, and knowledge shall be increased' come the end times, had long been taken to mean that, the more that scientific understanding, trade and exploration grew, the closer we were to Doomsday and the subsequent resurrection of the righteous in a state of Heaven on Earth. Advances like gunpowder and the printing press were taken as signs of the imminent Second Coming, as was the discovery of new lands by Columbus in his new, super-advanced sailing ships. To a man like the English poet John Milton (1608–74), such developments were a clear signal that mankind would soon progress from 'command of the Earth and seas to dominion over the stars', a necessary progress if we were to prepare ourselves properly for the return of our godlike powers in Eden. Indeed, if man was going to conquer the stars, then perhaps he might find that Eden was actually located in outer space; in his *Apology for Galileo*, the Italian mystic Tommaso Campanella (1568–1639) suggested that Adam had actually lived on the moon, which was high enough to have survived the Great Flood. Campanella speculated the Moon-Eden must have a warm climate, because its original two inhabitants had been able to walk around up there naked! Bacon himself genuinely believed in the Daniel prophecy, and reprinted it in some of his books, showing just how millenarian his own outlook on the world was; indeed, he specifically described his book *The Great Instauration* as an 'apocalypse'.[102]

The New Adam

Exploiting technology to make man 'fall back upwards' into an Adam-like state was the obvious next step. Bacon's disciples began to think that, by creating new scientific instruments, mankind would be able to restore the perfect physical and intellectual faculties he had once enjoyed before the Fall. Even better, maybe technology would one day allow us to *exceed* Adam's old powers?[103] When the Royal Society's chief dog-dissector Robert Hooke got hold of a new-fangled microscope during the 1660s, for instance, he argued that this was not an advance at all, simply a return back towards the superb level of eyesight which Adam himself had once possessed.[104] Another Royal Society man, the enthusiastic ghost-hunter Joseph Glanvill (1636–80), agreed. In Eden, Glanvill explained:

> the senses, the Soul's windows, were without any spot or opacity ... Adam needed no Spectacles. The acuteness of his natural optics showed him most of the celestial magnificence and bravery [stars and planets] without a Galileo's tube [telescope] ... While man knew no sin, he was ignorant of nothing else.[105]

What, exactly, would this new Adam, this new-fangled thing we would later come to term a 'scientist', look like, though? What precise powers, alongside superb eyesight, would he possess? The Rosicrucian in Comenius' allegory claimed such people's skills would be amazing indeed:

> To make gold, he said, was one of the smallest of their hundred feats, for all Nature was bared and revealed to them; they were able to give to, or take from, each creature whatever shape they chose, according to their pleasure ... He said that they had the [Philosopher's] Stone, and could by means of it entirely heal all illnesses and confer long life. For [their leader] was already 562 years old, and his colleagues were not much younger ... they knew that a reformation would shortly befall the whole world; therefore openly showing themselves, they were ready to share their precious secrets with everyone whom they should consider worthy.[106]

Healing all illnesses? Adding features to, or subtracting them from, all living creatures as they see fit? Doesn't that sound familiar to us from somewhere else? The boastful Rosicrucian claims that his Brothers have the power to work such miracles because they have found the famous Philosopher's Stone, a magical wonder-working substance alleged to have the power to impart immortality and cure all ills; modern-day biologists also claim the imminent ability to do something similar using maps of the human genome. In 1655, another Baconian, the Cambridge-educated German émigré Samuel Hartlib (1600–62), wrote his *Chymical Address*, in which he proposed that the new science of his day should ultimately be considered a search for 'the true universal Medicine', or a quest for the immortality Adam had once enjoyed in Eden, by which means 'all men and all flesh shall be delivered from death'.[107] It seems the hunt still continues.

Utopia Is No Place

It has been aptly said that the Baconian philosophy was really just the English equivalent of what continental Europeans termed Rosicrucianism, and vice versa.[108] Thus, when Francis Bacon himself encountered imported Rosicrucian texts, we must presume he was easily able to recognise the basic core of what they were getting at; and that, when European sympathisers with Rosicrucianism read things like *New Atlantis*, they must have been instantly able to understand them, too. Bacon was very specific in his other writings, in particular 1605's *The Advancement of Learning*, that there should be a universal brotherhood in knowledge, that science should be without borders, and that fraternities of like-minded men should be quickly established so as to share their findings with one another.[109] Given that these opinions of Bacon's were well known, it is little surprise that people with a Rosicrucian-type background like Comenius were predisposed to think the Royal Society was effectively Salomon's House made flesh.

Comenius was well aware of the admiration many English intellectuals had for Bacon, seeing as in 1641 he had been invited over to England with the tacit approval of many Members of Parliament, to help try and build a genuinely green and pleasant land for the nation's people to live in. These were the years of the so-called 'Long Parliament', which sat from 1640 to 1660 and whose MPs were given to openly defying their sovereign, King Charles I, something which was to contribute to the outbreak of the English Civil War in 1642. This was a heady time, if soon to be a bloody one, when both religious enthusiasts and some previously level-headed men, optimistic that the tyranny of an era of absolute monarchy was about to end, began speculating in all sincerity that England now stood revealed as being a blessed land, chosen by God to be the real-life location for Francis Bacon's 'Great Instauration' of humanity. In such an atmosphere, Comenius and his pansophist philosophy were taken seriously not as utopian dreaming, but as practical plans for improving the lot of man; in 1640, a sermon was preached in Parliament openly praising Comenius and saying his proposals for a better world ought to be followed. When word of this reached Comenius, then living in Poland, he was, according to Frances Yates, 'overjoyed', misinterpreting the sermon as meaning that he now had an official 'mandate from Parliament to build Bacon's New Atlantis in England.' The only trouble was, of course, that he had no such mandate at all ...[110]

The word 'utopia' had its origins in a book of that same name, published in 1516 by Sir Thomas More (1478–1535), Henry VIII's one-time Lord Chancellor. It is a portmanteau word derived from Greek which means, literally, 'not-place', or 'no-place'. In other words, it is a place which both does not, and cannot, exist – a fantasy.[111] To think otherwise would be to fall victim to hubris, something Comenius would soon discover for himself.

Initially, the Bohemian mystic found a warm welcome in England, being afforded the honour of a splendid banquet laid on by the Bishop of Lincoln. It was during this time, in 1641, that Comenius wrote the book which was eventually published in 1668 as *The Way of Light*, in whose preface he had fulsomely praised the new Royal Society. In this text, the proposals of which he had once hoped would be enacted by Parliament, Comenius laid out his plans for ensuring that, in England's new utopia, 'an Art of Arts, a Science of Sciences, a Wisdom of Wisdom, a Light of Light' would be created. First of all, a series of 'universal books' (textbooks, basically) would be published, and used to teach all men. Then, a 'Book of Pansophia' would need to be written, which would contain all possible knowledge and thus presumably be rather long. A universal language would be required, too, to solve the problem of misunderstandings between nations, and some kind of actual bricks-and-mortar version of Bacon's Salomon's House should also be built and staffed with the very wisest of wise men. The Sacred Order who dwelled within would then direct lesser men's lives, 'so that each of them may know what he has to do, and for whom and when and with what assistance ... for the public benefit'. Thus, a cynic might say, the old tyranny of the king would be replaced with a new and equally comprehensive tyranny of the scientific priesthood. Eventually King Charles fought back against such utopian dreams, of course, and in 1642 Comenius thought it best to leave England to seek safety in Sweden.[112]

Back to the Future

England's Civil War lasted until 1651, and was so brutal that, long before its end, many persons who had previously been utopian optimists began to think that, instead of a New Eden, the realm had become a kind of Hell on Earth. Bacon's 'Great Instauration' seemed as far away as ever – which makes sense, seeing as it was obviously wholly unachievable. You might have thought that Comenius would have realised this fact, given that his utopian fantasies had previously collapsed once before, following the defeat of the Protestant monarch Frederick V (1596–1632), Elector Palatine, by Catholic forces at the Battle of White Mountain in 1620. Frederick had been sympathetic towards the ideals of Rosicrucian-tinged science, and seemed likely to lend patronage and protection to its leading figures, thus ushering in a new golden age across central Europe; but, alas, his defeat put a stop to such fantasies, and the dreadful Thirty Years' War soon brought ruin to much of the Continent. Naturally, Comenius was left despondent by such developments, and in his fable *The Labyrinth of the World*, where earlier on we saw him praising the wonders promised by the Rosicrucians, he saw fit also to record the ultimate emptiness of their promises of universal enlightenment, peace and social well-being. In this latter part of the tale, the Rosicrucians set up a stall in the marketplace and start selling the townsfolk special boxes which, they say, contain all of their scientific wonders – allegories for their mystical books. However, explained Comenius, there was a catch:

> Now everyone who purchased was forbidden to open his box; for it was said that the force of this secret wisdom was such that it worked by penetrating through the cover; but if the box was opened it would evaporate and vanish. Nonetheless, some of those who were more forward could not refrain from opening them, and finding them quite empty, showed this to others ... They then cried 'Fraud! Fraud!' and spoke furiously to him who sold the wares; but he calmed them, saying that these were the most secret of secret things, and that they were invisible to all but 'Filiae Scientiae' (that is, the sons of science); therefore if but one out of a thousand obtained anything, this was no fault of his.[113]

It seemed that, when writing these words in 1623 at least, Comenius felt he had been duped. Evidently at some point he changed his mind, however, as by 1641 he was happy to sail over to England and begin selling quite a few empty boxes to the gullible himself. According to the old Baconian ideal, if an experiment doesn't work then you are supposed to admit the fact and move on, not continue endlessly repeating the same thing forever in the expectation of one day getting a different result; as Einstein was allegedly to point out several centuries later, to do that would be the very definition of madness.[114] There have always been people like Comenius, though, and there always will be. As science advances, some overly optimistic souls will forever see in its wonders cause to acclaim the dawning of an imminent utopia. This is how Comenius described the response of some townspeople to the Rosicrucians' absurd yet appealing promises:

> And behold, some rejoiced exceedingly, not knowing for joy where to go. They pitied their ancestors because, during their lifetime, nothing such had happened. They congratulated themselves because perfect philosophy had been fully given unto them. Thus could they, without error, know everything; without want have sufficient of everything; live for several hundred years without sickness and grey hair, if they only wished it. And they ever repeated: 'Happy, verily happy, is our age.'[115]

That sounds like a pretty good description of modern, twenty-first-century man. Maybe Bacon was right; history does progress along in cycles after all. Like Comenius, humanity endlessly repeats its mistakes – and, given the current scientifically inspired myth of progress as being one long straight line from which U-turns are entirely banned, we are now actively given direct social sanction to do so. 'Happy, verily happy, is our age' is a cry which is supposed to be getting louder with each passing generation. But is it really?

Evolutionary Dead-Ends: Fairies Playing at the Bottom of Darwin's Garden

Creation took place on the twenty-third of October, 4004 BC, at nine o'clock in the morning.

Dr John Lightfoot[1]

In his book *Straw Dogs*, John Gray argues that the theories of Charles Darwin (1809–92) are the single greatest unconscious basis of our modern, secular belief in progress. Western society, it is generally thought, is evolving all the time, thanks to progress in science, politics and ethics – an unacknowledged reflection of the Darwinist creed. However, as Gray shows, there is a hidden but 'exquisite' irony in this mode of thought, seeing as Darwin's work implied, above all else, that mankind was simply another form of animal, descended from a common ancestor shared with apes. In this way, Darwinism appeared to remove mankind's special place in the universe. And yet, as Gray pointed out, 'Darwinism is now the central prop of the humanist faith that we can transcend our animal natures and rule the Earth' through our mastery of science; or in other words, it turns out we *are* special after all![2] We are now so used to Western scientists and politicians speaking and acting as if this was true that the idea of mankind's special nature in the scheme of things has become an issue of unthinking acceptance among us. However, this outlook, while clearly couched in secular and rationalist terms, is ultimately derived from the sphere of religion, not science. It is the Bible which teaches man he has a special place in Creation, not Darwin. The essentially mystical or theological basis of our current myth of incessant and eternal social progress can be made clearer for us when it is expressed in unfamiliar terms by somebody from without mainstream Western currents of thought, a person at whom many of us would probably laugh.

For example, there is a very interesting lady named Svetlana Semenova, a philosopher, literary critic and member of the Russian Academy of Sciences, who has some rather unorthodox opinions about the topic of evolution. Semenova is interested in the issue of 'cephalisation', a genuine evolutionary trend in which, over countless generations, nervous tissue becomes concentrated in one end of an organism until, eventually, this extremity becomes a head with a functioning brain. To Semenova, this phenomenon implies that evolution has a distinct purpose to it; the ultimate goal of increasing the consciousness of living things. Just as interesting is the argument she then builds upon this belief. Seeing as human beings are the creatures in whom cephalisation is most advanced, Semenova says that we thus have a unique responsibility over the future evolutionary development of our universe. Rather than thinking of man in conventional Darwinian terms as just another animal, Semenova prefers to accord us a special place in the cosmos, on account of our equally special brains. She is very critical of the modern environmentalist movement, whom she labels 'eco-sophists' and accuses of excessively romanticising Mother Nature, endlessly protesting against any and all human-driven alterations of our environment, from building dams to redirecting rivers. To Semenova, such processes are a positive thing, a deliberate re-engineering by man of our planet up towards the next stage in its evolutionary path. Eco-sophists, says Semenova, 'invite man to accept the position of absolutely equal rights among all living beings', a weak and defeatist manner of thought which would lead only to 'suicide' for humanity.[3]

At first, this may seem like naive romanticism – yet, examined more closely, her mindset is actually uncomfortably similar to that of many of our own secular Western priests of progress. For example, GM crops, far from the dangerous 'Franken-Food' of popular myth, are constantly (and rightly) being praised by scientists as a more plausible solution to feeding the world's poor than any international aid programmes. Using our highly developed brains and the science which these organs have spawned, mankind is now able to interfere in the evolutionary development of certain crops in a way which will, hopefully, prove to be for our future benefit. This seems to us like a triumph of rationality – but to Svetlana Semenova, the triumph would seem equally to be a spiritual one.

The Darwin Diet

As well as projecting Darwinian theory outwards onto Western society as a whole, however, we have also applied the idea of 'survival of the fittest' to the path followed by science itself; just as with the combat between Drs Beddoes and Jenner, we are told that competing scientific theories are in the uncontrollable habit of fighting one another, with the best always winning out in the end, leading us ever onwards and upwards towards a state of final and ultimate truth and worldly perfection. Thanks to science, we are now supposed to be rapidly approaching the point at which we can begin directing both our own species' evolution and that of other species, too. This is meant to be the 'century of biology', if you believe the hype, and there is often good reason to do so. Biologists really can now alter living creatures' genes to make them behave in amazing new ways; one particularly startling example is that of the so-called 'transgenic' goats which have been produced in America. By splicing their genes with those of golden orb-weaver spiders, scientists have been able to seed the goats' milk with a new form of spider-silk protein, which they then extract and spin into a super-strong fibrous substance with elastic properties termed 'biosteel', which it is hoped will one day be used to make bulletproof vests and artificial tendons for transplant into humans.[4] Even Christian Rosenkreutz would have been impressed by that. My point is not that scientists will not one day be able to guide our evolution to some degree, then, simply that the idea of there being some kind of perfect end point to such a process is nothing but a dream. If dream it be, though, then there have certainly been plenty of people throughout history who have dreamed it. Take Alexander Sukhovo-Kobylin (1817–1903), a Russian playwright who retreated from public life after being falsely accused of murdering a lover. Immensely rich, Sukhovo-Kobylin devoted his new free time to translating the German philosopher Georg Wilhelm Friedrich Hegel (1770–1831) into Russian.

Hegel may be little read now, but his ideas were once – and perhaps still are, unconsciously – hugely influential. Hegel projected a kind of proto-evolutionary theory out onto the realm of human thought, which he termed the 'dialectic'. Simply put, over time people would inevitably come up with certain concepts, in philosophy, science or politics, each of which he termed a 'thesis'. Then, this thesis would be opposed by people who believed the opposite, an 'antithesis'. These two ideas would collide until, eventually, whatever was found to be most true in each camp of thought was extracted and merged into a new and improved idea, termed a 'synthesis'. This synthesis of the two old ideas would still contain errors, though, and eventually ossify into a thesis itself, only to be opposed by new dissenters. The newest thesis and antithesis would then collide and combine into another synthesis, and so on forever, until a state of final and ultimate truth comparable to that contained within the mind of God was reached, which Hegel termed 'the Absolute Idea'. This quasi-evolutionary process, said

Hegel, applied both to politics and science, meaning that progress in society must ultimately go hand in hand with progress in knowledge and invention. Just like Francis Fukuyama, Hegel predicted that one day a kind of 'end of history' would arrive, beyond which no further advance would be either necessary or conceivable. In an amusing example of hubris, Hegel then boldly stated that this state of total perfection had already been achieved, in the authoritarian Prussian state of his own day – a 'utopian' society within which, by a fortuitous quirk of history, he just happened to live himself.[5]

Alexander Sukhovo-Kobylin decided to add to his hero Hegel's philosophy by predicting an end to history of his own. In his view, man's physical and social evolution would only truly conclude when we became a kind of Absolute Idea ourselves – by shedding our corporeal bodies and becoming 'human angels', or pure thought-forms. There were three basic stages in mankind's forthcoming cosmic evolution, said Sukhovo-Kobylin: the 'telluric' stage, when we were confined to planet Earth; the 'solar' stage, when we ventured out into space; and finally the 'sidereal' stage, when we achieved a state of ultimate synthesis with the universe by inhabiting it all at once. By this, he did not mean humans having a base on every planet; instead, he meant it literally, saying each human angel would be able to spread its wings and dwell in all places simultaneously, like gods. To conquer outer space, however, we would first have to conquer actual physical space by no longer inhabiting three dimensions at all. To do this, explained Sukhovo-Kobylin, humanity would just have to start getting smaller and smaller until, eventually, we all simply disappeared. Becoming vegetarian would be a good start, he thought, seeing as eating only lettuce and grass would make us thinner and lighter, thus setting us out on the long road to spacelessness. During the aeons prior to this coming to pass, however, Sukhovo-Kobylin argued that we needed temporarily to use technology to make us less constrained by space. The bicycle, for example, was a new form of 'horizontal flight', and 'a person flying horizontally on a bicycle' a precursor towards our growing actual wings which would one day allow us to fly into outer-space.

We would do well, he said, to observe the humble housefly; because they were so small, they were also 'wonderfully mobile', and, he calculated, could fly one hundred times their own body length in a mere second. Over time, as humanity ate less and less, 'consuming our own spaciousness' while we simultaneously consumed less meat, eventually we too would become capable of such superb feats, developing insect-like wings and so becoming able to 'move through space with the velocity of a cannon-ball'. Eventually, we would shrink so small that we no longer had any dimensional existence whatsoever,

Sukhovo-Kobylin confidently predicted, at which point we would become entirely free from all spatial constraints, in our ultimate evolutionary end form of human angels. Finally, mankind would have achieved the greatest Hegelian synthesis of all – a state of total unity with Creation itself. Eating more lettuce, it seemed, would turn humanity into God. That's quite a diet plan.[6]

Marx and Spencer

All diet plans, no matter how daft, attract their acolytes. Today, Sukhovo-Kobylin's vision of human spacelessness has been unconsciously reconfigured by a breed of technological evangelists called 'transhumanists', who fantasise about one day being able to upload their brains into a computer-simulated universe, achieving non-corporeality electronically. And don't we all, at heart, still believe in the process of Hegelian synthesis, even if nine-tenths of us have never even heard of Hegel? If we do, then it might explain why our society currently so idolises Darwin above any other scientist there has ever been. Not for nothing is his picture on our banknotes; capitalism itself is supposed to be a kind of Darwinism in action. It is now the prime scientific myth of the modern West. A Darwinian nonbeliever, the mystical writer John Michell (1933–2009), once opined that Darwinism has become 'a form of secular religion ... a cosmosgony or creation myth ... [with] the same magical, fairy-tale appeal as all traditional cosmogonies ... These various myths both reflect and influence the cultures they belong to, and determine the ways in which people think and relate to their surroundings. The myth sanctifies the action.'[7]

While I do not agree with Michell that Darwin was wrong in his actual theory, I do agree with the other part of his critique. Darwinism has been remade many times over, adapted and tweaked to serve the interests of many different societies and groups. Nowadays, it is the unconscious motivating myth not only of right-leaning free-marketeers who project the idea of survival of the fittest out onto the companies of the stock exchange, but also of those vast legions of left-leaning politicians and campaigners who proclaim loudly that their policies are 'progressive' in nature, a truly empty word in that context if ever there was one. Do they never stop to ask, 'Progressing towards *what*?' Darwin never claimed evolution had any clear end point to it; neither does society, no matter how much socialists wish to enter into their much-promised (but curiously always absent) New Jerusalem. We ignore that Darwinism's true governing idea is not one of progress towards perfection, but of adaption towards an environment. If the environment any given civilisation occupies were to change in a negative way, then its members could quite easily regress back down to savages; how many liberal humanists do you think are now living under the rule of Islamic State, for instance? The Islamists too have their own motivating myth of 'progress', though. It is very much like a twisted version of that possessed by Francis Bacon, namely a kind of 'Great Instauration' of the world as it (supposedly) once was during the early days of Islam. For jihadis, as for Bacon, evolution is a deliberate movement backwards.

Communism also adopted an evolutionary myth as its underpinning intellectual basis; Karl Marx's (1818–83) famous 'dialectic of history', a clear response to Hegel, promised there were several distinct and unavoidable stages all human society had to pass through, from feudalism to capitalism and then on to full-blown socialism, in just the same way that mankind's earliest ancestors had once had to morph from single-celled organisms into sea creatures, then grow legs and lungs and drag themselves up out of the primordial slime towards the Promised Land of *terra firma*. Marx considered himself in some sense a scientist, and admired Darwin, though he apparently misread him, making the common mistake that his theories implied progress, not adaptation. There is a story that Marx asked Darwin's permission to dedicate a volume of *Das Kapital* to him but was turned down. The tale was untrue, but you can see how it might have arisen. At Marx's funeral in 1883, his colleague Friedrich Engels (1820–95), co-author of *The Communist Manifesto*, gave a eulogy stating that 'as Darwin discovered the law of evolution in organic Nature, so Marx discovered the law of evolution in human history'.[8] If so, then presumably the fall of the Berlin Wall represented some kind of dialectical mass-extinction event.

There is a common misperception that the term 'survival of the fittest' was coined by Darwin. In fact it was invented by the Victorian philosopher Herbert Spencer (1820–1903), who was thought worth a minute's silence by the Italian Parliament when he died, but is nowadays little read. The very definition of an autodidact, Spencer wrote millions of words on many topics. The titles of some of his books are very telling. 1862's *First Principles* spoke eloquently of his absurdly ambitious desire to systematise all human knowledge; it was followed by endless sequels, including *The Principles of Biology* and *The Principles of Psychology*. Such was Spencer's claimed expertise in everything from metaphysics to town planning that, when complimented upon the lack of wrinkles in his forehead one day, he is supposed to have replied it was because 'nothing had ever really puzzled' him. Perhaps this was because, whenever he encountered somebody with strong ideas of their own, he often simply refused to listen to them – Spencer carried around a pair of earplugs which he used to block out the speech of interesting persons lest it should overstimulate his brain and prevent him from sleeping at night. His own ideas were often enough to keep anyone from a good night's rest. As well as a writer and philosopher, Spencer was an inventor; he made a velocimeter for measuring the speed of steam engines, and a new kind of staple for binding magazines. He also claimed to have designed his own flying machine, but could never persuade anyone to build the thing. Onwards and upwards was the way he thought things were going, in society in general, and in science and industry in particular.[9] We shall return to Herbert Spencer later.

Cereal Killer

Spencer never got around to writing a book called *The Principles of Progress*, but perhaps he should have done. He buttressed his unshakable faith in his progressive creed with evolutionary theory, although less that of Darwin than of his now discredited predecessor Jean-Baptiste Lamarck (1744–1829), a Frenchman who believed in the inheritance of acquired characteristics; that, for example, by constantly straining upwards to reach leaves on trees, giraffes would slightly but permanently stretch their necks, leading to the next generation of animals being born with longer ones. Lamarck's views had many adherents, including Darwin himself to some degree. Darwin minimised the idea of acquired characteristics in evolution, but did not entirely eliminate it.[10]

A later fan of Lamarck was Josef Stalin, who found in him 'proof' that evolution could be planned and controlled; if giraffes could strive for longer necks, then so too could all good Communists strive to be supermen. In short, Lamarck recapitulated both Marx and Hegel; the rival theory of purely genetic inheritance increasingly espoused by Western thinkers did not fit the desired Marxist model quite so well. Stalin found an evolutionary ally in the shape of Trofim Lysenko (1898–1976), Russia's most influential pseudoscientist. In 1935, Lysenko had absurdly declared that 'in our Soviet Union, Comrades, people are not born ... but people are created!'[11] If so, then Lysenko himself should have been returned to the tractor factory immediately on grounds of faulty workmanship.

Born to a family of Ukrainian peasant farmers, Lysenko's humble background proved of distinct political advantage, allowing him to be promoted as a genius of the proletariat. Attending agricultural college, Lysenko came under the influence of Ivan Michurin (1855–1935), a horticulturalist specialising in the hybridisation of fruit, known for his proclamation that 'we cannot wait for favours from Nature; we must wrest them from her' through selective breeding. In 1927, Lysenko began research into increasing wheat yields. There was famine brewing across the USSR, and desperate need for more crops.

Lysenko's solution was something called 'vernalisation'. This is a real phenomenon which had already been described as long ago as 1854. However, Lysenko allegedly did not know this, as he was not very well read even in his own field! By treating winter wheat seeds with moisture and exposure to cold before sowing them in the springtime, Lysenko (re-)discovered that the crop would mature more rapidly and produce greater yields. Lysenko's father put his son's findings to the test on his own farm and found they worked. When this was announced in newspapers, Lysenko became a Party hero, the so-called 'barefoot professor'.[12]

However, Lysenko completely misunderstood why he had succeeded in increasing wheat yields. Imagine you are growing a potted plant at home. By giving the seed good soil, lots of water and fertiliser, and exposing it to plenty of warmth, you produce an excellent flower. Eventually, the plant produces seeds of its own. You take one and give it to a friend who treats it poorly, putting it in average soil with only average amounts of water, heat and fertiliser. It will not grow to be as good as the original plant. For Lysenko, though, by exposing the original plant seed to excellent care and conditions, you would have altered its very nature; the healthy, vigorous flower you grew originally would be a kind of *überpflanze*, bearing an inherent seal of quality which would inevitably then be passed down to its equally superb and prize-winning offspring. Like Lamarck, Lysenko felt that acquired characteristics could be inherited, but this is not so – at least not on the scale he imagined. Whilst Lysenko's idea of vernalisation was a genuine process, it couldn't be passed on from parent plants to their offspring; it would have to be repeated anew with every fresh set of seeds. Dye your skin purple and you won't have purple babies.

Lysenko tried every trick in the book to appear right. He preferred to announce new breakthroughs in the popular press, not actual scientific journals, describing his working methods in deliberately opaque terms – a good job, seeing as some of his assistants apparently didn't understand such elementary things as the need for plant pots to have drainage holes. If it was impossible to work out precisely what he had done in his experiments, then it would also be impossible for others to repeat them to test if they really worked. Denouncing the statistical analysis of data as 'harmful' to science, Lysenko announced his trials as successes before they had even started, then simply rigged the results. For example, when there were two wheat fields, one filled with ordinary wheat, the other with his superwheat, Lysenko fraudulently ensured that not all the wheat from the first field would be collected for weighing. Thus, it would appear that the vernalised field had produced way more crops than the other had done. A. A. Sapegin, the director at the institute where these trials were conducted, discovered the fraud and denounced Lysenko. Soon after, he was thrown into prison. He would not be the last.[13]

Cruel as a Cucumber

Denouncing his opponents as 'fly-lovers and people-haters' (a reference to contemporary work on genetics involving fruit flies), Lysenko had been happy for Stalin to eliminate his rivals. Exile, execution and banishment to labour camps followed. N. I. Vavilov (1887–1943), Russia's most distinguished geneticist, was declared a British spy after falling out with Lysenko, and sent to freeze to death in Siberia. Another anti-Lysenko scientist was surprised to be told one day that he had bravely 'volunteered' to go and fight on the front line in the Spanish Civil War. He packed his things and fled.[14] Becoming Stalin's favourite scientist, Lysenko was appointed Deputy Head of the Supreme Soviet in 1938, and acclaimed 'Hero of Socialist Labour' in 1945.[15] Now, even trivial slights against him could have severe consequences. Vadim Birstein, a Russian-American geneticist, tells the tale of three Russian students who stole a portrait of Lysenko and hung it in a women's lavatory as a harmless prank in 1962. Once apprehended, the culprits were treated with 'mercy'. Declared 'insurgents', two were expelled, while the third, shown even more compassion, was put in a mental home. Scientists quickly realised they had to imitate Lysenko in all matters if they wanted to survive. Birstein recalls being lectured himself by a fanatical professor dressed in Russian folk costume who would write the phrase 'DNA' on a blackboard then angrily cross it out with chalk while screaming 'This *does not* exist!' at his students in an insane manner while claiming that Lysenko had the power to perform 'miracles'.[16] 'Don't tangle with Lysenko,' the high-ranking official Andrei Zhdanov (1896–1948) once warned his son. 'He'll cross you with a cucumber.'[17]

Speaking out against Lysenko became increasingly impossible, which was unfortunate as at some point he appeared to become insane. By the 1950s, he was claiming he could transmute plant species into one another; he said he could make rye become wheat, and grow spruce trees from pines, the rough equivalent, as one critic observed, of a cat giving birth to a lion. He also bizarrely claimed he had observed a warbler lay eggs containing cuckoos as a result of eating too many caterpillars.[18] Asked to describe his new theory of fertilisation to visiting British scientists in 1945, Lysenko explained that sex cells entered into a 'love-based marriage' in which one cell 'eats' another. The cell doing the eating passed on its characteristics to the next generation; any rejected characteristics from the cannibalised partner were simply burped out via a process of 'belching'. Claiming that 'nobody knows what a species is', Lysenko proclaimed that 'everything written in books is not true', something which certainly applied to his own publications. His literacy levels were so poor that in one essay he appeared to accidentally claim that sheep were made from grass![19]

Here in the West Lysenko was mocked, with the English novelist John Wyndham (1903–69) suggesting in his 1951 sci-fi fable *The Day of the Triffids* that the titular flesh-eating plants on legs were the result of Lysenko's mad experiments. So why in Russia was he almost deified? It was because Lysenko's theories were compatible not with the genuine science of genetics, but with the pretend 'science' of Marxism. The idea of genes, apparently, was simply 'bourgeois idealism' and 'an image of the [Western capitalist] ruling class as it sees itself'. Communists were obliged to accept Darwin's teachings because Karl Marx had too, and Darwin was not entirely against Lamarck. As such, it was the subsequent idea of genes – about which Darwin knew nothing – against which the Soviets rebelled. While Stalin didn't interfere in the work of nuclear physicists, trying to impose a 'Marxist' method of splitting the atom, he did feel qualified to comment upon biology, thinking incorrectly that evolution itself was actively left-wing in nature. In the Soviet media, Stalin was praised as the lord of 'progressive materialistic biology' as if he was a scientist himself. But how can biology be 'progressive'?[20]

Lysenko's idea of turning Nature into a progressive force was to replace 'competition' with 'co-operation'. In reality, animals within the same species compete with one another for food, shelter and mates. The stronger or better adapted ones live, the weaker or less adapted die. Lysenko denied this. Instead, individuals within any given species helped one another out, like good comrades. There could be no such thing as 'class struggle' among plants and animals. Decadent capitalists may have competed against one another, but ideologically pure Marxists were all working towards the same goal: the perfection of humanity itself, and the creation of Paradise on Earth. It was the environment that caused people to inherit their qualities, not the 'fake' Western concept of genes. Thus, just like with plants given good soil, the ideologically correct way to make mankind evolve was to improve his environment. People, like cereal crops, were to be 'vernalised', and made to grow better. Perhaps one day children might even inherit Communist beliefs at birth?[21] The critic of pseudoscience Martin Gardner summarised the ideology extracted from Lysenko's ideas thus:

> Heredity is transmitted with every particle of the body (just as every worker in Russia contributes toward the future of the State). When a plant is suddenly given new environmental conditioning,

> there is a 'shattering' of its heredity (like a political revolution) ... [The Western genetic theory of evolution] is a slow process which operates by means of random, purposeless mutations ... Lysenkoism offers a more immediately attractive vision. Humanity becomes plastic – capable of being moulded quickly by new conditions and individual efforts. Russian children can be taught that the Revolution has "shattered" the hereditary structure of the Soviet people – that each new generation growing up in the new environment will be a finer stock than the last.[22]

Another key to Lysenkoism's success was its compatibility with Stalin's obsession with collectivising agriculture. In 1928, Stalin had ordered that Russia's traditional, family-owned farms be reorganised into gigantic, collectively owned ones in the name of 'efficiency'. The result was mass starvation. But so what? According to Lysenko, collective farms were 'the one and only scientific guiding principle' inspiring his work, a principle 'which Comrade Stalin teaches us daily'. In 1935, Lysenko gave a speech to the Congress of Collective Farm Shock-Workers which caused Stalin to shout, 'Bravo, Comrade Lysenko, Bravo!' such was its affinity with his own hubristic doctrines. Lysenko had compared his own opponents to opponents of collectivisation and thus implied both alike were enemies of Communism. In 1948, in front of the Soviet Agricultural Academy, Lysenko gave another speech which was even better-received by his patron – because Stalin had written it himself. Despite Stalin having no scientific qualifications, Lysenko said he still 'in detail explained to me his corrections, [and] provided me with directions as to how to write certain passages in the paper'. Following this speech, dissent from Lysenkoism was officially outlawed; it was now 'the only correct theory' and its opponents 'fascists'. The ideology was directing the science, not the other way around.[23] Not that Stalin cared. As he so famously said, 'One death is a tragedy; a million is a statistic.'[24] In 1921/2, during an earlier period of needless, ideologically created famine, Maxim Gorky, the Russian novelist and fanatical Marxist, had openly proclaimed, 'I assume that most of the 35 million affected by the famine will die ... The half-savage, stupid, difficult people of the Russian village will die out ... and their place will be taken by a new tribe of the literate, the intelligent, and the vigorous.'[25]

Or: let's kill off the people in the name of the People. Now *that's* what I call progress!

The Fake's Progress

One of the strangest tales from this period in Russian history was that of Paul Kammerer (1880–1926), a Viennese biologist, writer and committed socialist, whose 1924 book *The Inheritance of Acquired Characteristics* was full of Marx disguised as Lamarck:

> If acquired characteristics cannot be passed on ... then no true organic progress is possible. Man lives and suffers in vain. Whatever he might have acquired in the course of a lifetime dies with him. His children and his children's children must ever and again start again from the bottom.

If Lamarck was right, however, then instead of 'slaves of the past' Kammerer said we would become 'captains of our future', able to influence our own evolution like new Rosicrucians and so 'ascend into higher and ever higher strata of development':

> Education and civilisation, hygiene and social endeavours, are achievements which are not alone benefitting the single individual, for every action, every word, aye, even every thought, may possibly leave an imprint on the [future] generation.[26]

A manic-depressive, Kammerer possessed a number of other unusual ideas, notably his concept of 'seriality'. As expounded in a 1919 book, seriality was an extremely complex attempt to explain the idea of coincidence via the concept of mathematical 'sets'. Einstein was a fan of the notion, calling it 'by no means absurd', but to most it was basically incomprehensible. One recent reassessment defined seriality as being a 'Law of the Conservation of Information', the idea that everything in the universe could be reclassified as information somehow, and that this information was never actually lost, it just *appeared* to be. Everything in existence was connected, said Kammerer, and the world was made up of so much information that the patterns were beyond human comprehension or analysis. There was, in theory, a mathematical reason why, say, a person might suddenly think of someone they hadn't seen for ten years and then bump into him in the street two minutes later, but it was too difficult to be accurately calculated.[27]

What, though, were the chances of *this* happening? On 23 September 1926, the body of a man, well dressed in a dark suit, was found propped up on the slopes of an Austrian mountainside, the gun he had apparently used to shoot himself still clutched in his right hand – even though in life he had been left-handed. He left a suicide note (typed, not handwritten) willing his body to medical science, and his library to the University of Moscow, where he had been due to take up a post teaching biology. The corpse was that of Paul Kammerer, who was deemed to have killed himself in despair following the recent ruin of his reputation.[28]

The scandal centred around one of Nature's more obscure creatures, *Alytes obstreticus*, or the 'midwife toad', an amphibian so named because of its odd mating cycle. Unlike most toads, this species mates on dry land, with the male mounting the female and clinging there, sometimes for weeks, until she lays her spawn, which the male then fertilises. Then, he wraps the spawn in a string around his legs and waits for them to hatch. In order to keep a grip on each other, though, the midwife toads' skin has to be dry. Other toad species can follow a similar routine, but in water, where their skin becomes slippery. In order to maintain a grip on their mate, these water-breeding male toads develop small blackened spines along their hands at breeding-time, called 'nuptial pads'. Knowing this, Kammerer put some ordinary midwife toads in glass tanks and forced them to breed in water like their peers. He did this, generation after generation, from 1906 to 1919, until eventually, he said, some males were born who developed pigmented spines for grip. It seemed that, like giraffes wanting longer necks, Kammerer's toads had wanted spines for better grip, and so had somehow influenced their own bodies to develop them. Lamarck had been proved right after all!

Except ... it turned out the toads were fakes, or at least that the main preserved specimen being hawked around as proof of Kammerer's idea was. In 1926, G. K. Noble (1894–1940), of the American Museum of Natural History, examined the animal and concluded the black 'pigmentation' had been caused by an injection of Indian ink. (This is strangely reminiscent of the 1974 case of American dermatologist William Summerlin (b. 1938) who, in a desperate struggle for research funds, coloured a white mouse in black with a felt-tip then claimed to have successfully carried out a skin transplant on it![29]) The nuptial pads had by now disappeared too, although seeing as plenty of other people had seen them previously, they may simply have worn down through constant handling. Had the ink been injected by Kammerer to spruce up an old specimen, or had it been a fake all along? Evidence is inconclusive. As Arthur Koestler showed in his 1971 book *The Case of the Midwife Toad*, it may not even have been Kammerer who injected the ink at all. It could have been a well-meaning assistant, or even an unknown loon who had been going around Vienna's Institute of Biological Research playing malicious pranks for some time. The hoax took place not long before Kammerer was due to leave Vienna for Moscow, with the Russians being keen to get a prominent Western Lamarckian into their ranks. Had someone deliberately tampered with his toad to prevent him going? In later years, that became the Soviet take on affairs. Following Lysenko's rise, the Russians made a film glorifying Kammerer and claiming he had been framed by sinister capitalists with a hatred for all those who would seek to undermine Western science.[30]

Evolutionary Economics

Outside Russia Lamarck's ideas rapidly lost favour, with experiments performed by August Weismann (1834–1914), who chopped the tails off twenty-two successive generations of mice only to find that their young were still born with such appendages, seeming to undermine the theory. Weismann was the leading proponent of the idea of 'germ-plasm' (chromosomes, basically), which he saw as being impervious to environmental change during an organism's lifetime.[31] As has been pointed out by several critics, both seriously and in jest, if the idea of Lamarckian inheritance was really true, then Jews would surely be being born without foreskins by now – but they aren't! The last significant Western Lamarckian was William McDougall (1871–1938), a Fellow of the Royal Society who taught at Harvard during the 1920s, where he performed a highly flawed series of experiments purporting to prove that rats could inherit the ability to find their way through a maze from their parents. By 1928, with his work falling out of favour, McDougall moved to North Carolina's Duke University, whose president, William Preston Few (1867–1940), was specifically looking out for 'facts incompatible with materialism'. This quest led to McDougall eventually speculating that perhaps the secret ingredient in Lamarckian inheritance was telepathy, proposing that not only rats, but also pigeons, seals and penguins might well be highly psychic.[32]

Surprisingly, a kind of variant neo-Lamarckism is now making something of a comeback, with a 2015 study led by Rachel Yehuda, of New York's Mount Sinai hospital, concluding that Holocaust survivors might somehow have passed on the trauma they experienced to their children through their genes, a process known as 'epigenetic inheritance'. According to Dr Yehuda, such survivors' children are more likely to have various stress disorders themselves than a sample of Jewish children whose parents were not interned by the Nazis. They may not have been born without foreskins, Yehuda says, but these particular Jews appear to have inherited their parents' mental traumas nonetheless.[33] Such ideas are controversial; it doesn't take long to see that there could be other explanations for Holocaust-survivors' children having high stress-levels, after all. Despite this, epigenetics is a growing field and, whilst giraffes can't grow longer necks through willpower alone, it does appear that the process of inheritance is a bit more complex than was once thought.

One man who would surely have appreciated this unexpected comeback for Lamarck was Herbert Spencer, the inventor of the term 'survival of the fittest'. As a thoroughly self-made man, Spencer thought it was possible not only for a man to pull himself up by the bootstraps, but in doing so to pull his offspring further up the evolutionary ladder, too. By striving for a better tomorrow, like a short-necked proto-giraffe reaching up towards distant leaves, perhaps that better tomorrow might eventually come? Spencer believed that each race inherited the character traits of its forebears, attributing his own cussed and stubborn nature to his (supposedly) being descended from Huguenot Nonconformists and Bohemian rebels. So, together with things like skin colour, perhaps people might be able to inherit gumption and good character from their ancestors, too? In western Europe (and particularly Britain), said Spencer, evolution had simply worked more quickly than it had in places like Africa, something he linked with long-standing habits of trade and free enterprise. A man had to fight to succeed in life, and those who were most evolved succeeded the best. Furthermore, as individuals evolved, so did their businesses and occupations, even the whole cities and countries in which they traded. There was, said Spencer, a general law of evolution, which applied not only to animals and plants, but to all human endeavour. Manchester, for example, the world's first true industrialised city, had recently evolved, and now stood in relation to more primitive pre-industrial towns as man did to ape. As one industry or town failed to adapt to new conditions and went extinct, another inevitably evolved to take its place, one more fitting to the new commercial climate. Capitalism, then, was pure evolution in action. Britain, the famous 'nation of shopkeepers', had the racial inheritance best fitted to exploit this fact, and its citizens could, via a process of continual and deliberate self-improvement, ensure that their offspring would be even better adapted to help rule the world through commerce and empire.[34]

Some of Spencer's ideas may now seem silly, but their basic thrust – that society is evolving, in an upwards fashion – is still the prevailing outlook of our own time. Few mainstream scientists now believe in the inheritance of acquired characteristics on the scale Lamarck imagined, but our own intellectual inheritance has certainly acquired more than a tinge of the Spencerian outlook to it nonetheless. Darwin himself was related to one of Britain's most successful pre-Victorian capitalists, Josiah Wedgwood (1730–95), of pottery fame, and it has been speculated that his great idea was actively influenced by both his family background and the prevailing, thoroughly Spencerian, attitudes of his day. A. N. Wilson, for instance, whose book *God's Funeral* features much interesting discussion of the nexus between Darwin, Spencer and Marx, notes how (as we shall soon see) Darwin's conclusions were arrived at independently by another man at almost exactly the same time, and says that this curious fact should make us ask how much his famous theory was really 'a scientific discovery … and how much the expression of a metaphysical idea having much in common with *laissez-faire* economics, competition, progress, and other ideas of the time'.[35] Maybe theories of evolution, themselves, tend to evolve?

How to Get A-Head

Evolutionary theory did not begin with Darwin. Perhaps the first known evolutionary theorist was the Greek philosopher Anaximander (*c.* 610 BC–*c.* 546 BC), who felt that all life emerged from muddy moisture as it evaporated in the sun, leading him to conclude (not entirely incorrectly) that man evolved from fish. He also thought the Earth was a giant cylinder, and is traditionally said to have been the first man to make a map – though presumably not a very accurate one.[36] The best ancient Greek evolutionary theory came from Empedocles (*c.* 495 BC–*c.* 430 BC), who thought that, in aeons past, disembodied limbs, heads, eyes and breasts wandered the globe, looking for mates. Breeding randomly, they sired a range of bizarre offspring whose forms were quite impractical: big bundles of living hands, people with back-to-front faces, humans with ox-heads and so forth. Eventually, via a process of natural selection, advantageous forms – heads with two eyes, arms connected to shoulders – won out, joining with other advantageous forms until we ended up with the familiar and more practical fauna we see living on Earth today.[37]

There has, then, been plenty of time for people to invent evolution-based myths; whatever your pre-existing cast of mind, you can create your own personal pseudo-evolutionary creed to fit it. A 2001 book, *Strange Creations*, by Donna Kossy, gives an amazing catalogue of nutty ideas about the evolutionary process, some of which are worth citing here. There is, for instance, the curious phenomenon of people who compensate for there being no Divine Creator by taking the more 'scientific' route of substituting aliens to stand in for Him instead. Several fringe thinkers have trodden this path, but none more strangely than the British author David Barclay, whose 1995 *Aliens: The Final Answer?* posited the unlikely theory that some dinosaurs were really super-intelligent lizard men, who once kept a species of less intelligent, smaller dinosaurs as pets. These pets, said Barclay, were us – through a long process of selective breeding, the lizard people gradually bred humans out of an original root race of 'dog-like domestic dinosaurs'. Eventually the lizard people developed space flight, claimed Barclay, and left Earth in flying saucers, vehicles in which they can still be seen whizzing through our skies today. Even odder, Barclay also argued that Biblical accounts of people encountering heavenly 'pillars of fire' and shining angels were primitive man's mistaken accounts of reptilian UFOs. Thus, 'God' does exist after all, but is really a dinosaur![38]

Another common reason for creating your own personal evolutionary myth involves racial prejudice, a context in which it is impossible not to mention the Nazis. Everyone knows what the dream of Germany becoming the home of a 'pure' Aryan master race ultimately led to, but fewer know just how peculiar some of the claims made in support of such an idea were. Take the views of Professor Hans F. K. Günther (1891–1968), one of the Nazis' favourite anthropologists, who went to great lengths to prove that Nordic types were quite simply so much cleaner than all other races were. Ignoring the well-established historical fact that soap was an Arab invention, Günther claimed to have unearthed proof that it was actually an Aryan creation, as was the hairbrush and, presumably, all other personal grooming products. Furthermore, pure Germans naturally preferred the 'clean' colours blue and green to any others, apparently – so what about the brownshirts? Most ridiculously, Günther 'discovered' that German females had an inbuilt genetic predisposition towards modesty. By closely examining women's legs on public transport, he said he had gathered indisputable proof that the frauleins of the Fatherland sat with their legs held closer together than the shameless and 'diabolically alluring' women of inferior races, thus making it harder for him to see up their skirts in the name of science.[39]

The Ape of God

Particularly worth dwelling upon were claims that Jesus Christ was really Aryan, an assertion made, for instance, by the Nazi philosopher Alfred Rosenberg (1893–1946) in his Party-approved *The Myth of the Twentieth Century*.[40] Actually, the legend of Jesus' Aryan nature predated the Nazi era, as can be seen from an examination of one of the looniest books of all time, 1905's *Theo-Zoology: Or, the Lore of the Sodom-Apelings and the Electron of the Gods* by Jörg Lanz von Liebenfels (1874–1954). Lanz was a fan of the *völkisch* (folkish) ideas of the Viennese mystic Guido von List (1848–1919), who had invented a romantic, and wholly false, 'history' for the Germanic race, involving ancient priest-kings, rune-casting wizards and other such *Hobbit*-like nonsense. Inspired, Lanz himself then founded another *völkisch* cult, the vaguely homoerotic Order of the New Templars, which met in a ruined castle bedecked with swastikas, and which was allegedly a partial influence upon Himmler's later development of the SS.

Lanz's main contribution to German racism, though, was his frankly insane doctrine of 'Theo-Zoology'. In his 1905 book, Lanz claimed that the blonde-haired, blue-eyed Aryan race were literally on the side of God, and all others – blacks, Jews, Asians and

'Mediterraneanoids' – on the side of Satan. Indeed, Aryans were the only genuine human beings left on Earth, said Lanz, as all other races were the result of immoral interbreeding between racially pure ancient Aryan ladies and a race of genetically inferior half-monkey midgets he called the *Buhlzwerge*, or 'love-pygmies'. Interpreting the images on certain eighth-century BC Assyrian reliefs in a totally deluded way, Lanz concluded that almost the entire purpose of ancient civilisation had been to breed such semi-simians en masse for deviant sexual pleasure. The tale of the Garden of Eden was really a metaphor for mankind's genuine original Fall – the dark day when a weak Aryan female named Eve first gave in to temptation and had sex with a tiny monkey called Adam. The result of this perverted act was the birth of the blacks, Orientals and Jews, whose existence Lanz saw as being a kind of 'de-evolution'. The true purpose of the Old Testament – which Lanz had studied in-depth, being a former monk – was thus to warn any remaining Aryans not to sleep with hairy dwarfs. By obsessively reinterpreting common words like 'bread' and 'water' in the text as really being secret codes for 'beast-man', Lanz patiently developed his theory.

The early Aryans, said Lanz, had been gods on Earth – *Theozoa*. Adam and Eve's half-monkey offspring, though, were merely devolved humanoid animals – *Anthropozoa*. Early Aryan man had been partially 'electronic' in nature, having powers of telepathy and clairvoyance which had since atrophied due to rampant pygmy abuse. However, some relatively pure Aryans still existed in places like Germany and Scandinavia, and could be specially interbred to restore these skills in their children, Lanz proposed. A demonstration of what kind of powers these super-beings of the future might hope to possess could be found in the study of the life of Our Lord Jesus Christ (real Aryan name: 'Frauja'), who had been a kind of godsent genetic throwback to the days of the *Theozoa*. Jesus, said Lanz, was highly electronic in nature, hence accounting for his miracles like raising the dead and walking on water. In 1896 Henri Becquerel (1852–1908) had announced his famous discovery of radioactivity to the world, an event Lanz took note of. Clearly, thought Lanz, Jesus – and by implication, all ancient Aryans – had been radioactive and psychically able to control the emission of their body's electrons, thereby allowing them to work wonders.

Lanz's obscene interpretation of the life of Christ was truly unique. Sent down to Earth by God to warn His 'chosen people' of the dangers of monkey love, Jesus was ultimately defeated by the Satan-worshipping descendents of Adam, who tried to negate his threat by crucifying him. They then released a group of evil apelings to gang-rape Jesus on the cross, hoping thereby to pollute his racial purity. Christ had somehow resisted, however, and thereby provided an admirable model of sexual abstinence for all modern-day Aryans; but his wider message of 'sexo-racist gnosis' had since been wholly misinterpreted. Christ did not preach compassion for the weak at all, said Lanz, and so all inferior non-Aryan races should be destroyed immediately. After being sterilised, these modern apelings could be deported to Madagascar, used as humanoid slave-labour, or burned alive as a sacrifice to God. The purest Aryan women, meanwhile, should be immediately imprisoned in special convents and used as sex-slaves by the noblest Aryan men, in order thereby to create generation after generation of increasingly electronic children. Thereafter, Germany would one day conquer the entire world and 'enchain the apes of Sodom' with a new breed of holy electronic knights, a sacred and highly radioactive SS-style warrior-priesthood designed to worship, in Lanz's words, 'the electron and the Holy Grail ... and transform the Earth into the Isles of the Blessed'.

It would be easy to over-exaggerate how much influence Lanz's thinking had on Nazism, however. Lanz himself spent the Second World War hiding in Switzerland, but after the war had finished he claimed that Adolf Hitler himself had been a reader of his popular monkey-cult magazine, *Ostara*, which had once had a print run of 100,000. He even said that Adolf had visited him at his editorial office one day in 1909, seeking back-issues to complete his collection! Whether this is true or not, there were certainly some similarities between Lanz's ideas and the later Nazi schemes to breed a new generation of Aryan supermen – one of the earliest SS proposals for cleansing the Reich was to send all Jews off to Madagascar, for instance, a direct echo of Lanz's thinking.[41]

Monkey Madness

Black supremacists, too, have invented their own perversions of Darwin, most notably the infamous Elijah Muhammad (1897–1975), leader of America's Nation of Islam sect for over thirty years, populariser of racist terms like 'blue-eyed devil' and author of 1965's *Message to the Blackman in America*. But what was this message? According to Muhammad, it was one relayed to him direct by Allah Himself, whom he specifically identified as being a 'scientist'. Muhammad's tale itself is certainly science-fiction. Sixty-six trillion years ago, said the author, Allah tried in vain to teach all Earth-men to speak the same tongue, but failed, became angry, and caused an explosion so huge it split the planet in two, thus creating the moon. Only one human tribe, that of Shabazz, survived. The people of Shabazz were black and produced various great scientists, including one who, 50,000 years ago, gave them curly hair 'to make us all tough and hard' and thus able to battle wild beasts in Africa's primeval jungles.

However, the tribe also had one particularly evil scientist, named Yakub – apparently a rip-off of the Biblical Jacob, who subjected sheep to selective breeding in the Book of Genesis. Yakub, though, went one further. Putting a black man's sperm under a microscope and examining it carefully, he discovered it contained the germ of two tiny men, one black, the other brown. Potentially, this meant that a new and artificial white race could be created. Being evil, Yakub went into eugenics, killing all the blackest babies born on the island where he lived with his 59,999 followers and saving the lightest, brown-skinned ones, thus gradually creating a white master-race to rule over the people of Shabazz. After some 600 years Yakub finally succeeded in manufacturing wholly white babies – the 'blue-eyed devils'. Confined in Europe away from the superior Shabazz race, some of these devils desperately tried to breed themselves black again, giving birth to gorillas. After 2,000 years of striving, the unrepentant majority of whites managed to create an advanced European civilisation and used its technology to enslave the godly tribe of Allah. It was Elijah Muhammad's desire to inform all black men of this 'truth', convert them back to Islam, and thereby foment the overthrow of white Western civilisation – the world's first sci-fi jihad! Amazingly, Muhammad gained many followers – including Muhammad Ali and Malcolm X – and his cult still goes strong today.[42]

Another implausible evolutionary cult was created by the Polish-American artist Stanislaw Szukalski (1893–1987), founder of the theory of Zermatism, which taught that the violent thugs who staffed all dictatorships were the result of their distant ancestors having been sexually abused by apes. In an odd echo of Elijah Muhammad, Szukalski was also interested in the idea of an original universal language, which he called 'Protong', and whose details he revealed in a 1980 tract, *Behold!!! The Protong*. This book also contained information about his theory that, many years ago, lustful gorillas had gone around raping humans, causing them to give birth to a series of hybrid sub-species called the Yetinsyny, or Manapes. These Manapes passed for human beings at first sight but were given away by their being really quite ugly, with pot-bellies, long lips, short arms and occasionally even secret tails. Their sense of morality was even less evolved. Full of hatred and animal desire, Manapes often became rapists, criminals and hooligans. Sometimes, they even achieved political power through violent revolution, becoming dictators – and, particularly, Communists.

The true purpose of Communism, said Szulkalski, was to allow ugly Manapes like Lenin and Stalin to kill all the handsome non-ape males in purges, thus giving them a better chance with women: 'Politically, [Yetinsyny] forever remain ... potential Communists, because they cannot forgive [true] humankind its popularity with the opposite sex.' Some particularly prominent examples of 'politically dangerous Yetis' given by Szukalski included Karl Marx, Peter Kropotkin and Chairman Mao. Maybe he was bitter that, after first being forced to flee his homeland by Nazi bombs, Poland had then been quickly overtaken by the Soviets? Or, then again, maybe a commissar just stole his girlfriend. Generously, though, Szukalski did admit that some Manapes could make excellent artists, if encouraged to channel their sexual frustrations into painting instead of politics. Szukalski certainly channelled his own frustration into such an arena; during his many years of Californian exile, he obsessively produced some 40,000 drawings of Communist monkey men and their hairy-legged female counterparts.[43]

Sick in the Brain

Not all fringe evolutionary theories are ideologically or racially motivated. Some are simply the product of diseased minds – in the case of the Hungarian writer Oscar Kiss Maerth (1914–90), quite literally so. Of independent means, Maerth devoted his life to developing the thesis that mankind had only evolved into an intelligent species in the first place because of its early habit of eating its own brains. His 1973 book *The Beginning Was the End* explained that, at the dawn of time, our ape ancestors discovered that eating other monkeys' brains could stimulate their sexual appetite. Maerth could vouch for this, he said, as he had eaten a monkey's brain himself as research and it gave him an erection. Early man could not have eaten such a thing for its flavour alone, he continued, as the one he sampled had 'the consistency of rubber'. However, 'about 28 hours' after chewing it, Maerth began to experience 'increased sexual impulses'. Early humans couldn't get enough of this source of free Viagra either, so began guzzling each other's brains too – as do remote jungle tribes even today.

However, swallowing so much grey matter had the inadvertent side effect of making prehistoric man super-intelligent. Man thus lost his fur and started to walk upright and use tools, but also lost all sense of morality in his insane lust for sex. The results were disastrous – human 'civilisation', filled with war, violence and crime by our cannibalism-corrupted minds, made us worse off than the monkeys who, Maerth assured us, had 'fewer worries', enjoyed 'better health' and could communicate telepathically with other planets. Ultimately mankind renounced brain-feasting as it caused epilepsy, but this didn't mean that 'progress' was anything other than an illusion. After all, only human beings, due to their distance from Nature, had to work to survive, said Maerth, something which led to our species developing the disgraceful genetic defect of smelly armpits:

> Most people earn their money by the sweat of their brow, but some do so by the sweat of other people's armpits. This is called the 'damages of civilisation', but for its origins one must look not in the armpits but in the head.

Looking inside Maerth's own head was certainly a rewardingly comic experience. The Biblical fruit of knowledge eaten by Adam and Eve, he thought, was really a symbolic representation of a human brain, and a warning that eating such a forbidden foodstuff led only to expulsion from Eden. Maerth did suggest a solution to make us all de-evolve back into happy, non-sweaty psychic monkeys by going back to Nature, performing yoga and artificially extending our skulls to look like big domed eggs, but the idea found few fans.[44]

Naturally, we laugh at stupid ideas like these, and think we are immune to them; but we are not. The evolutionary myths our own societies currently promulgate may not be as overtly lunatic as saying that God is a dinosaur, but they are myths nonetheless. We have even managed to apply evolutionary theory to the realm of competing ideas these days, with the once popular 1990s notion of 'memetics', which tried to claim that all human beliefs should be viewed as some kind of 'thought-animals' competing with one another for dominance of their own natural environment, our minds. According to this fanciful notion, the best ideas – or 'memes' – always win out in the end via a kind of 'survival of the fittest', something which is palpably untrue. We are still stuck in the trap of recapitulating Hegel. Science itself has now been remodelled in Darwinian terms, with our automatic presumption being that the best and most truthful theories will always win out in the end, and progress will thus inevitably occur. According to this view of history, the chief opponent science has had to combat down the years has been religion. We all know the story of Galileo (1564–1642) being forced to recant his opinion that the Earth orbited the sun by an oppressive Church, and are asked to take these things as being typical, a Darwinian struggle for survival from which science, as the truer philosophy, finally emerged victorious. The trouble is that this is nothing more than a crude cartoon.

I Depict a Riot

As an outstanding instance of the myth of science automatically representing progress towards a state of anti-religious human enlightenment, consider the so-called 'Priestley Riots' of 1791. These took the form of violent public disturbances perpetrated by a

Birmingham mob protesting about a dinner held in the locality by a group called 'the Revolution Society'. Members of this society liked to present themselves as friends of progress – the social experiment which followed upon the French Revolution of 1789 seemed to them to be the very embodiment of this ideal, with old and undemocratic forms of government being swept away and replaced with exciting new ones, deemed to be more rational in style. Of course, the new cult of 'rationality' in France ended up expressing itself in bloodshed more than anything – even the guillotine was celebrated as a uniquely 'rational' and 'humane' form of execution – but it also involved a bizarre campaign of trying to dethrone Christianity and replace it with a new, state-sanctioned religion of reason of which Auguste Comte would surely have approved. Notre Dame was deemed a cathedral no longer, but a 'Temple of Reason' in which abstract concepts like Logic, Nation and Brotherhood were to be worshipped, while everywhere Catholic priests were defrocked and persecuted.[45]

Though many of the Jacobin revolutionaries did believe in a Supreme Being, it was one cast very much in their own image, a God of *liberté*, *égalité*, and *fraternité*. To many more conservative persons in England, it looked uncannily like a new form of anti-religion had been instituted over in France, one in which man had usurped the place of God. Thus, when on 14 July 1791 – Bastille Day, one of the new secularised 'holy days' in the remodelled French calendar – Birmingham's Revolution Society had held a dinner of celebration, there was outrage. Slogans in favour of God and the king were scrawled over walls, and an orgy of arson, rioting and drunkenness erupted, with the homes of dissenters of any kind, religious or social, picked out for particular attention. One man targeted was Joseph Priestley (1733–1804), a chemist of genius, and a prominent member of the Revolution Society, who fled from his home on the outskirts of the city when he got wind the mob were rising. It was a good job he did – reputedly, the rabble had brought along a spit and intended to roast the chemist alive. Finding him absent from his Fair Hill home, they settled instead for creating an effigy of him, hanging it and then setting it alight. Then, they set about destroying and torching his laboratory, perhaps the best-equipped in England, whose valuable glassware and instruments were quickly reduced to smithereens, as was Fair Hill itself; so thoroughly did the rioters destroy Priestley's house that one was actually killed by falling timber. It was like a real-life version of the torch-wielding village folk destroying Castle Frankenstein – but what, precisely, had Priestley done to incur the citizenry's wrath?

It seemed his main sin, like the Jacobins over in France, was his very public proclamation of the new religion of reason, and his seeming attempts to play at being God. Priestley and some of the other scientists with whom he associated were very open about the idea that their new discoveries were the scientific counterpart of their belief in revolutionary politics; their findings were broadcast as tearing down the old order of things and putting a new worldview in its place. Over in France, Antoine Lavoisier (1743–94), a chemist of even greater genius, had been officially appointed as director of the Gunpowder and Saltpetre Administration, an apt metaphor for the supposedly explosive social effects of the new ideas in chemistry; scientific experimentation and social experimentation were now seen to go together. Erasmus Darwin, studying Lavoisier's findings at the same time as the upheaval in France, tellingly declared, 'I feel myself becoming all French, both in chemistry and in politics.' So did Joseph Priestley.

In 1790, in his book *Experiments and Observations on Different Kinds of Air*, Priestley had provided a classic example of scientific hubris, stating that his new chemical discoveries would become the means of mankind 'extirpating all error and prejudice, and of putting an end to all undue and usurped authority', saying that the established social order had just reason to 'tremble' at the new inventions and innovations coming their way. Apparently, he was half-right and half-wrong in this claim. The forces of reaction had indeed begun to tremble at such talk, so much so that in the end Priestley had his house and laboratory destroyed – but this fact hardly said much for his discoveries 'extirpating all error and prejudice' from the world.[46]

The Mind of God

The moral we are meant to draw is obvious. The scientist is the hero of progress, and ignorant religious and conservative types are his enemy – and the hero, as we know, always wins out in the end. Well, up to a point. To portray Priestley as having been diametrically opposed to the forces of religion and conservatism would be a gross oversimplification. Priestley was a deeply religious man, one of the founders of Unitarianism, and saw his scientific enquiries as an attempt to unravel the mind of God; a sacred quest, not a secularising one. He believed in the literal reality of the bizarre predictions about the end times found in the Book of Revelation, studying Hebrew, Syriac, Arabic and Chaldee to understand them better, and one of the main reasons why he celebrated the French Revolution was in fact because he thought it indicated the Second Coming of Christ was at hand, not because of its focus upon rationality and equality. As he put it, 'I take it that the ten horns of the great Beast in Revelations mean the ten crowned heads of Europe, and that the execution of the King of France is the falling off of the first of these horns; and that the nine [remaining] monarchies of Europe will fall ... in the same way.'[47]

Priestley's idea of unravelling the mind of God through science is not dead. Francis Collins (b. 1950), for example, one-time Director of the Human Genome Project, is a born-again Christian. Collins has proved open about his beliefs, claiming belief in the literal resurrection of Christ, the existence of Heaven and the reality of miracles. In 1996, he gave a speech describing how working on the Genome Project, for 'a scientist who has the joy of being a Christian', was a delight because 'a work of discovery ... can also be a form of worship'. The point, he said, was 'to understand something that no human understood before – but God did'.[48] Even scientists who do not actually believe in God sometimes acknowledge that they still act upon an essentially religious impulse. Robert Sinsheimer, another early mover in the Genome Project, wrote in 1994 that

> throughout history, some have sought to live in contact with the eternal. In an earlier era they sought such as monks and nuns ... Today, they seek such a contact through science, through the search for understanding of the laws and structure of the universe ... I am a scientist, a member of a most fortunate species. The lives of most people are filled with ephemera ... But a happy few of us have the privilege to live with and explore the eternal.[49]

Just so long as you don't call 'the eternal' God, that is.

Maths Problems

Many of its modern-day proselytisers may not like to admit it, but science itself to a very great degree grew from out of both religion and the occult, as with the Royal Society. Opinions like those of Francis Collins were once *de rigueur*. I wouldn't exactly say that such facts have been covered up in the standard narrative we are given today, but attention is certainly not drawn to them. Consider the word 'theory', which now seems about as stereotypically scientific a word as you can get. And yet, its roots are religious. Originally, it came from the ancient Greek mystery-religion of Orphism, and meant something like 'passionate contemplation', a state in which the adept considers his god so deeply that he seems to merge with him. The word was later adopted by Pythagoras (*c.* 570 BC–*c.* 495 BC), whose famous theorem is still used when calculating the angles of triangles. Pythagoras founded a bizarre-sounding cult centring on mathematics, reincarnation, making your bed properly in the morning, and a taboo upon eating beans. When Pythagoras went into a state of 'passionate contemplation' the holy revelations he received were mathematical in nature, and it is from his own use of the word 'theory' that we get our own modern meaning of the term. Sometimes, Pythagoras would achieve sudden 'Eureka!'-type moments during his meditations, and give birth to a new mathematical theory which helped further elucidate the mind of God. Pythagoras believed that Creation itself was made of numbers, and was so disturbed by his discovery of so-called 'irrational numbers' like Pi, whose value can only be calculated approximately, that he is alleged to have drowned one of his students, Hipparsus of Tarentum, after he blasphemously revealed this fact to outsiders.[50]

Aristotle (384 BC–322 BC) too felt that when man made use of mathematics he was gaining temporary access to the mind of God, as he explained in his *Nicomachean Ethics*. For him, man possessed both a 'rational soul' and an 'irrational soul', the irrational soul being linked to bodily tastes, appetites and opinions, whereas the rational soul dealt with things eternally and universally true. For instance, two men's irrational souls might make them differ in taste of music, but their rational souls would make both agree that two plus two is four. In this sense, said Aristotle, 'reason is divine'.[51] This is not quite as we are taught to view the matter nowadays. Reason is still presented to us by science as being divine, but only in the sense that it has now *replaced* God, rather than giving us contact with Him. Mathematics can still, to those who love it, facilitate moments of sudden joy and revelation as a problem is solved, but to most of us it now seems joyless and dead. Modern scientists and mathematicians who speak about their work bringing them closer to the mind of God usually do so only metaphorically, using such language to try and communicate the sense of awe it makes them feel about the universe. Whenever one of their number attempts to return to the old way of looking at things, the new religion of rationalism often does its best to excommunicate him.

Take the case of Pavel Florensky (1882–1937), who was both a priest and an occultist, and a mathematician and an electrical engineer, an unusual-sounding combination to modern ears. Despite priests being deemed enemies of the people in post-revolutionary Russia, Florensky continued to wear his cassock and bother men about God, even while he taught mathematics to the workers and supervised Soviet electrification projects in the name of Communism. He managed to justify this apparent contradiction by developing a 'mathematics of discontinuity' which, essentially, held that two opposing things could be true simultaneously; instead of designating a component of an equation 'A' and another 'B', he would rather unify both possibilities at once by writing '(A + B)'. So, if 'A' was 'religion' and 'B' was 'science', then both could be united in '(A+B)', or 'Communism', which, being a greater truth than either, contained a synthesis of both possibilities within it – an echo of Lenin's famous statement that 'Communism equals Soviet power plus the electrification of the entire country'.[52] Thus, the Soviet engineer defiantly kept his cassock. During this period, Florensky managed somehow to find the time to keep abreast of the latest developments in physics, seeking some esoteric fusion of advanced maths and Christian theology which might have appealed more to Pythagoras than it did to the People's Commissars. Eventually he was arrested after penning a paper about Einstein's theory of relativity which tried to prove the existence of God by claiming that certain imaginary numbers which applied to a theoretical body moving faster than the speed of light represented 'the Geometry of the Kingdom of the Divine'. Deported to the Siberian gulag, Florensky tried to make the best of a bad situation by conducting intensive research into the nature of permafrost, his papers on the subject being so good that they were actually published by the very same authorities who were simultaneously imprisoning him. In 1937, he was executed and dumped in a mass grave near Leningrad, his number finally having come up for good.[53]

For sinning against Pythagoras' religion of rationality by revealing the imperfect and 'irrational' value of Pi to the world, Hipparsus of Tarentum was drowned in a pool; for sinning against Lenin's own religion of rationality by saying that maths was really God, Florensky was shot and buried. One was killed for saying that mathematics undermined the existence of a deity; the other for saying mathematics revealed it. Both were deemed unreasonable and dangerous by different observers.

The Science of the Supernatural

It would be easy to give a long list of scientists who have been religious, mystically inclined, or even outright occultists. Famously, Isaac Newton (1642–1727) was an alchemist who sought out hidden codes in the Bible; born upon Christmas Day, he believed himself to be a kind of Messiah and Prophet, something Seventh-Day Adventists still believe about him now.[54] Even Galileo believed in God, contrary to the automatic assumption many might hold today. Robert Boyle actually thought he had *met* God, undergoing a mystical experience while studying in Geneva during the 1630s. Seeing a storm so severe he thought the end times were at hand, he pleaded for salvation, and was apparently answered. Awed by this encounter, Boyle questioned why he had been saved, concluding he had been given the divine mission of advancing knowledge to demonstrate thereby the existence and mercy of Jehovah. In Boyle's view, scientists were sacred servants, 'born the Priests of Nature', whose role was to 'mediate between God and Creation'. So devout was Boyle, he learned Hebrew to read the Bible in God's original language, just so he could be certain it hadn't been mistranslated. It is hard to believe that the celebrated 'father of chemistry' wrote a tract explaining the resurrection of the dead on Judgement Day in terms of chemical reactions, but he did.[55] As well as worshipping God through his discoveries, Boyle also belonged to an early paranormal research-type group, which met at Warwickshire's Ragley Hall from 1665 onwards, to discuss apparitions, faith healing and witchcraft. Boyle's most celebrated book may have been called *The Sceptical Chymist*, but he was not one himself. In 1658, he even helped arrange the publication of *The Devil of Mascon*, a book detailing a French poltergeist haunting, so impressed was he by the case when he heard about it.[56]

Other Fellows of the Royal Society who have taken an interest in the occult – albeit later restyled as 'psychical research' – would include R. J. Tillyard (1881–1937), Lord Rayleigh (1842–1919) and Julian Huxley (1887–1975). One Fellow, Edward Heron-Allen (1861–1943), even wrote a book about as flaky a topic as palm-reading, 1885's *A Manual of Cheirosophy*. Yet another, Major Edward Moor (1771–1848) published a book about alleged incidents of ghostly bell-ringing in his Suffolk mansion, 1841's *Bealing Bells*. The Royal Society's President throughout the Second World War, Sir Richard Gregory (1864–1952), was also intrigued by such subjects, an interest reflected in the pages of *Nature*, of which journal he was editor for some years.[57] Britain's Society for Psychical Research (SPR), founded in 1882, was particularly rich in such figures. During its first century of existence, as SPR luminary Archie E. Roy (1924–2012) – himself a former Professor Emeritus of Astronomy at the University of Glasgow – has pointed out, the SPR numbered among its fifty-one presidents 'nineteen professors, ten Fellows of the Royal Society, five Fellows of the British Academy, four holders of the Order of Merit and one Nobel Prize-winner'. Among its more notable members have been J. J. Thomson (1856–1940), father of atomic theory, the great physicists Sir Oliver Lodge (1851–1940) and Sir William Barrett (1844–1925), Charles Richet (1850–1935), the Nobel-winning physiologist Marie Curie (1867–1934), discoverer of radium, and F. J. M. Stratton (1881–1960), President of the Royal Astronomical Society. Some were sceptics, but others genuinely believed.[58]

My point isn't necessarily that such people discovered indisputable (as opposed to merely suggestive) proof of the existence of ghosts or psychic powers. Indeed, some of the SPR's most brilliant scientists were remarkably credulous in their investigations, such as the celebrated physicist and chemist Sir William Crookes (1832–1919). His work on vacuum tubes and cathode rays led ultimately to the invention of television, as well as the detection of the electron and x-rays, and in 1861 he himself discovered the element thallium, achievements for which he was knighted in 1897. Nonetheless, beginning in 1873, Crookes also investigated the obviously fraudulent activities of a teenage medium named Florence Cook (1856–1904), who would vanish behind a curtain at séances then re-emerge a few moments later disguised as a 'spirit' named Katie King, to the amazement of the gullible folk who attended – including Sir William. In spite of the blatant falsity of Cook's claims, Crookes concluded her conjuring was genuine, saying it 'beggared belief' that a mere 'innocent schoolgirl' should be able to 'conceive and then successfully carry out ... so gigantic an imposture'. As it turned out, she was entirely capable![59]

Bad Apples

Whether such eminent SPR-men were correct to believe in ghosts is beside the point. What I am trying to get at is only that they were *interested* in such subjects, and didn't simply treat them with automatic contempt. Compare that to the situation today, where for most scientists to admit to any similar curiosity would be tantamount to career suicide. In his book *The Science Delusion*, Rupert Sheldrake gives a fascinating example of this kind of attitude at work. Sheldrake notes the common piece of laboratory folklore that there exists such a thing as an 'experimenter effect', in which the presence of a certain person has either a negative or a positive influence upon the outcome of an experiment. The most famous example was the quantum physicist Wolfgang Pauli (1900–58), who was convinced he had the inadvertent quality of causing expensive equipment to fail in his presence – he honestly believed he once caused a cyclotron to catch fire simply by being nearby. Sheldrake gives the further case of a necessarily anonymous 'professor of biochemistry from a major US university' he knows, who claims to be able to achieve superior separations of protein molecules to his lab partners simply by standing in the room with his apparatus, willing the molecules to split apart, and chanting the word 'separate'![60]

Sheldrake took this unlikely-sounding idea seriously and suggested a simple experiment – get two samples of the same mixture of proteins, leave one alone in a room and put the other in the same room as the psychic scientist, and when the separation process was over, see if there was any difference in the results. My guess is that there would not be (and in any case, two experiments is too small a sample from which to draw conclusions), but it is hard to see what damage would come of the test. It would be cheap and easy to arrange and, if nothing else, would be a bit of harmless fun. Nonetheless, said Sheldrake, the scientist would not perform the trial – 'although he was curious, he could not risk the potential damage to his credibility and career'.[61] That such an action could plausibly ruin a man's livelihood seems to me rather stranger even than the idea that you can get proteins to split simply by shouting at them. Whatever would have become of the likes of Sir Thomas Browne and Dr Thomas Beddoes today?

Presumably they would have been derided as pseudoscientists, a label often attached to a very appropriately named Japanese researcher called Dr Masaru Emoto (1943–2014), who also attempted to influence substances in a laboratory using only the power of his mind. By pouring ordinary water into two containers, one marked 'LOVE', the other 'HATE', and then beaming positive or negative thoughts at them before introducing a freezing process, Emoto claimed to have produced amazing results. Examining the resultant ice crystals under a microscope, Emoto said the water subjected to brainwaves of love formed beautiful symmetrical shapes, while the water given nothing but psychic hatred developed ugly, malformed crystals. Whatever you think of Emoto's findings, this does sound like exactly the kind of thing Sir Thomas Browne would have done!

You would think it would be easy for scientists to repeat such simple experiments in order to prove or disprove (presumably the latter) Emoto's idea. It may be an amusing thing to do one idle lunchtime. It may even win you an IgNobel Prize. Instead, the field is left to enthusiastic amateurs, the most recent being a self-improvement guru from Kent called Nikki Owen, whose book *Charismatic to the Core* lays out her unusual beliefs about fruit. Owen has recently repeated Emoto's trials, but with apple halves in sealed jars instead of frozen water. She says that the loved apples stayed fresh, whereas the hated ones quickly shrivelled and dried. Impressed by Owen's claims, many members of the public performed experiments themselves to back her up, and provided important testimony about their innovative research methods. 'Several times a day I would sing romantic songs to the "love" apple,' said one bold citizen experimenter. 'Billy Joel's "Just the Way You Are" was a particular favourite.' The 'hate' apple, however, was subjected to open abuse: 'I shouted at it and said it was to blame when I got a parking ticket, and I started putting it on a "naughty apple" step.' Asked about the science behind her findings, Owen was quite open in her honesty. 'There isn't any,' she told a reporter from the *Daily Mail* in 2015.[62]

One-Eyed CSICOPS

Most people probably think all parapsychology is equally as laughable, but this is not the case. Whether or not psychic powers are real, parapsychology does produce some genuinely anomalous results, yet this fact is often simply brushed aside. Maybe such studies are flawed. Maybe they are not. But surely they still deserve to be properly investigated? If nothing else, their claims can sometimes conclusively be disproved. The trouble is that sometimes they cannot; and so perhaps it is better not really to look into them fully at all. The most obvious contemporary example of this kind of thinking is an organisation called CSICOP – the Committee for the Scientific Investigation of Claims of the Paranormal (recently rechristened CSI). Founded in 1976, the organisation publishes a magazine, *The Skeptical Inquirer*, and has as its aim the scrutiny and debunking of anything they deem to be pseudoscience; things like psychical research, for instance. Its membership backlist is just as impressive as that of the SPR, with multiple Nobel laureates, and such famed scientific figures as Carl Sagan and Richard Dawkins. There is little doubt that CSICOP has helped expose any number of woeful frauds and examples of dubious thinking; but one of the most dubious of all pieces of thought it has ever been involved with was one of its own.

CSICOP's origins lay in a 1975 campaign by the philosopher Paul Kurtz (1925–2012), editor of *The Humanist* magazine, to denounce astrology as nonsense. Kurtz's crusade gained much media publicity and provided the initial impetus for the founding of CSICOP. Soon, the astrologer Michel Gauquelin (1928–91) challenged Kurtz and his colleagues to investigate certain astrological findings he claimed to have made, a test which was accepted. However, the approach CSICOP adopted towards data analysis was flawed, as an astronomer sitting on their executive council, Dennis Rawlins (b. 1937), tried to point out. Nonetheless, Rawlins' warnings were ignored, and CSICOP, amusingly, ended up producing a study which appeared to provide statistical proof that a person's level of sporting ability was determined by the position of Mars in the sky at the time of their birth. Kurtz refused to publish the findings. Rawlins, while sensibly not believing the study's conclusions, pointed out that this was hardly in the spirit of scientific openness. Eventually, he was forced out of CSICOP for his troubles, taking some other members with him in protest. In 1981, Rawlins published an exposé in an American paranormal magazine, after which CSICOP adopted a formal policy of not conducting any further such research in future – presumably on the grounds that they might not like the findings it produced.[63] Astrology is not a science. But neither is the approach adopted by science's supposed defenders at CSICOP!

CSICOP's true main function is to ridicule the paranormal in the media and thus promote a doctrine of materialism among the general public. As far as I can tell, its standard tactic for denouncing psychical research is to argue that 'this specific example of psychic powers cannot exist, because psychic powers themselves do not exist' – even when this shameless tautology is in (apparent) contradiction with actual research findings. One of CSICOP's critics, the American parapsychologist George P. Hansen, summarises its attitude thus:

> ... scientific investigation is inappropriate for [CSICOP]. If a serious, sustained effort were undertaken to investigate the paranormal, that by itself would confer status upon the topic. It would signal the paranormal to be worthy of study. Instead, the Committee belittles such efforts, and its magazine carries cartoons and caricatures that ridicule researchers.[64]

Some such cartoons are funny; in one a fortune teller looks at a customer's palm and says, 'I see you are very gullible.'[65] Believers in the supernatural, however, hold no monopoly upon bizarre and unlikely ideas; Paul Kurtz himself once wrote a book in which it was suggested that Jesus and Lazarus were in a gay relationship together.[66] Predictably, given such Church-baiting notions, CSICOP has been formally allied with various atheist groups down the years, for a while sharing a building, equipment and personnel with CODESH, the Council for Democratic and Secular Humanism – whose head was also Paul Kurtz.[67] An exception to this rule was the CSICOP founding member Martin Gardner (1914–2010), upon whose 1957 study *Fads & Fallacies in the Name of Science* I have drawn in writing this present book. Gardner devoted a fair portion of his life to debunking pseudoscience and the paranormal, and often did so very well (and very entertainingly) indeed, becoming lauded as 'the godfather of the skeptical movement'. Naturally, his motivation was one of materialism; wasn't it? Not so. In fact, as George P. Hansen had confirmed through personal correspondence, Gardner was a religious man, believing in both an afterlife and the power of prayer. His distaste for parapsychology was partly based upon a belief it was an 'insult to God', and the use of psychic abilities a blasphemous attempt to test His divine powers.[68]

The Uncertainty Principle

Gardner was an exception, though, and CSICOP's purpose overall seems less to *disprove* the paranormal than to propagandise upon behalf of materialism. Science in this outlook becomes hubristically redefined as 'what *is* real' as opposed to simply 'what we *know* to be real' (the word comes from the Latin *scientia*, meaning 'to know'). By conflating the two ideas, organisations like CSICOP try to establish science, and its current body of knowledge, as being the only arbiter and definition of reality, which clearly cannot be true. Presumably there is indeed a form of ultimate underlying reality to the universe, but, strictly speaking, it must lie outside of science, which will always remain at least somewhat incomplete. Two hundred years ago, the very notion of subatomic particles lay outside of science, for instance, and yet they still existed. 'It is the theory which decides what we can observe,' said Einstein,[69] but it is equally the implicit assumptions and methods underlying science which decide what we can observe, too.

There are many aspects of lab-based parapsychology whose sins seem not only to go against known physical laws but challenge science in other ways, too. Many of its results are non-repeatable, and there are also problems relating to which hypothetical psychic force is being made use of in any given experiment; for example, in a test based around guessing the images printed on a series of cards, does the successful sitter read the minds of the supervising scientists through telepathy, view the images themselves through clairvoyance, or sense the future cards to come through precognition?[70] There is also the problem that psychic powers may be 'goal-oriented', a term coined by the German physicist Helmut Schmidt (1928–2011) in 1974 to describe the way that the effectiveness of psychic processes seem to be unaffected by the complexity of the task at hand. For instance, say in one experiment a psychic was asked to influence a die to fall on six, and in a second experiment asked to guess the correct number to be targeted first before then influencing the fall of the die, the level of success might not be altered. Thus, the psychic would be equally as good performing a one-stage task as a two-stage task.[71] Schmidt's conclusion was that, instead of viewing psychic force as 'some mechanism by which the mind interferes with the [process] in question', maybe we should see it as one that 'aims successfully at a final event, no matter how intricate the intermediate steps'.[72] Therefore, rather than viewing what went on as the sitter beaming waves of psychic force out of his head and using this to alter the die's fall, perhaps no energetic processes would be involved in such a feat at all? If this were so, then it is hard to see how the precise mechanism involved could ever be successfully deciphered by science, seeing as all observations must by definition be based ultimately upon the observation of energies at work or bodies in motion; maybe the term 'mechanism' is not really appropriate to describe such a hypothetical process at all?

Some other experiments produce even more bizarre implications. For example, in 1976 Schmidt set up a random-number generator (RNG) which would make a click whenever it produced a certain number. He then asked sitters to try and increase the rate at which clicks occurred; apparently, they could do so. Intrigued, Schmidt recorded the RNG at work without the psychics being present, then played half of the tape back to them and asked them to increase the rate of clicks. The data showed that the first half of such tapes showed a higher click rate than the second halves, which the psychics had left alone. Schmidt's conclusion was that, somehow, the psychics had influenced past events from a point in the future! (That is to say, a point in the future relative to the time at which the tapes had been recorded.)[73] Other researchers replicated such results. Another German, Elmar Gruber, set up photoelectric sensors which produced a click on a tape every time someone entered a certain supermarket, or a car passed through a certain tunnel. Then, he played parts of the tapes to psychics and asked them to increase the click rate; Gruber said they succeeded, these sections of the recording having more clicks than the other, control sections.[74] Presuming that these results were accurate (and they do strain credulity), then the problems which could potentially arise for researchers are profound; because, as George P. Hansen put it, 'if people in the future can influence the past ... how can we know whether an experiment is ever finished? If a study is conducted and published, could someone later read it and alter the outcome?'[75]

These seem like comical, made-up problems fit only to be tackled by Doctor Who. But Schmidt and Gruber do appear to have produced some genuinely anomalous results. Maybe their methods were flawed in some obscure way that nobody has yet managed to determine. Possibly the data admit of other interpretations. If what Schmidt and Gruber were saying *was* true, though, then such findings would act to undermine the ultimate validity of the scientific method itself. It is this, I think, that really makes such topics so incredibly taboo, and leads to people like Sheldrake's protein-separating psychic professor fearing for their careers should they dare to try and investigate them. If there are elements of the world which science cannot fully investigate and understand, then it would have to drop its stance of being the only arbiter of reality. Science would once more mean 'what we can *know* to be real', not 'what *is* real'. Many would find such a notion to be profoundly disturbing.

Weighing the Evidence

Nonetheless, the underlying assumptions of materialistic science and its methods are so ingrained within the modern Western mindset by now that even parapsychologists seek to imitate it in their experiments. The general public may be surprised to hear how incredibly dull most parapsychology experiments are. A man sat at a desk spending half an hour trying to will a die to fall on 'six', or shift a tiny cube one millimetre to the left, are not the stuff of *Carrie* or *The Sixth Sense*. In fact, one of the main alleged problems with lab-based psychical research is that the tests are so overwhelmingly tedious that the psychics involved get so bored that their results start to tail off! So keen are many parapsychologists to conform to the standards of conventional science that in some respects they actually outdo their more mainstream counterparts; one 1999 study performed by Rupert Sheldrake, for instance, found that 85.2 per cent of parapsychology experiments were performed 'blind', as opposed to merely 24.2 per cent of medical trials, and an astonishing 0 per cent in the physical sciences.[76]

Some of the most comical experiments in psychical research have come when their inventors, having internalised the implications of materialism, have sought to provide proof of matters immaterial in avowedly material terms. For example, several attempts have been made over the years to weigh the souls of both animals and human beings. Joseph Priestley himself once set up a trial in which he weighed mice just before and after they died – finding no loss in weight he could attribute to their spirits vacating the body.[77] This didn't stop people trying to perform similar experiments with humans, though. When the idea of doing so was first put forward in 1854 by the German anatomist Rudolph Wagner (1805–64), he was mocked by facetious suggestions that it might also be possible to flash-freeze the soul at the moment of death and then bottle it for display purposes. By 1896, however, Dr Duncan MacDougall (1866–1920), a surgeon and member of the American SPR, was making serious plans to weigh a dying human immediately before and after he shuffled off this mortal coil. In 1901 he claimed to have registered a sudden 21-gram weight loss in a TB patient at the point of his demise. A second patient of MacDougall's then registered a mere 14-gram weight loss after death, this particular individual evidently having somewhat less soul to spare than the first. Another patient, though, recorded none at all; perhaps he had no soul to shed? Extending the trial to dogs, MacDougall anaesthetised fifteen animals then killed them on an industrial beam-scale; they also registered no weight loss, leading the doctor to conclude that there was no Doggy Heaven after all. Eventually, hospital authorities put a stop to MacDougall's (deeply flawed) tests, and they seem never to have been repeated, having been weighed in the balance and found wanting.[78]

I have no idea if there is such a thing as a human (or rodent) soul. But, if there is, I somehow doubt it would prove to be measurable by weight. Absence of evidence is not, however, evidence of absence; while nothing can be found to suggest the existence of the soul as a material body, that would by definition only be a problem for someone who was inherently wedded to the idea of it actually *being* a material body. The insistence upon reducing absolutely everything down to materialist terms and then dismissing anything which is not material as automatically not real is characteristic of a movement which has been termed scient*ism*, not science as such. Scientism, roughly speaking, is what you get when science pokes its nose into areas where it does not truly belong, such as religion or metaphysics. By insisting only that which can be analysed via scientific methods is truly real, scientism is a school of thought that reduces reality down only to that which can be measured – making love into the mere result of chemical imbalances in the brain, or an inevitable consequence of the evolutionary imperative for species to reproduce, for instance.[79]

Spiritual matters were once largely left alone by science, and deemed a separate sphere of experience – the two areas were, as the palaeontologist Stephen Jay Gould (1941–2002) famously put it in a 1997 essay, 'non-overlapping magisteria'. In the view of Gould, science deals with facts whereas religion deals with spiritual experience and values, two legitimate but separate domains which 'do not overlap'. As he put it:

> Science tries to document the factual character of the natural world and to develop theories that coordinate and explain these facts. Religion, on the other hand, operates in the equally important, but utterly different, realm of human purposes, meanings, and values – subjects that the factual realm of science might illuminate, but can never resolve.[80]

If only his critics had agreed with him. Gould's most prominent opponent was the evolutionary biologist Richard Dawkins, probably the best-known militant atheist in the world today. Dawkins and Jay Gould already had differing views as to how precisely evolution worked, but their opinions of science and religion's proper place in the world were, if anything, even more divergent, and the two men ended up having something of a feud over the matter. According to Dawkins, 'religions make existence claims, [i.e. that God exists] and this means scientific claims'.[81] As such, he saw no reason why he should not undertake to step in and subject God to rigorous scientific analysis. But is such a thing truly possible? Repeatedly, people of Dawkins' mindset construct religious straw men and then demolish them (and I speak here as someone who possesses no religious beliefs whatsoever). In his bestselling 2006 book *The God Delusion*, for instance, while Dawkins admits that he cannot actually disprove the existence of God,[82] he nonetheless sets out to do so; but the God which he tries to eliminate is basically a cartoon which few actual theologians would consider fit to be believed in by anyone beyond small children and abject simpletons.

Good Fences Make Good Neighbours

Some of the pronouncements of the chief prophets of scientism can be quite amusing in nature, in the way that they singularly fail to see the point of the things they airily dismiss. Far odder than his diatribes against organised religion, for example, Dawkins has also in the past preached anathema upon such heinous and 'harmful' practices as telling your children fairy tales or letting them believe in Father Christmas. According to Dawkins, his super-intelligent infant self had been able to work out that the 'Santa Claus' who visited his childhood home at Christmas-time was simply a family friend wearing cheap disguise, something he seemed very pleased to relate. Deeming it harmful to teach small children to believe in such fictional festive figures, in a 2014 speech Dawkins moved on to explaining why it was also 'rather pernicious' to tell toddlers fairy tales, as they act 'to inculcate into a child a view of the world which includes supernaturalism'. Specifically, Dawkins objected to the much-loved tale of the frog prince, and its message that human beings might potentially be transformed into amphibians via magic. Rather than reading this unscientific tosh to their kids before bedtime, Dawkins said that conscientious parents should instead overload their minds with mathematical data, by telling them that 'there's a very interesting reason why a prince could not turn into a frog – it's statistically too improbable'.[83] To be fair, such an approach would certainly get little ones quickly nodding off to sleep ...

This is all despite the fact that Dawkins himself once penned his own scientific fairy tale in his 2004 book *The Ancestor's Tale*, which attempted to resituate the story of humankind's evolution within a pastiche of the framework of Chaucer's *Canterbury Tales*. Even his 1976 masterpiece *The Selfish Gene*, which recasts all living creatures as, in his words, 'gigantic lumbering robots' designed to ensure the survival of their genes, relies to a great degree upon metaphor to get its message across and could be seen as being a kind of engaging 'origins of life'-type fable every bit as much as it is a scientific theory. At one point, Dawkins actually says that DNA 'moves in mysterious ways' just like God is meant to.[84] What genes actually do, though, is to control the sequence of amino acids in protein molecules. They have no deliberate 'purpose' to them at all; they do not actually 'desire' to replicate themselves through the offspring of their 'lumbering robots', nor are they actually 'selfish', and much of Dawkins' persuasive power – and, being an excellent writer, he *is* persuasive – lies in the metaphor he lays onto fact, rather than upon the underlying facts themselves *per se*, at least for the lay reader.[85]

The book is, in other words, a kind of scientific fairy tale (which is not to say that Dawkins' theory *itself* is actually untrue, of course).

Such techniques of narrative have a longer history than you might presume. A 2015 book, *Science in Wonderland* by Melanie Keene, has explored the curious Victorian phenomenon of 'scientific fairy tales', a genre of children's stories in which fantastic characters with names like The Fairy of Hydrochloric Acid tried to explain the wonders of the world in scientific terms, in an attempt to wean readers off any lingering infant belief in magic. The *real* magic in the world, such tales implied, lay in advancing Victorian science, not in the supernatural. Or, as one reviewer of Keene's book put it, while Shakespeare's fairy Puck 'claimed to be able to put a girdle round the Earth ... submarine cables could do it so much better'.[86] Few little girls, however, go around dressing up as cable engineers ...

Another wrong-headed expression of such attitudes can be seen in the occasional dismissal that is made of deliberately far-fetched but enjoyable TV programmes purely on the grounds of their scientific inaccuracy. A particular materialist bugbear was the hugely popular 1990s American series *The X-Files*, in which a pair of FBI agents investigated paranormal phenomena, frequently finding evidence they might be genuine. In a 1996 lecture, Richard Dawkins criticised the show as being far too partisan. Every week, he said, its storylines 'ram home the same prejudice or bias'; namely, that things like UFOs, ghosts and Bigfoot were real, not imaginary. Absurdly, Dawkins compared this to the idea of a programme deliberately designed to reinforce racial prejudices:

> Imagine a crime series in which, every week, there is a white suspect and a black suspect. And every week, lo and behold, the black one turns out to have done it. Unpardonable, of course. And my point is that you could not defend it by saying, 'But it's only fiction, only entertainment.'[87]

Carl Sagan, meanwhile, another *X-Files* critic, chided the show as providing less of a rationalist 'public service' than did old *Scooby-Doo* cartoons, in which the ghost or monster is always unmasked as a disguised human criminal come the end of the episode. Far more realistic, Sagan said, would be a drama series 'in which paranormal claims are systematically investigated and every case is found to be explicable in prosaic terms'. More realistic, maybe – but more entertaining? Sagan also had an issue with *Star Trek*, specifically the fact that the character Mr Spock, who hailed from the Planet Vulcan, had evolved to so closely resemble a human being with two arms, legs and a head. Admittedly, said Sagan, an alien of this kind would be easier to represent on-screen, but nonetheless such a thing still 'flies in the face of the stochastic [statistically unpredictable] nature of the evolutionary process'.[88] Heaven forfend!

What are scientists doing, making mad pronouncements upon the evolutionary accuracy of fictional TV shows, or what stories parents should read their children in bed at night? The answer, surely, is 'interfering outside of their proper sphere'. But then, if science is increasingly coming to be defined *as* reality by such people, then I suppose they might counter that there is actually *nothing* which lies outside of their proper sphere, not even Mr Spock's body shape. Ironically, this was just the attitude once adopted by scientism's chief and eternal bogeyman, the Catholic Church. That institution, too, once tried to claim there was nothing which lay beyond its own personal purview, an outlook famously contested by scientism's chief hero, Galileo Galilei. In his 1615 *Letter to the Grand Duchess Christina*, Galileo protested eloquently against the traditional Church view of only accepting as true what was said in the Bible. Were there not, he said, other sources of knowledge open to humanity? Who, he queried, 'will say that geometry, astronomy, music and medicine are more excellently contained in the Bible than they are in the books of Archimedes, Ptolemy, Boethius and Galen'? If the Church's professors of theology had 'no regard' for such subjects, deriding them as comparatively unimportant 'because they are not concerned with blessedness', he said, then they

> should not arrogate to themselves the authority to decide on controversies in professions which they have neither studied nor practised. Why, this would be as if an absolute despot, being neither a physician nor an architect, but knowing himself free to command, should undertake to administer medicines and erect buildings according to his whim – at grave peril of his poor patients' lives, and the speedy collapse of his edifices.[89]

The likes of Richard Dawkins would do well to apply such words to themselves. Science, in the fundamentalist guise of scientism at least, cannot admit to the existence of spheres outside its scope, as that would be tantamount to dropping its current status as the only uncontested arbiter of reality. It was once precisely the same with organised religion. Are the implications of this parallel really so difficult for materialist hard-liners to grasp?

The Spirit of the Age

The fact that so many of the great scientists of the Victorian era, often thought of as one of the high points of philosophical materialism, were interested in spiritual matters may now seem surprising, but this is only because the stereotypical image of the scientist we are continually presented with today is itself a purely materialist one. Historically-speaking, this is not an accurate picture to present. Many of the nineteenth century's greatest men of science were in fact highly religious, with figures like James Clerk-Maxwell (1831–79), Lord Kelvin (1824–1907), James Joule (1818–89), Michael Faraday (1791–1867) and Sir Charles Lyell (1797–1875) all being fixed believers in a deity of some kind.[90] Some of the very greatest scientists of the day were members of religious orders. Gregor Mendel (1842–84), for example, whose famous experiments with peas proved the influence of dominant and recessive genes in heredity, was an Augustinian monk who liked to rise before daylight and sing a hymn to the dawn, giving voice to the opinion that it was God 'who hath made us and not we ourselves'.[91]

Even some of the fathers of evolutionary theory were not always quite as materialist as we are led to believe. Darwin himself was set on becoming an Anglican minister during his early days, before losing his faith through the close study of what he was later to conclude was not really God's Creation at all,[92] and, following the death of his beloved daughter Annie, even once saw fit to attend a spiritualist séance. Here, a number of chairs supposedly flew through the air, over the sitters' heads, and then landed on a table – frolics Darwin himself sadly missed, having gone upstairs partway through the ritual.[93] Darwin ultimately came to dismiss the spiritualist craze, but one other prominent naturalist who did not was Alfred Russel Wallace (1823–1913), the co-discoverer of the theory of evolution. Darwin, nervous about the implications of his ideas, had sat on them for several years, and was disturbed in 1858 to find that Wallace had sent him a paper demonstrating that he, too, had come to the exact same conclusions. Not wishing to be scooped, on 1 July that year Darwin agreed to have a joint paper by himself and Wallace read before the Linnean Society, in which the pair's ideas were finally made public.[94] An issue which Darwin might have preferred Wallace not to have publicised, though, was his own frequent attendance at séances. These began in July 1865 after Wallace's return from a scientific expedition to the Dutch East Indies, when he attended a sitting at which a table moved, rapping noises were heard, and he purportedly received a message from his dead brother. At first Wallace was sceptical, but soon changed his mind, claiming to have observed fresh flowers falling down from the ceiling, and the levitation of a heavy table with a woman sat on top. Once, during a sitting with the famous Victorian medium D. D. Home (1833–86), Wallace even claimed to have witnessed a disembodied ghost hand playing an accordion beneath a table![95]

Before long, he was writing books with titles like *The Scientific Aspect of the Supernatural* and *Miracles and Modern Spiritualism*, and questioning the implications of his experiences for the theory he and Darwin had introduced to the world. His ultimate conclusion was one still routinely trotted out by some religious types today; namely, that the laws of evolution were indeed real, but had been created by an 'Overruling Intelligence' which 'has watched over the action of those laws, so directing their variations, so determining their accumulation, as finally to produce an organisation sufficiently perfect to admit of, and even to aid in, the indefinite advancement of our mental and moral nature'.[96]

The idea, so common among people like Richard Dawkins today, that Darwin has somehow 'disproved' the existence of God, and 'proved' the truth of materialism, is nothing but a myth. It's certainly a reasonable assumption, but not an unarguable one, clearly, if even one of the co-discoverers of evolutionary theory himself could feel moved to reject it. Even at the end of *On the Origin of Species*, Darwin put forward the suggestion that evolutionary processes could simply be viewed as being 'laws impressed on matter by the Creator'.[97] Darwin didn't genuinely believe this, however, he just didn't wish to offend public sensibilities, actually considering Wallace's conclusions to be a 'little heresy'.[98] If so, though, then it was a 'little heresy' which many persons were also to make in the years to come – and, indeed, that some had already been making for some time previously.

The Wonders of Creation

We often think of Creationism as being a modern pseudoscience, which sprang up in direct response to Darwin's theories, but its intellectual ancestors actually appeared many years beforehand. Fossils predated Darwin, and religious people had to account for them somehow. One way of managing the situation was to suppress the ideas of early evolutionary theorists. When a French potter named Bernard Palissy (*c.* 1510–90) suggested fossils might be the bones of ancient animals, he was charged with heresy and thrown into the Bastille, where he died in 1590. In 1708, Dr Johann Scheuchzer (1672–1733) said the same thing and was treated more kindly – he was simply ridiculed, and told fossils were just rocks which *happened* to look like animal bones. That arch-rationalist Voltaire (1694–1778) had a more nuanced view. When fossilised fish were found on mountaintops, he agreed they were indeed fossilised fish. He disagreed, though, that this meant that the mountains themselves had once been underwater. Instead, he said that travellers had eaten fish suppers on the mountains aeons ago and tossed away the remains. Karl Linnaeus (1707–78), the great taxonomist – he who gave us 'Latin names' for plants and animals – refused to believe new species could emerge or old ones become extinct. Instead, he thought fossils were simply oddly shaped minerals.[99] Other ideas about the 'true' nature of fossils included that they grew from seeds or sea sperm, were produced by the action of salts, or that they were simply Nature's little jokes. The Chinese thought they were dragon bones.[100]

Some scholars, like Dr Scheuchzer, tried to reconcile the fossil record with the Bible. Maybe there were once many animals which are now extinct, this extinction being caused not by evolution but by the biblical Great Flood? Scheuchzer found a large skeleton which he claimed to be of a now-extinct pre-Flood type of human that he called *Homo diluvii testis*. It later turned out to be the fossil of a giant salamander.[101] Scheuchzer did have allies, though, such as the Frenchmen Comte Buffon (1707–88) and Georges Cuvier (1769–1832), who thought that God had 'wiped clean the slate' with a series of catastrophic Great Floods, getting rid of many prehistoric animals to make way for a new and improved batch of beings. Creation, Cuvier thought, must have taken place in a series of separate stages, not all at once.[102] Further speculation came from Dr Johann Beringer (1667–1740), who in 1726 published a remarkable book, *Lithographia Wirceburgensis*, featuring a wealth of illustrations of fossils he had found littering a mountainside near Würzberg in Germany. Some showed dead animals caught in remarkable poses: wasps drinking from flowers, spiders sat in webs, frogs having sex. Others showed pictures of comets and stars. Most peculiar were stones filled with fossilised Hebrew letters spelling out the name of God. Were these fossils God's early practice attempts at Creation, signed by Himself as if by an artist? No. After his book was published, Beringer found another stone, bearing his own name. It turned out that his academic rivals and a gang of rebellious students had carved the fossils themselves and then planted them on the mountain purely to make him look stupid![103]

Perhaps the most ingenious attempt to reconcile the poetic testimony of the Bible with actual geological fact came from the Victorian zoologist Philip Gosse (1810–88). Gosse was a well-respected naturalist, who was both the inventor of the aquarium and the world's leading expert on the genitalia of butterflies (I suppose someone had to be), so the outrageous claims he made in his controversial 1857 book *Omphalos: An Attempt to Untie the Geological Knot* shocked the world of Victorian science. In this much-mocked title, Gosse made the extraordinary assertion that all fossils and time-worn geological features were simply fakes which had been placed there by God to test our faith. The bones of extinct animals, the remnants of vanished volcanoes, the telltale signs that our landscapes have been formed over millions of years by the actions of glaciers and rivers – all were created by Jehovah purely in order to make our planet seem more realistic. Just as a set designer might leave food and dishes from a meal which never really took place scattered across a table on stage so as not to break the illusion of reality for theatregoers, so God left the fossilised remains of animals like woolly mammoths which never really existed lying buried within rock strata and sediments. Such was the Deity's commitment to verisimilitude, Gosse explained, that He even went so far as to scatter the globe with fossilised dinosaur turds! The strange thing about Gosse's theory, however, is that it is completely beyond proving either way – while clearly ridiculous, it is actually just as impossible to show that it is definitely false as that it is definitely true. As he himself pointed out, God might have created the world as it is now only ten minutes ago, planting false memories in all our minds about every detail of our 'lives' up to this point, and we would have no way of knowing the fact.[104]

A modern theory every bit as bizarre as Gosse's can be found in the American TV host and so-called 'Young Earth Creationist' Carl Baugh's (b. 1936) 1987 book *Dinosaur: Scientific Evidence That Dinosaurs and Men Walked Together*. Baugh organised fossil-hunting expeditions in Texas which apparently uncovered several human footprints existing alongside fossilised dinosaur tracks, something he cited as proof that humans didn't evolve after dinosaurs after all, because there is no such thing as evolution, seeing as it isn't mentioned in the Bible (neither are dinosaurs ...). Furthermore, Baugh also professed to have discovered a fossilised human finger from the cretaceous period, 145–66 million years ago – but, regrettably, it appears to be just a small, elongated rock which he has simply said is a finger, because he wants it to be. Baugh's weird conclusion is that ancient man used giant herbivorous dinosaurs as living, domesticated combine-harvesters; without their 'beastly friends' eating or harvesting all the vines, ferns and trees of the primeval jungle, humans would have been overgrown by rampant vegetation and died, he said. In interviews, he has openly compared prehistory to an episode of *The Flintstones*, a hypothesis he appears to think is genuinely plausible. Subsequent investigation of Baugh's supposedly 'human' footprints, however, has shown them to be nothing of the kind, so the involvement of dinosaurs with early agriculture must remain as yet considered unproven. As an aside, Baugh also believes in the existence of something called 'hexagonal water' which has amazing, God-given curative properties, but the less said about that particular fantasy, the better.[105]

Sensitive Souls

Other people have accepted the reality of evolution, but tried to reclaim it as a spiritual phenomenon, notably the French philosopher Henri

Bergson (1859–1941), who spoke of an *élan vital*, or 'life force', which controlled the process. Bergson described evolution in a somewhat personified way, as if it itself was God; in his 1911 book *Creative Evolution*, Bergson spoke of evolution as a path of 'creation that goes on forever in virtue of an initial movement'. Rather than having an overall goal or end point in mind, evolution for Bergson was a kind of ongoing creativity which would proceed forever. In a sense, God was creating *Himself* through the new life forms spawned by the *élan vital*, he suggested. Bergson's thought could be viewed as being a mystical variant of Lamarckism, with a certain species having some vague desire in mind before the *élan vital* steps in and makes their wish come true somehow. Imagine, for instance, early sightless animals having the desire to see. There is no such thing as an eye yet, but the vital life force steps in and gradually invents one by influencing the species' development, an act of quasi-artistic creativity.[106] Obviously, there are problems with this idea. As far as we know, a race of seafloor-dwelling creatures called trilobites were some of the first prehistoric animals to develop sophisticated eyes. Why them? Did they want to see their world more than any other species did? And why are some creatures living under similar conditions, like sea cucumbers, still blind today? Did they signal to the *élan vital* to leave them as they were? If so, why? Are they all luddites? Do they feel that ignorance is bliss? Bergson's ideas are fascinating, but essentially have to be taken on faith.

Several people did embrace Bergson's faith, though, like his translator and disciple T. E. Hulme (1883–1917), an English poet who described evolution as being 'the gradual insertion of more and more freedom into matter'. In an amoeba, said Hulme, the *élan vital* had 'manufactured a small leak, through which free activity could be inserted into the world, and the process of evolution has been the gradual enlargement of this leak'.[107] Such ideas have proved as appealing to those with an interest in the supernatural as to those with an interest in science, however – little surprise, seeing as Bergson himself was once President of the SPR, and his sister was married to a prominent occultist, the bizarre fake Scotsman MacGregor Mathers (1854–1918). This, for example, is a passage from a book by a Celtic scholar and mystic named W. Y. Evans-Wentz (1878–1965), *The Fairy-Faith in Celtic Countries*, published in 1911, the same year as Bergson's *Creative Evolution*:

> A view of evolution is rapidly developing in the scientific world ... regarding all evolutionary processes, reaching from the lowest to the highest organisms, as illustrating a gradual unfolding ... of a pre-existing psychical power through an ever-increasing complexity of specialised structures, this complexity being brought about by

> natural selection ... [therefore] there must have been life before its physical manifestation or its physical evolution began. We may regard this psychical power as like a vast reservoir of consciousness ever trying to force itself through matter, the walls of the reservoir. Through the microscopic body of an amoeba there has percolated a very minute drop from the reservoir. As evolution advances, the walls of the reservoir become more and more porous, and little by little the drop advances to a tiny rivulet. Through the higher animals, the tiny rivulet flows as a brook. Through man as he is, the brook flows as a deep and broad river. Throughout the completely evolved man of the far distant future, the deep and broad river will have overflowed all its banks [and he will become psychic].[108]

Living prior to modern knowledge about genetics, Evans-Wentz considered 'the theory of the mechanical transmission of acquired characteristics' to have failed. As such, he said, there had to be some equivalent to Bergson's *élan vital* in existence which used the genes of any given species 'merely as a physical basis for its manifestation', using them to 'build up a body suited to its further evolutionary needs'.[109] An expert upon what he called 'the Celtic Esoteric Doctrine of Re-Birth', Evans-Wentz went on to imply that ancient ideas of reincarnation were really just a primitive attempt to formulate a basic theory of evolution. Rather than any individual person or animal being reincarnated, Evans-Wentz proposed, the ancient Celts actually believed that it was the 'life force' inhabiting them which was reborn, with each new baby creature thus being 'the bearer of all evolutionary gains' made by the *élan vital* in its previous earthly incarnations, reaching right back to the first single-celled organisms. He tried to prove this by virtue of the fact that, while in the womb, a foetus undergoes various developmental stages derived from humanity's previous, non-human forms, temporarily growing gill slits in an echo of the days when our distant ancestors were water dwellers, for instance. Life itself, Evans-Wentz says, is thus a 'permanent evolving principle', just as Bergson argued.[110] Therefore, 'the Druids and other ancients ... seem not only to have anticipated Darwin by thousands of years, but also to have quite surpassed him'.[111]

While he cites Bergson, Evans-Wentz doesn't actually use the term *élan vital* himself, preferring the phrase 'soul-stuff', taken from the American founder of psychology, William James (1842–1910) – another prominent SPR man. In a 1909 essay, James defined this soul-stuff as being 'a cosmic environment of [disembodied] other consciousness of some sort', imagining a vast sea of it extending out across the universe, within which each individual human mind participated. Our minds may be, he said, 'like trees in the forest' which 'commingle their roots in the darkness underground', with everyone part of a collective unconscious lying inherent within Creation.[112] Perhaps this was the medium through which trilobites could ask for their future eyes and sea-cucumbers politely decline them? If so, then Evans-Wentz was part of a growing esoteric movement which was also petitioning evolution for a new kind of eye – a third eye, which would allow humanity one day to participate within the psychic realm, seeing spirits and, in particular, fairies. His 1911 book is a record of his wanderings around the Celtic Fringe hunting both fairies and fairy-seers, and Evans-Wentz is quite clear that one day he expects mankind to develop the ability to perceive the Hidden Folk. Man's future evolution, he says, will allow his individual mind to merge with James' soul-stuff, thus allowing the super-being of tomorrow to see whatever spiritual entities he so desires.[113]

Fairy Family Trees

Other people, too, were hunting fairies in the name of Darwin around this time. Theosophy, for instance, an esoteric movement founded in 1875, involved a deliberate recasting of Darwinian theory in mystical terms. One bizarre result of this was detailed in a 1913 book by the clergyman turned Theosophist Charles Webster Leadbeater (1854–1934), called *The Hidden Side of Things*. For Leadbeater, evolution was an expression of a divine force, immanent in nature, which explained not only the existence of all physical life on Earth, but all spiritual life, too. The end result of this belief was the creation of a new branch of the evolutionary tree – for fairies.

Leadbeater was quite sincere in this endeavour, and unlike Darwin in that he viewed evolution as something with a pre-defined goal at the end – namely, the progression of all things from a state of so-called 'Mineral Life' up to the status of 'Solar Spirits', angelic beings next only to God. It appears Leadbeater thought all life came ultimately from the sun, or Solar Deity, and was slowly returning back up to it through a process of evolution. With each stage of this evolution, living things became more and more conscious, with men being more fully alive than apes, and both being of a higher mentality than rodents. This same pattern held true for fairies, said Leadbeater, illustrating his concept by referring to the alleged existence of two classes of primitive nature-spirit known as 'Detachable Gnomes' and 'Undetachable Gnomes'. Both of these curious races, said Leadbeater, had evolved from out of certain minerals, but one species was more evolutionarily advanced than the other. The Undetachable Gnomes had evolved first, and lived within subterranean rocks from which they were unable to separate their so-called 'etheric bodies'. They were trapped underground forever – until, over time, they evolved into Detachable Gnomes, which lived in rocks closer to the Earth's surface from which they were occasionally able to free themselves and walk around for a little time through the grass and flowers. Eventually, the most advanced of these Detachable Gnomes then evolved into something dubbed 'Surface Fairies' – the kind of free-range Little Folk that W. Y. Evans-Wentz would like to have seen skipping through the fields himself.[114]

Fairies leave behind few fossils, however, so Leadbeater was unable to provide any actual physical proof of these assertions. Nonetheless, he continued to elaborate his theory of 'evolvement', as he called it, showing how there were several distinct lines of fairy evolution in Nature. One parallel path of fairy evolution, for instance, led to the creation of the 'Tiny Creatures', spirits resembling miniature hummingbirds which, he said, were constantly hovering around flowers and had evolved from grass. According to Leadbeater:

> These beautiful little creatures will never become human, because they are not in the same line of evolution as we are. The life which is now animating them has come up through grasses and cereals, such as wheat and oats, when it was in the vegetable-kingdom, and afterwards through ants and bees when it was in the animal-kingdom. Now it has reached the level of these tiny Nature-spirits.[115]

Darwin might have been surprised to learn that bees evolved from ants, and that ants evolved from cereal, but there you go – Mr Leadbeater was full of such little-known wisdom. There was, he said, a 'ladder of evolution' in the unseen world, and fairies and nature-spirits could only ever climb upwards, not descend downwards, upon its rungs. For example, 'the life which has ensouled one of our great forest trees,' said Leadbeater, could never deign to 'descend to animate a swarm of mosquitoes, nor even a family of rats or mice'. Instead, these horrid vermin would provide appropriate vehicles only for that lowly 'part of the life-wave which had left the vegetable kingdom at the level of the daisy or the dandelion'.[116] During the gradual evolution of incorporeal nature-spirits into divine Solar Spirits, there were many grotesque and 'undeveloped tribes' through which they first had to pass, he explained. There were, for instance, the 'shapeless masses with huge red gaping mouths' which float around corpses and suck the 'etheric emanations of blood and decaying flesh' out of them, the 'rapacious red-brown crustacean creatures' which hover over brothels, and, last but not least, the 'savage octopus-like monsters' which 'gloat over the orgies of the drunkard' and feed off the fumes of his alcohol.[117]

Once these incorporeal beasts had evolved into fairies, Leadbeater clarified, they quickly became much more pleasant to observe – and observe them he did, helpfully laying out their different genera and species just like Darwin or Linnaeus would have done. According to Leadbeater, the different fairy races have

> colours of their own, marking the difference between their tribe or species, just as the birds have difference of plumage ... They are on the whole distributed much as are the other kingdoms of Nature; like the birds, from whom some of them have evolved, some varieties are peculiar to one country, others are common in one country and rare elsewhere, while others again are to be found almost anywhere. Again like the birds, it is broadly true that the most brilliantly coloured orders are to be found in tropical countries ... In England the emerald-green variety is probably the commonest, and I have seen it also in the woods of France and Belgium, in Massachusetts and on the banks of the Niagara River ... Java seems especially prolific in these graceful creatures ... A striking local variety is gaudily ringed with alternate bars of green and yellow, like a football jersey ...[Found in volcanic regions across the world] is a curious variety which looks as though cast out of bronze and burnished ... Several indications seem to point to the conclusion that this is a survival of a primitive type, and represents a sort of intermediate stage between the [detachable] gnome and the fairy ... A curious fact is that altitude above the sea-level seems to affect their distribution, those who belong to the mountains scarcely ever intermingling with those of the plains. I well remember, when climbing Slieve-na-mon, one of the traditionally sacred hills of Ireland, noticing the very definite lines of demarcation between the different types.[118]

Sensitive Flowers

As well as demonstrating the processes of evolution, though, some of these fairies living at the bottom of Darwin's garden also actively *intervened* in it. Alongside the hummingbird-like Tiny Creatures which hovered around flowers, said Leadbeater, were temporary thought-forms created by highly evolved 'Great Ones' – angels which were 'in charge of the evolution of the vegetable kingdom'. According to him, whenever a Great One had a nice idea for a new development in the plant species under its charge, it created an impermanent thought-form which then 'hangs around the plant or the flower all through the time that the buds are forming, and gradually builds them into the shape or colour of which the angel has thought' before it then 'just simply dissolves, because the will to do that piece of work was the only soul that it had'.[119]

Surprisingly, you can see (wholly inadvertent) echoes of Leadbeater's odd ideas in the thinking of certain actual scientists, in the way that some have occasionally dared to speculate about the possible existence of a psychic component lurking somewhere within the evolutionary process. In a series of lectures published in 1965 as *The Living Stream*, for instance, Sir Alister Hardy (1896–1985), the eminent marine zoologist, Antarctic explorer and Oxford don, spoke of his investigations into a type of flatworm named *Microstomum* which had evolved a bizarre defence-system involving the eating of polyps of the species *Hydra*. These *Hydra* polyps contain stinging capsules called nematocysts, whose presence is detected in the flatworm's stomach and then transferred directly to the surface of the flatworm's skin. Here, they remain mounted like cannon, ready to fire off their stings at any threats the flatworm should encounter. Strangely, the flatworms only ever eat these polyps as a means of replacing any 'cannon' whose ammo has already been fired; they never just eat them for food, and when their armoury is full leave them alone completely. This, said Hardy, caused problems for the standard Darwinian narrative; it was difficult for him to see how a process of random mutations and accidental selection could account for the development of such an apparently purposefully created feature. Hardy, while he certainly didn't reject Darwin, was stumped by the puzzle, and saw fit to propose the idea of a kind of 'group mind' among the animals of species such as *Microstomum*, which directs the development of its genes to a certain specific end. He didn't actually call it *élan vital*, but he may as well have done.[120]

Hardy was a very interesting figure. One of his students at Oxford was a young Richard Dawkins, but it appears that not many of his wilder ideas rubbed off on the future scourge of supernaturalism. Quite apart from being a leading proponent of the 'Aquatic Ape' theory of human origins, which held that mankind's early ancestors lived much of their lives in the water, Hardy was also something of a nature-mystic who openly admitted in later life that he was much more interested in religious and spiritual matters than he was in zoology, setting up Oxford's Religious Experience Research Centre in 1969 and joining the SPR. As a child, Hardy had been so overwhelmed by the beauty of Nature that he had literally got down on his knees in the middle of rambles and given prayers of thanks to God for the world's loveliness, and it seems his decision to pursue a career in zoology was to some extent an attempt by him to reconcile God with Darwin.[121]

One of the most speculative chapters in Hardy's book *The Living Stream* was entitled 'Biology and Telepathy' and, as a former military camouflage expert, he had a particular interest in the strange way that certain species of insect managed to disguise themselves in a fashion which at first could seem almost paranormal in nature. One particularly striking example is that of the African flattid bug, or 'planthopper', which comes in a variety of colours, from green, pink and coral to half one shade, half another. These clever insects, in order to evade predators, are able to arrange themselves on a twig so that they look like a lovely blooming flower with a green tip. This is not the same as the evolution of an insect which on its own resembles a single leaf, a resemblance facilitated purely by random mutations; there, a few lucky animals that look a bit leaf-like survive predation, pass their genes on to offspring, the most leaf-like of which again survive while the least leaf-like get eaten by predators, and so on forever. The flattid bugs, though, seem different; how can they just 'learn', quite accidentally, to swarm together collectively in such a way as to convincingly imitate a flower?[122] C. W. Leadbeater would have had little trouble in attributing the process to the intervention of a Great One; but would science's own Great One, Charles Darwin, be quite so easily able to explain the bugs' behaviour? My point is not that either the flattid bugs or the flatworms mentioned above actually evolved in a supernatural fashion via the use of group minds, but simply that some reputable scientists such as Sir Alister Hardy have occasionally speculated that they *might* have done.

Creation Myths

The other modern theory which is often spoken of as 'disproving' the existence of God is the Big Bang, a notion which would have come as some surprise to the man who first properly formulated it, Father Georges Lemaître (1894–1966), an ordained Catholic priest. Erasmus Darwin had proposed a similar idea over a century beforehand, speculating that a gigantic explosion at the beginning of time might have dispersed whirling gases throughout space which had gradually condensed into stars and planets,[123] but Lemaître was the first man to lay out the basic theory as we know it today back in 1927. Despite this, Lemaître's idea wasn't fully accepted by science until the late 1960s, one of the reasons for which being that it could be construed as providing potential evidence for the existence of God. After all, if the universe had a beginning, then who had set the whole thing in motion? Pope Pius XII (1876–1958) had no doubt that the answer was 'God', telling an audience at the Pontifical Academy of Sciences in 1951 that science had now managed to peer back through time to the moment of Divine Creation itself.[124]

One of the most vociferous opponents of the Big Bang idea was the astronomer Fred Hoyle (1915–2001), a Yorkshireman who sometimes seemed to gain pleasure from opposing the scientific mainstream just for the sheer, cussed fun of it. Among other things, in 1986 he claimed that archaeopteryx, a 160 million-year-old feathered reptile hailed as the evolutionary 'missing link' between dinosaurs and birds, was a fake, whose fossils had been forged in the 1860s by conmen keen to cash in on Darwin's new theories.[125] Hoyle objected to orthodox Darwinism, and in his 1971 book *Lifecloud*, co-written with his fellow anti-Darwinian Chandre Wickramasinghe (b. 1939), put forward the idea of 'panspermia', claiming that life on Earth was 'seeded' by comets which crashed into our planet, bringing our world's first primitive bacteria down with them. Furthermore, Hoyle said, maybe passing comets still periodically dropped down virus storms onto us, leading to new illnesses, and new genetic mutations and mass-extinction events in our flora and fauna. He even suggested that AIDS could have come from outer space! You might presume that, being somewhat opposed to Darwin, Hoyle had religious feelings – but not a bit of it. Rather than seeing mankind as having any special place in the cosmos, Hoyle felt his theory of panspermia helped prove the opposite:

> The idea that in the whole universe life is unique to the Earth is essentially pre-Copernican ... Just as no one country has been the centre of the Earth, so the Earth is not the centre of the universe.[126]

Hoyle's major objection to the Big Bang idea (a name he coined himself, intending it to be dismissive) was that it was inherently 'Judaeo-Christian' in nature, and proposed his own Steady-State theory as an alternative. Steady-State held that the universe was infinite and eternal, having no beginning and no end, with galaxies forever moving apart from one another and new ones springing up to fill the gaps in-between, with new matter and energy being continuously created from out of a hypothetical entity he called the C-Field. By 1965, though, the phenomenon of cosmic background radiation – presumed to be the 'echo' of the Big Bang itself – was discovered, and Lemaître's theory thus considered proven.[127] It is not really clear, though, why at one point the Big Bang was thought dangerously suggestive of the existence of God, and the next held up as triumphant proof of His *non*-existence. It might undermine hard-line Biblical literalists' belief in God creating the Earth in six days, or the Irishman Bishop Ussher's (1581–1656) idea that it was only 4,004 years old when Christ was born (a dating he largely derived from adding up the ages of all the people mentioned in the Bible),[128] but clearly Lemaître's notion was considered wholly compatible with religious belief by people like Pius XII. Rupert Sheldrake likes to quote his friend and fellow radical theorist Terence McKenna (1946–2000) on this point, who observed that, with the Big Bang, 'modern science is based on one principle: "Give us one free miracle and we'll explain the rest." The one free miracle is the appearance of all the mass and energy in the universe and all the laws that govern it in a single instant from nothing.'[129]

Or, in other words, 'What caused the Big Bang?' Religious people might like to claim that it was God. Scientists might try and push the answer back a bit by invoking various prior conjectures, but how far back can they go? Every cause would presumably have to have a preceding cause, and you could go on like this forever. Whichever ultimate answer you come up with for the universe's existence, whether scientific or religious in nature, must remain ultimately somewhat arbitrary. In the end, the fact of existence is a complete mystery, and the Big Bang every bit as much of a creation myth as some of its old religious counterparts. It is a myth based upon a reality – the Big Bang did actually occur – but the subsequent projection of various materialist assumptions onto it is really just yet another scientific fairy tale.

Revolutionary Thinking

The biggest scientific fairy tale of all, though, is the straight Darwinian path of development which we have projected out onto the discipline. To say that modern science emerged from a survival-of-the-fittest-style struggle with religion and reactionary conservative social forces, and won out purely because its ideas were true and correct, is, ironically, very far indeed from being true and correct. It is not the *truth* of science which ultimately persuaded Western society of its value, but the *usefulness* of that truth, the things which science allowed us to do. Furthermore, ideas which best fit in with the prevailing spirit of the times can stand a better chance of being accepted than those which do not, as with global warming. When faced with irrefutable evidence that something is true, then it would be foolish to deny that it is so. But when faced with somewhat disputable or uncertain evidence that something is true, the most socially acceptable theory will often tend to win out, at least temporarily. Even the heliocentric model of the solar system proposed by Nicolaus Copernicus (1473–1543) and Galileo eventually found favour not simply because it was true, but because as time passed it began to fit in with wider social and intellectual movements and currents of prevailing thought.[130]

The most famous attempt to chronicle the way that scientific endeavour actually works was the 1962 book *The Structure of Scientific Revolutions* by Thomas S. Kuhn (1922–96). Kuhn taught that science in any given era was governed by what he termed 'paradigms', presiding ideas which determined how most scientists thought about the world. For instance, in astronomy for a long time the ruling paradigm was that of geocentricism, the notion that all heavenly bodies orbited around the Earth. While the paradigm rules, most scientists engage in what Kuhn called 'normal science', defined as a kind of elaborate puzzle-solving exercise, or 'mopping-up operation' where knowledge of a kind which does not upset the ruling paradigm is sought; for instance, trying to get data which would allow you to predict the paths of the stars more accurately. The old geocentric model developed by the Greco-Egyptian astronomer Ptolemy (*c.* AD 90–*c.* AD 168) was actually quite good in this respect. It was not necessary for ancient astronomers to deviate from it in order to get better results, so they didn't. According to Kuhn, during the era of normal science, no genuine novelty is sought in terms of discoveries. Instead, it can be viewed as 'an attempt to force Nature into ... the [governing] paradigm'. Many phenomena which will not fit into such a box, says Kuhn, 'are often not seen at all', or simply ignored. Most ordinary scientists do not 'normally aim to invent new theories' and are 'often intolerant of those invented by others'.[131] Instead, the puzzle solving during such an era is best seen as an exercise in 'achieving the anticipated in a new way', something which gives the impression of rapid progress; seeing as most normal scientists concentrate upon attacking puzzles they think have definite solutions and are confident of solving, these puzzles normally do get solved. For instance, if an astronomer has successfully charted the path of Mars through the night sky, then he should really be able to do the same for Jupiter and Venus, too, and could easily do so without upsetting Ptolemy's ruling paradigm.[132]

However, more and more anomalies, niggling little things that just don't work or make sense within the terms of the prevailing paradigm, begin to accumulate, until eventually a state of crisis is reached, in which it becomes obvious there is something seriously wrong. The old paradigm can't just be instantly abandoned, though, for 'to reject one paradigm without simultaneously substituting another is to reject science itself'.[133] Eventually, a new governing idea is put forward – the heliocentric model of Copernicus and Galileo in which Earth orbits the sun, for example – which seems to solve many of the puzzling anomalies bedevilling the old one. This new paradigm gains followers, and what Kuhn calls a 'paradigm-shift' finally occurs in which the new idea becomes the new ruling paradigm in place of the old one.[134] Often, the people who create such new paradigms are young or new to the field, so see things with eyes left unjaundiced by years of groupthink.[135]

Of course, some of these brave dissenters will always simply be cranks. I think in particular of George Francis Gillette (1875–?), an American engineer with an obsessive hatred of the theories of Albert Einstein. Deriding Einstein's ideas as the 'moronic brain-child of mental colic', Gillette published a book called *Orthodox Oxen* in 1929, declaring all mainstream scientists who accepted them to be nothing but idiots. There is 'no ox so dumb as the orthodox', he wrote, and explained to his readers in plain language mercifully free of 'hi-de-hi mathematics' the hitherto unknown existence of such non-Einsteinian wonders as all-cosmos doughnuts, maximotes, ultimotes, unimotes, supraunimotes, allplane velocities, interscrewed sub-units and, of course, laminated solid-solid-solid-solids.[136] In a classic explanation of his basic theory about how the universe really worked, Gillette wrote the following:

> All motions ever strive to go straight – until they bump ... Nothing else ever happens at all. That's all there is ... In all the cosmos there is naught but straight-flying bumping, caroming and again straight-flying. Phenomena are but lumps, jumps and bumps. A mass-unit's career is but lumping, jumping, bumping, rejumping, rebumping, and finally unlumping.[137]

I bet you didn't know that. Nor do most current-day scientists. Some proposed new paradigms, needless to say, never catch on at all – often for very good reasons.

However, other new paradigms turn out to be entirely true but are still resisted nonetheless. Such resistance may well be futile, but it comes in many forms. Kuhn cites the esteemed physicist Lord Rayleigh having a paper discussing certain anomalies of electrodynamics rejected by the British Association as 'the work of some paradoxer' after accidentally leaving his name off it, for example. When they discovered its author, however, the body quickly accepted it with 'profuse apologies' – the scientific equivalent of the old story about Charlie Chaplin getting third place in a Charlie Chaplin lookalike contest, maybe.[138] When x-rays were accidentally discovered by Wilhelm Roentgen (1845–1923) in 1895, meanwhile, the more established physicist Lord Kelvin famously dismissed them as an elaborate hoax. One problem was that the cathode-ray laboratory equipment of Europe had not been shielded with lead to prevent interference from x-rays – they were unknown, so why would they be? – and Roentgen's discovery thus necessitated much cost in updating apparatus, and potentially invalidated many previous experiments which would now have to be repeated. No wonder so many scientists initially objected to Roentgen's discovery; it was a massive damned nuisance![139]

It Came from Beyond the Stars

The list of perfectly genuine theories and inventions which were once denounced as heresy or fraud is a long one. In 1838, two months before the successful public demonstration of his new technique, Louis Daguerre

(1787–1851), inventor of photography, was denounced by his wife to the respected chemist Jean-Baptiste Dumás (1800–84) as a madman for having the apparently insane belief he could fix his shadow onto a metallic plate.[140] This was similar to the occasion, on 11 March 1878, when Thomas Edison's (1847–1931) new phonograph was demonstrated to a gathering of French scientists. Most were pleased by the device; but not one professor. He rushed towards the machine's demonstrator, grabbed him by the lapels, and accused him of being a ventriloquist![141]

The classic example is the long-standing scientific denial of the reality of meteorites. For centuries, many authorities simply refused to believe they existed. An interesting report into the topic was produced in 1769 by Antoine Lavoisier, the great French chemist, at the behest of his country's *Académie des Sciences*. Lavoisier looked into one particular instance of a meteorite which was seen by several persons to have fallen from the sky in September 1768. Reading Lavoisier's description of this event, it seems quite clear that this was a genuine event. As Lavoisier describes it:

> On September 13, 1768, at half-past four in the afternoon, a stormcloud appeared in the direction of the castle of La Chevallerie, near Lucé, a small village of Maine, and a sharp thunderclap was heard which resembled the report of a gun. Then, over a space of some two leagues and a half, a considerable whistling sound was heard in the air, without any appearance of fire. It resembled the lowing of a cow so closely that several people were deceived by it. Several crofters who were harvesting in the parish of Périgué, about three hours from Lucé, having heard the same noise, looked up and saw an opaque body describing a curve, and falling on a meadow near the main road to Le Mans, beside which they were working. They all ran up to the spot and found a sort of stone about half buried in the earth, but it was so hot and burning that it was impossible to touch it. Then they were all seized with terror and ran away.[142]

Clearly, the natives were superstitious – but so was Lavoisier, in a different kind of way. The great genius did not deny that the stone existed. Indeed, he examined a sample of it in his laboratory, and concluded that it was just an ordinary stone, nothing more. It contained a large amount of iron, and its surface seemed to have been fused, apparently by great heat. Lavoisier's conclusion was that the stone had been there in the field all along, buried beneath soil. When the storm came along, a lightning bolt must have found itself attracted to the iron in the rock and struck it, fusing its surface and blasting away the earth which had previously covered it up. Then, when the country folk went to the point of impact and saw the smoking stone lying there in a spot where, as far as they knew, there had been no stone before, they drew the mistaken conclusion that it had fallen down from the sky together with the lightning. This conclusion, though, said Lavoisier could not be true – after all, according to the prevailing paradigm of his day, there were no stones in the sky which could fall down to Earth in the first place.[143]

Nowadays, we know better. The paradigm has shifted, and meteorites are a piece of common knowledge. While still operating under the influence of the previous prevailing paradigm, however, scientists could be blind not only to the evidence of other people's eyes, but even of their own. In 1627, a meteorite weighing some 66 pounds fell from a clear sky in Provence, right in front of the early French scientist and astronomer Pierre Gassendi (1592–1655). Not only did Gassendi see the thing fall, he went up to its impact site, touched and thoroughly examined it. He had to admit that the stone had fallen from the sky – but he also felt he had to maintain the fiction (to him, a truth) that there were no stones in the sky to fall. His solution was to say that the stone must have been propelled into the heavens from a volcanic eruption. The fact that there are no active volcanoes in France was an unimportant detail. It was not until as late as 1803 that the French *Académie* finally admitted that meteorites were a real phenomenon. Old paradigms, it seems, die hard.[144]

Anomalous Ideas

According to Kuhn, the way that science is structured and taught to new, trainee scientists aids in this whole process, acting to 'render [past scientific] revolutions invisible'.[145] Students are taught only about the paradigms that prevail today, and textbooks misrepresent the great scientists of the past as 'having worked upon the same set of fixed problems in accordance with the same set of fixed canons' as scientists now do.[146] This, however, isn't true; for instance, someone studying magnetism today may well be taught about William Gilbert (1540–1603), who first saw that the Earth itself could be conceived as a giant lodestone with magnetic poles, something we now know to be true. They are less likely, though, to be taught that Gilbert, being a child of his age, thought that magnetic force was actually alive, the 'soul of the Earth', and that magnets attracted iron filings because this ghostly spirit 'reaches out like an arm' to the iron and 'draw[s] it to itself'.[147] Simply saying 'Gilbert developed the theory of the magnet' doesn't really give the full picture of his world view, and is at best a sin by omission which implies he thought like us, and was trying to solve the same problems about the world as we are.

Kuhn shows how this kind of thinking leads to the natural impression that scientific knowledge is simply cumulative, that 'one by one, in a process often compared to the addition of bricks to a building', all scientists of the past have been steadily supplying a never-ending conveyor belt of facts and theories which together encompass the prevailing paradigms which rule today.[148] In fact, someone like Gilbert was laying the bricks of an entirely different type of building; his science was a cathedral, illuminating the mind of God. Our science is more like the French Revolutionaries' Temple of Reason, denying it. As any good Lego fan knows, you can build a lot of different things with the same boxful of bricks. For Kuhn, scientific knowledge was only properly cumulative *within* paradigms, not *between* them. However, the modern science student is taught to see his subject's past as 'leading in a straight line to the discipline's present vantage ... In short, he comes to see it as progress'. However, 'there are losses as well as gains' in paradigm shifts, 'and scientists tend to be peculiarly blind to the former'.[149] Furthermore, while Kuhn admits that the history of science demonstrates 'a process of evolution *from* primitive beginnings', this does not necessarily make it 'a process of evolution *toward* anything'. There is no state of ultimate graspable truth towards which science is capable of progressing; that would be yet again to mistake Darwin for Hegel.[150]

The most famous expression of such hubris was Lord Kelvin's (alleged) statement of 1900 that 'there is nothing new to be discovered in physics now. All that remains is more and more precise measurement.'[151] Try telling that to Einstein. Around the turn of the millennium, back when Fukuyama still had his disciples, proponents of hubris were at it again, with John Horgan's (b. 1953) 1997 book *The End of Science: Facing the Limits of Knowledge in the Twilight of the Scientific Age* being essentially a book-length treatment of Lord Kelvin's supposed quip. That kind of perspective makes sense only if you ignore the possibility of future paradigm shifts taking place. How do we know that there will not be any? We cannot. Niggling little anomalies which seem of little importance now may prove of immense consequence in the future. In 1827, the Scottish botanist Robert Brown (1773–1858) observed through a microscope that tiny pollen grains, floating on the surface of perfectly undisturbed water, were continually jittering about. This 'Brownian motion' made no sense, and was ignored as a trivial minor anomaly. Then, in 1905, Einstein interpreted this as being due to these grains colliding with individual water molecules, all of which, he argued, were in a state of constant and random motion. The Frenchman Jean Perrin (1870–1942) confirmed this experimentally, providing final proof of the reality of atomic theory, which was still being resisted by some at the time.[152] As William James once put it:

> Anyone will renovate his science who will steadily look after the irregular [anomalous] phenomena, and when the science is renewed, its new formulas often have more of the voice of the exception in them than of what were supposed to be the rules.[153]

Anomalies in all fields continue to accumulate – readers wishing to see this fact proved before their very eyes in a highly humorous fashion should look into the works of Charles Hoy Fort (1874–1932), an American writer and philosopher who spent several decades of his life scouring through neglected records in the public libraries of London and New York in search of bizarre nuggets of data before later reprinting them in his books in an attempt to embarrass scientists. He stitched together accounts of weird anomalies like giant hailstones and alleged rains of blood in order to create a series of deliberately ridiculous explanatory theories to account for such events, thereby hoping to parody the way that scientists' minds worked. Fort's excellent books, with evocative titles like *The Book of the Damned* (the 'Damned' of the title being obscure and paradigm-defying data which mainstream science chooses to ignore and suppress) still have a large following today, and a highly entertaining magazine devoted to his memory called *Fortean Times* has been continuing his good work for the past forty years.[154]

Another man who devoted his life to cataloguing such anomalies was the American researcher William R. Corliss (1926–2011), who was in many ways Fort's modern-day equivalent, albeit without the satirical edge. Corliss ran something called the *Sourcebook Project*, searching through various publications in search of strange data and then presenting them in a kind of ongoing anomalistic encyclopaedia. A randomly chosen selection might include the finding of a new bone-bed in Florida in which fossils from various different environments were surprisingly found jumbled together, a series of strange 'sky-flashes' observed by Canadian astronomers, and the mysterious discovery that the venom of the Australian funnel-web spider is only effective against a group of animals which it does not usually prey on – like humans and monkeys – rather than on those it does.[155] Perhaps these anomalies really are trivial, and digestible into prevailing paradigms. Possibly they have been solved by papers in obscure journals already (they date from 1984/5). Or, on the other hand, maybe they will have to wait for new paradigms to have been created to be fully explained. Who knows? According to Corliss, anomalies 'reveal Nature as it really is: complex, chaotic, possibly even unplumbable'.[156] That, though, is not how we are usually taught to look at the matter today. Maybe, as Rupert Sheldrake once put it, what we really need now is 'an enlightenment of the Enlightenment'.[157] Our faith in the inherent rationality of our own society is, I think, in itself a demonstrably irrational thing. Far from evolving, in so many respects we now seem instead to be going backwards.

Two Little Boys: Perpetual Motion, Meat Machines and Science as a Spiritual Quest

The value of science as metaphysic ... belongs with religion and art and love.

Bertrand Russell[1]

There were once two little boys who had two little toys – each had a home-made perpetual motion machine. The dream of creating such a device – a kind of engine which can run forever without exhausting its energy source or, better yet, which creates more energy than it consumes – is one as old as science itself. One of the earliest known attempts was the idea put forward in 1150 by the Indian philosopher Bhaskara (1114–85) that, if a certain weight was placed on the rim of a wheel at a certain point and in a certain way, it might leave it in a state of perpetual imbalance which would cause it to keep on turning around under the weight's influence *ad infinitum*. If you then used such a wheel to turn components in an engine then this engine would, in theory, keep on running forever, thus making it a genuine perpetual motion machine. The trouble was, of course, that Bhaskara's miraculous wheel *was* just a theory, and that no such device has ever been successfully built.[2]

The dream of perpetual motion is just that – a dream, an unrealisable impossibility. Machines which purport to achieve this goal violate the First and Second Laws of Thermodynamics. Think, for a moment, of a wheel held on top of a vertical rod which rotates when you spin it; no matter how hard you push, eventually it will come to a stop. The energy you imparted to it will dissipate out into the world around it, a process known as entropy, and the wheel will stop spinning. This represents the Second Law of Thermodynamics, namely that entropy always increases – energy or heat always dissipates, however slowly, outwards. If, however, this wheel had somehow been manufactured to be so efficient in its spin that the energy you had imparted to it through your initial shove never did manage to disperse outwards via entropy, then it would create a kind of closed loop which would enable the wheel to go on turning until the end of time, and the Second Law would have been broken. Imagine, on the other hand, that you gave the wheel a weak and tiny push, and it then magically went on and on spinning forever at a very great speed indeed, like an everlasting Catherine wheel. A tiny shove should not produce anything like such a reaction, and would mean that, once the initial push had been given to the wheel, it had somehow managed to begin generating its own source of energy from out of nowhere, thus breaking the First Law of Thermodynamics. This states that the total amount of potential energy available in the universe remains a constant, meaning you can't get more energy out of any process than you put into it in the first place – so a feeble push on a wheel should produce only an equally feeble spin.[3] Perpetual motion is, then, an absolute impossibility, at least as far as anyone can tell – and yet to the imagination of a small child, as they say, nothing seems impossible ...

Curiosity Skinned the Cat

Our two infant inventors took wildly different routes towards manufacturing their own personal impossible dreams. The first was inspired by a desire not only for knowledge, but also for money. Something of a child prodigy, this particular eleven-year-old genius was a voracious reader, of everything from adventure stories to poetry, from theology to a practical guide for German midwives (though he himself was Swedish). Clad in a dressing gown and nightcap, smoking a grown-up's tobacco pipe and with his fingers shoved tightly into his ears to block out all distractions, whenever one of his brothers appeared on the scene and tried to disturb him for a joke, the lad would jump up and threaten to beat his sibling to a pulp for disturbing both his reading and his tobacco puffing. Particularly loved by the child were books on geometry and chemistry.

When he had first encountered the former subject at school, it had profoundly disturbed the lad, the idea of there being some kind of hidden, invisible mathematical laws underpinning the visible world he saw around him seeming positively terrifying to his infant mind. Then, however, the tiny bookworm had gotten hold of a textbook upon surveying, and suddenly found himself entranced by the idea of being able to measure every object he could find, and discovering the secret geometry which underlay them all. Buildings, trees, streets – he measured them all, then calculated the distances and volumes of features in the landscape, and made little cardboard scale models of them. The curious child never lost this mania for measuring things, even as an adult. When his wife had taunted him about the supposedly small size of his manhood one day, he hired an obliging doctor and then bundled him off with him to a brothel in Geneva. Here, 'provoked to the very root of my testicles' by his wife's cruel words, he hired a prostitute, had her induce a state of tumescence in his rapidly swelling organ and then handed it over for examination to the waiting doctor, who evidently had a tape measure conveniently to hand. The angry man's penis measured some 16 centimetres in length by 4 centimetres in circumference, dimensions the medic pronounced to be entirely normal. After taking advantage of the prostitute's more customary services and noting her statement that his performance in the matter was entirely satisfactory, the irate husband then returned home to his wife and provided her with the 'scientific' evidence that she was talking absolute rot, as usual. Could she really doubt the empirical, first-hand testimony of both a trained medic *and* an experienced professional whore? To make the true state of affairs absolutely clear to her, he then made use of a handy metaphor from the world of engineering which he had taken the time to prepare during his journey homewards. 'The screw,' he said, 'is not necessarily too small just because the nut is too big.' The biggest nut in this particular marriage, however, was undoubtedly the husband, and the union later ended in divorce.

It is not possible for parents to divorce their child, however, and so during his youth the boy's mother and father simply had to put up with him, no matter where his burgeoning enthusiasm for the sciences took him. Taking apart the world around him by surveying it and reducing it down to its fundamental geometrical components was not enough. Soon, the child was physically dismantling all kinds of household objects, from toys to clocks, to see what made them tick. Sometimes he later reassembled these items back correctly in perfect working order, and sometimes he did not. Getting hold of chemistry books, he thought it not enough merely to reproduce the experiments he found detailed within their pages, but also to invent many of his own. Finding an old galvanic battery lying in the attic, he removed the sulphuric acid from the appliance and performed a number of ingenious tests with it which succeeded in confirming conclusively that applying such a caustic substance to piles of the family's clean linen would completely ruin them. The child was often beaten.

His many home-made inventions, meanwhile, were quite elaborate, though not always entirely functional, such as the primitive electrical generator he attempted to create by taking apart his mother's spinning wheel and then fusing its components together with the spokes of an umbrella and some stolen violin strings. One day, he found a dead cat lying in the road, which he skinned with a knife and then transformed into a Leyden Jar. Leyden Jars were primitive batteries which stored up small amounts of electrical charge, and the fur from a dead cat, when faintly warmed, can be used to build up just such a charge in a few amenable substances like resin or vulcanite, highly obscure facts which the budding scientist was evidently already aware of from his intense programme of tobacco-fuelled reading.[4]

It takes quite a bit of skill and learning to be able to turn a dead cat into a battery, though, and in an attempt to direct his son's annoying yet simultaneously impressive efforts towards something which might actually prove to be useful for once, the boy's father pointed out to him that the Swedish government was currently offering a large cash prize to anyone who could successfully create a genuine, workable perpetual motion machine. The boy was a bundle of perpetual motion himself, and setting him off on a fool's errand like this must have seemed like a sure-fire way to keep him harmlessly occupied for months if not years on end — it was, after all, a task which could never be completed. If this was his father's intention, however, then the plan did not work. Coming up with what he thought was an appropriate design to extract free energy from the ether, the boy decided that what he really needed to bring his blueprint to life was a large number of wooden planks. Therefore, he promptly smashed up his parents' bedroom chest of drawers, and made off with all the pieces. Then, he appropriated a birdcage, a coffee percolator, a hanging lamp and a bottle of soda water, took them all to bits, recombined them with the planks like Rube Goldberg made real, and set the whole contraption off in motion. It didn't work. The boy, in a fury, threw it against a wall, finding that there were some limits to his genius after all – a lesson which, as we shall soon see, he was immensely reluctant to take on board during his later adult life.[5]

A Bug's Life

Our second infant seeker after the secrets of perpetual motion was motivated not by money or fame, but a desire to improve the pitiable lot of humanity. Born in what is now Croatia in 1856, he claims to have come from a family whose capacity for invention and discovery was truly a thing of wonder. His elder brother, for instance, was 'one of those rare phenomena of mentality which biological investigation has failed to explain', he said, and would undoubtedly had grown up to be the greatest man of all time – had he not been accidentally kicked to death by the family's equally super-intelligent horse at an early age. His father was so clever, meanwhile, that, finding himself short of worthy opponents to argue various abstruse points of philosophy with, he would often simply shut himself away in a room, pretend to be two people at once, and then carry on heated discussions, half in his own voice, and half in an invented one. If the bone of contention was a particularly difficult one, he sometimes adopted more than two competing viewpoints, putting on so many accents that, in his son's words, 'a casual listener might have sworn that several people were in the room'. His mother, likewise, was 'a truly great woman', who came from 'one of the oldest families in the country and a line of inventors', who would surely have gone down in history alongside Leonardo da Vinci were it not for the fact that she had so much housework to do. According to the boy, his mother's fingers were so dextrous that, when she was gone sixty, she was still able to tie three knots in an eyelash.[6]

Placed next to these amazing beings, our second baby scientist himself seemed 'dull in comparison', and yet even he was far above the intelligence of most ordinary mortals. In his genuinely bizarre 1919 autobiography, published in serial form in a specialist journal that was *supposed* to be devoted entirely to the subject of electrical engineering, the grown-up inventor illustrated the power of his childhood imagination in words as unforgettable as they were frankly unbelievable. 'Like most children,' he wrote, 'I was fond of jumping and developed an intense desire to support myself in the air. Occasionally a strong wind richly charged with oxygen blew in from the mountains rendering my body as light as a cork, and then I would leap and float in space for a long time.'[7]

While it is debatable as to whether 'most children' really develop the belief that they can float, the boy in question really believed that *he* could – a highly imaginative notion which he did not fully disabuse himself of until experiencing a 'bad fall' one day after jumping off the top of a building whilst holding an open umbrella.[8] These were not the child's only strange habits. According to his own testimony:

> During that period [of early childhood] I contracted many strange likes, dislikes and habits, some of which I can trace to external impressions while others are unaccountable. I had a violent aversion against the earrings of women ... The sight of a pearl would almost give me a fit ... I would not touch the hair of other people except, perhaps, at the point of a revolver. I would get a fever by looking at a peach and if a piece of camphor was anywhere in the house it caused me the keenest discomfort. Even now [as an adult] I am not insensible to some of these upsetting impulses. When I drop little squares of paper in a dish filled with liquid, [he never explains *why* he does this ...] I always sense a peculiar and awful taste in my mouth. I counted the steps in my walks and calculated the cubical contents of soup-plates, coffee-cups and pieces of food – otherwise, my meal was unenjoyable. All repeated acts or operations I performed had to be divisible by three and if I mist [*sic*] I felt impelled to do it all over

> again, even if it took hours ... I was opprest [*sic*] by thoughts of pain in life and death and religious fear. I was swayed by superstitious belief and lived in constant dread of the spirit of evil, of ghosts and ogres and other unholy monsters of the dark.[9]

Clearly, this was a very nervous child indeed – and another youthful phobia he possessed was his intense dislike of insects. The trigger event for his dread of creepy-crawlies centred around the first of several attempts he was to make at creating a perpetual motion machine. Rather than designing some complex mechanical device based upon sensitively weighted wheels, magnets and pistons as others had tried before him, the young mastermind hit instead upon a far simpler and easier way to get free energy – namely, to enslave the world's insects and make them do his bidding. The particular species the boy chose to enchain in a state of eternal bondage was the 'May-Bug', a type of flying black beetle which was 'a veritable pest' in the boy's home village during the spring months. Annoying they might well have been, but these May-Bugs were also remarkably stupid, something the imaginative youth took advantage of by gluing them to the arms of a miniature home-made windmill. As he explained:

> I would attach as many of four of them to a cross-piece, rotably arranged on a thin spindle, and transmit the motion of the same to a large disc and so derive considerable 'power' [when they tried to fly away]. These creatures were remarkably efficient, for once they were started they had no sense to stop and continued whirling for hours and hours, and the hotter it was the harder they worked.[10]

Our second Solomon was beside himself with joy. He seemed to have successfully harnessed a source of free perpetual motion just waiting to be endlessly harnessed by a grateful world – during the month of May, at any rate. The entire plan was soon scuppered, however, by the arrival of a 'strange boy' in our hero's village. The son of a retired Austrian Army officer, he came along and saw the May-Bug machine in action. He, too, was delighted with it – because May-Bugs were his favourite food. Opening up a big jar filled with replacement slaves for when the current ones died, the boy reached in, produced a big handful of bugs, and shoved them right into his mouth, still alive and wriggling, and began munching away. This horrible sight, combined with the sickening sound of the insects' hard shells crunching up beneath the newcomer's teeth, caused the device's creator immediately to vomit, thus putting him off the idea of creating a mass insectoid labour force for life. As he later put it, not a little wistfully, 'that disgusting sight terminated my endeavours in this promising field', and he was never able to so much as touch an insect of any kind throughout his whole life, ever again, May-Bug or otherwise.[11]

Insects may now have been out of bounds as a means of creating perpetual motion, but other potential methods were not. After successfully manufacturing a kind of air-powered pop-gun and smashing several windows with it, the boy became abnormally interested in the principles of air pressure. Impressed by what he called 'the boundless possibilities of a vacuum', he began to long almost painfully to be able to harness what he saw as being 'this [source of] inexhaustible energy'. He actually thought he had succeeded, creating a complex wooden contraption which he described as being 'a cylinder freely rotatable on two bearings and partly surrounded by a rectangular trough which fits it perfectly' and divided into 'two compartments entirely separated from each other by air-tight sliding joints', to one end of which he then applied a vacuum pump before awaiting the inevitable result of perpetual motion. The thing moved only slightly, but this was a good enough start, he thought, and began making immediate plans to incorporate his air pump into the engine of a home-made flying machine, or primitive jetpack. It was only later on that he discovered the thing was turning slightly because there was a leak somewhere in it through which air was passing, thus meaning it moved only because it was faulty, and not because it actually worked.[12]

An Apple a Day

Who were these two perpetual motion fiends? Did they really grow up to become actual scientists or inventors of any note? The first one we met didn't, although he certainly *thought* he did. The boy who skinned a cat and threw his own unsuccessful perpetual motion device against a wall in fury was in fact none other than the celebrated Swedish playwright, novelist and occasional painter and musician August Strindberg (1849–1912). You may know him as the author of such masterpieces of early Scandinavian modernism as *Miss Julie* and *The Island of the Dead*, but there were certain times during his extraordinarily full life when he wanted to throw literature temporarily to one side and concentrate instead upon making his name in the sciences, a field in which he felt confident he was an unrecognised genius. The only problem was that his ideas were frequently ridiculous.

There was the time, for instance, that he decided that plants obviously had nervous systems, even though nobody, neither gardener nor botanist, had ever actually noticed them in any way, shape or form when cutting up and handling the things. Thinking this to be a ridiculous state of affairs, Strindberg took to carrying a syringe in his pocket during his early-morning walk and then injecting various pieces of vegetation he came across with morphine to see what would happen. Upon being spotted one day by a passing policeman who caught him red-handed in the act of injecting a low-hanging apple with drugs, Strindberg was arrested. It was only when the great playwright began to explain what he was actually doing that he was eventually set free, the constable reasoning that he was simply a harmless lunatic, not a sinister fruit poisoner.[13] You would have thought that a madman wielding a syringe was exactly the kind of person that *should* be immediately arrested, but Strindberg had a knack for convincing officers of the law that his various public experiments, no matter how bizarre, were absolutely vital contributions to the future of the human race, and that as such they simply *had* to be allowed to take place, no matter how much inconvenience they may have caused to others.

One night during 1892, for example, while staying in Berlin, Strindberg unexpectedly announced to two of his habitual drinking partners that he did not necessarily believe the world was round. Sure, people always *said* it was round, but why should he, the great poet-scientist August Strindberg, trust the word of others? All he needed to prove to his own personal satisfaction that the Earth was indeed spherical was a broom, he said. This being the middle of the night, it proved quite difficult to get hold of such an item, but eventually one was procured, and Strindberg ordered one of his fellow experimenters to lie down in the middle of the road and hold up the brush in question, with its bristle-end pointing skywards. Then, Strindberg and his other companion, Dr Carl Ludwig Schleich (1859–1922) – a pioneer in the medical use of cocaine as an anaesthetic (on people, not apples), and thus no mere uneducated mug – both lay down in the road themselves, some way away, and prepared to make their observations. It is not quite clear how exactly the men intended to test the Earth's rotundity by lying down on their backs and then staring very hard at a broom, but when a German policeman approached and implied they were nothing but a bunch of drunkards, Strindberg launched into yet another of his pseudoscientific spiels. Apparently, the explanation of his methodology which Strindberg gave was so convincing that the policeman was actually on the point of lying down in the road to take part in the investigation himself before it was noticed that the inebriated man with the broom had simply gotten bored and wandered off, thus rendering the whole exercise academic in any case.[14]

Strindberg's nocturnal drinking sessions with his fellow co-investigators in Berlin would often begin with him asking his friends to 'explain the riddle of the universe, my children',[15] something which he himself tried to do at every available opportunity. There was the period, for instance, while living in Paris, that he decided to walk around Montparnasse Cemetery waving test tubes around through the air in an attempt to capture the souls of the dead, before taking them back home to his 'laboratory' (read: bedroom) and subjecting them to chemical analysis. I am unaware of the precise results of these trials, but presumably he will have found that the chemical composition of the human soul was surprisingly similar to that of fresh air itself.[16] Another bright idea of Strindberg's was that all gorillas were not really apes at all, but the hybrid descendants of a shipwrecked sailor who had once been stranded alone on a desert island with an ordinary female monkey which he had taken as his wife after starting to get a bit lonely as the years passed by. His 'proof' for this assertion was that he for some reason possessed a close-up photograph of the palm of an elderly sailor which, he thought, looked extraordinarily like another photograph he inexplicably owned showing a close-up version of the hand of a gorilla. He enjoyed showing visitors these two pictures, and detailing his idea; apparently, people did agree with him that the two hands looked alike, but this was probably only because they were too scared not to.[17] Such was Strindberg's desire to be recognised as a titan of science that he even began claiming credit for the discoveries of others. Shortly after the first detection of x-rays by Wilhelm Roentgen in 1895, for example, he sat drinking absinthe in a café with a friend, before casually slipping into conversation the 'fact' that he had already discovered the same things himself ten years ago, but had never bothered to tell anyone.[18]

Occasionally, Strindberg liked to fantasise about creating his own mystical cult of scientist-monks, an 'order of superior beings seeking new paths', who would search out the meaning of life itself, a problem, he said, 'which must be solved'. The wonderful new inventions created by the members of this order would, he said, 'liberate the [human] spirit from its bonds and raise it over time and space'. A new kind of telescope, for instance, was in dire need of being invented, a telescope which would operate without the need for any lenses whatsoever, and which thus would in fact not really even *be* a telescope at all – shades of the early Royal Society men and their programme of reuniting fallen mankind with Adam's telescopic eyes. Once such impossible wonders had been created by his caste of scientist-monks, Strindberg had a dramatic plan for how to reveal their hitherto secret existence to the world: 'We'll build a white Viking ship painted in gold and other colours; dress ourselves in white ceremonial attire and row down the river Aisne [in Sweden] playing new instruments which I'll invent!'[19] It's an unusual vision of scientific triumphalism, it must be granted; but then August Strindberg wasn't *really* a scientist at all ...

Shocking Discoveries

But what of our other child genius, that strange little boy who attempted to harness the power of May-Bugs? Did he go on to invent or discover anything worthwhile? Well, he certainly kept on having the occasional pipe dream – in one case quite literally. During his adolescence he conceived a scheme for transporting mail across the Atlantic without any need for ships or aeroplanes. All that had to be done was to place letters and objects inside sealed spherical containers, then place these within a network of giant pipes laid out across the seabed, and turn on a high-pressure water jet to push the rotund parcels down through the pipe tunnels and straight through from Europe to America in a matter of hours – a proposal with just the one flaw that, as he himself later admitted, it wouldn't actually work. Another ingenious but aborted idea of his was that of building a gigantic artificial ring around the equator which would float in space but, rather than turning around in line with the rotation of the actual Earth, would be 'arrested ... by reactionary forces' so that it hovered entirely still while the globe spun beneath it. Thus, step into the ring somewhere above Central America, laze around for a few hours hovering in stasis above the planet, and then jump back out again a mere few hours later over the sunny savannahs of Africa, having travelled to your holiday destination at a speed far in excess of that possible by plane, boat or train. Time your descent incorrectly, though, and you would presumably end up drowning in the Pacific.[20]

Rather more useful than all these never-built devices, however, was another invention conceived by the same person; namely, the alternating-current or 'A/C' motor, an original means of generating and transmitting electrical energy which was based around the notion of using rotating magnetic fields to drive motors rather than solid mechanical parts, something which reduced friction and thereby massively increased their efficiency. An A/C current of this kind generated in a power plant is now something we all make use of every day, each time we so much as flick a switch to illuminate a lightbulb or turn our TV sets on. The computer upon which I am typing these words right now makes use of an A/C current, a much more practical and efficient idea than that of using a tiny windmill turned by insects to power the thing. The man behind the A/C system also has a good claim for having created the first viable means of transmitting and receiving radio waves and may even have beaten both Wilhelm Roentgen and even August Strindberg to the discovery of x-rays – something which, unlike the Swedish braggart, he really does seem to have kept puzzlingly quiet about.[21] Unless you live in a hollowed-out tree and survive by drinking rainwater and eating grass, then you live in a world which was, substantially, made possible by this man. So significant was his contribution to both science and industry that, in 1960, he was honoured by having the IS Unit for magnetic-flux density named after him. That unit is called the 'tesla', and our bug-collecting hero was, of course, none other than one of history's all-time great geniuses, Nikola Tesla (1856–1943), the esteemed Serbian-American electrical engineer and inventor.

The Fame Game

Tesla was well known in his day, one of the age's great celebrities in fact, but has a rather lower level of name recognition these days than many of history's other great scientists. Partially this is because the main form of recognition which Tesla sought for his work was simply that people made profitable use of it. He was offered the Nobel Prize in 1912 but is supposed to have turned it down flat,[22] and was much less adept at claiming responsibility for some of his own finds than others were – the Italian Guglielmo Marconi (1874–1937), for instance, is usually given credit as the inventor of radio, when Tesla's claim upon creating the medium is probably greater, or at least of equal strength.[23] Certainly, he was profoundly uninterested in the trappings of worldly success. He died virtually penniless, never owned his own home, preferring to live within rented hotel rooms, and, by the end of his life, counted his main friends as being the pigeons who lived in New York's Bryant Park. Women, too, never concerned him, it being his staunch belief that the true inventor, so as not to become distracted from his work, should never marry. Towards the end, Tesla somewhat breached this rule by falling in love with his favourite pigeon, but there can be few persons on this planet who would be content with that particular small reward for successfully electrifying the world. Even Rod Hull managed to bag himself an emu.

Maybe it is Tesla's incredible strangeness, however, which has acted to obscure his contribution most of all, leaving him much less known today than he deserves to be. Scientists nowadays aren't supposed to be like Nikola Tesla – or, indeed, like August Strindberg. The myth of science as a totally objective, almost dehumanised thing, which stands utterly apart from the (human!) scientists who practise it, will not allow it, and there is in general less space for the talented individual who tackles great problems alone and in isolation these days. Due to the institutionalised way in which science now works, many important results and findings are the result of collaboration among big teams toiling away in industrial research facilities like CERN, the famous underground Swiss home of the Large-Hadron Collider. Nobody can afford to build such facilities for their own personal use, as they cost billions, so scientists in many fields are forced to collaborate, thus becoming reduced down in the public mind to the level more of technicians than seekers of hidden wonders like Einstein. Doubtless this is a more efficient way to proceed, but when one recent particle physics paper ended up having over 3,000 listed contributors, a person could be forgiven for thinking that the era of individual genius is almost over. Often, to save having a list of names longer than the report itself, the 'lead author' on such papers is not listed as the person who made the greatest contribution, but simply the person who has the first available name in the alphabet; CERN, for instance, currently has some 350-plus papers credited to one 'Dr Georges Aad *et al*', not because Dr Aad is the next Newton, but simply because his name comes first in alphabetical order.[24]

As the idea of science being an utterly objective endeavour entirely separate from the human beings who pursued it really began to take root in the late nineteenth century, the standard practice (since somewhat abandoned) of writing in the passive voice started to spring up too – experimenters would say 'A culture was placed in a petri-dish' not 'I placed a culture in a petri-dish' to make it appear as if the process of their research simply unfolded logically step by step all by itself, rather than being based upon their own personal ideas and decisions. Recent surveys have shown that perhaps around 30–40 per cent of British schools still teach their pupils to write up their experiments in this way, perpetuating this obvious myth of the scientist-as-robot even further.[25] Given the prevalence of such ideas among the general public nowadays, it is perhaps no wonder that the fame of this chapter's twin heroes as scientists has all but faded from the average person's memory. A pair of genuinely massive loons who go around declaring their love for pigeons, skinning cats and trying to invent perpetual motion machines just *can't* be proper scientists – can they?

Of course they can; well, one of them could be, anyway. While Strindberg was a fantasist, Tesla was one of the foremost architects of modernity, and there have been many other deeply strange scientists throughout the years, as we have already seen. Arthur Koestler, in his 1959 book *The Sleepwalkers*, an interesting attempt to reclaim science as a subjectively motivated human activity rather than a dehumanised and entirely objective one, bemoaned the way that historians of science so frequently attempt to depict their heroes as 'reasoning-machines on austere marble pedestals ... probably on the assumption that in the case of a Philosopher of Nature, unlike that of a statesman or conqueror, character and personality are irrelevant', a conclusion Koestler viewed as being dubious at best. Koestler's book deals specifically with the history of cosmology, which he claimed 'may without exaggeration be called a history of collective obsessions and controlled schizophrenias'.[26] You could certainly say that about the work of Tesla and Strindberg!

The God Particle

Strindberg and Tesla might seem at first sight like very different people, their shared eccentricity aside, but there were in fact certain very real similarities between their weird quests. Both were in search of new scientific utopias – albeit of very different kinds – and both relied to some extent upon states of mystical revelation in order to get them there. Each of them, too, saw the potential of a kind of hidden, invisible realm located beyond the boundaries of the everyday world to transform humanity's lot on Earth. In Strindberg's case this transformation would be effected through a new form of scientific spirituality in which the boundaries between man and matter became erased, whereas for Tesla new forms of technology which tapped the secret sources of energy he was convinced lay all around us were the key to the coming change – but both were still convinced that change was coming, and that they were the only men who could bring it about.

Strindberg's most intense period of scientific research, the 1890s, coincided with the height of influence of a once-famous natural historian and philosopher named Ernst Haeckel (1834–1919), whose great idea was that Darwinism, rather than banishing God from the world, acted instead to prove His presence within it. Aiming to make a kind of scientific religion out of evolution, Haeckel became the most prominent proponent of a philosophy termed Monism, whose implications were well laid out by him in a short 1892 book, *Monism as Connecting Religion and Science*. Here, Haeckel explained his vision of 'Nature as a unity' and spoke of the need to 'establish a bond between religion and science, and thus contribute to the adjustment of the antithesis so needlessly maintained between these two highest spheres'. Haeckel's basic idea was, as mystics so often have it, that 'all is one'; for Haeckel, life itself evolved out of inorganic matter, and so all inanimate things could thus potentially at some far-off future point take on the quality of life to them. As he said, there is an 'essential unity of inorganic and organic Nature, the latter having been evolved from the former only at a relatively late period'. Thus, Haeckel explained, 'We cannot draw a sharp line of distinction between these two great divisions of Nature, any more than we can recognise an absolute distinction between the animal and the vegetable kingdom, or between the lower animals and man.'

This was Darwinism taken to its extreme; not only were men once monkeys, they were also once stones, chemicals and other such lifeless substances at some point during the distant past too, and ultimately would one day be so again, after death, when we rotted down into atoms. For Haeckel the idea of Monism had a real mystical quality to it, one which had in the past been anticipated and hymned by such great poets as Johann Wolfgang von Goethe (1749–1832), who had, he said, one 'fundamental thought ... that of the oneness of the cosmos, of the indissoluble connection between energy and matter ... or, as we may also say, between God and the world'. When Haeckel said 'God', though, he did not mean it in the traditional sense of a personal deity, an old man with a big beard living among the clouds. Indeed, he derided such notions as being childish and 'dualistic', seeing as they assumed a separation between God and His Creation; in Haeckel's view, God *was* Creation, the single unity of energy which underlay all matter. As he put it:

> Ever more clearly are we compelled by reflection to realise that God is not to be placed over against the material world as an external being, but must be placed as a 'divine power' or 'moving spirit' within the cosmos itself. Ever clearer does it become that all the wonderful phenomena of Nature are only various products of one and the same original force, various combinations of one and the same primitive matter.

This in itself was not exactly a new view – in variant form, it used to be called either 'deism' or 'panpsychism' – but what was new was the way Haeckel used the language of both Darwin and physics to back it up. God was not now merely 'Creation', but 'the infinite sum of all natural forces, the sum of all atomic forces, and all ether-vibrations'. Representing God as a kind of anthropomorphic being merely 'degrades' Him down to the level of a 'gaseous vertebrate', Haeckel said, and was an opinion fit only for small children and morons. There was no conscious God, and there was no special position for man in the universe, just as Darwin had truthfully implied, but there was nonetheless a consolation to this apparent downgrading of mankind's place in the scheme of things, as it also meant that he too now stood revealed as sharing in the essential nature of God, seeing as God was really all things. Haeckel quotes the great Renaissance mystic Giordano Bruno (1548–1600) to the effect that 'There is one spirit in all things, and no body is so small that it does not contain a part of the divine substance whereby it is animated', using these words to suggest that all atoms should be conceived 'not as dead masses, but as living elementary particles endowed with the power of attraction and repulsion', qualities we would now generally express in terms of positive and negative charges, but which Haeckel suggested should be renamed as either 'pleasure' and 'pain' or 'love' and 'hate' to make them seem more alive when speaking about them.

Elementary, My Dear Haeckel

Seeing as inorganic matter was thus somehow alive, it made sense to Haeckel to talk about its changes from one state to another in terms of Darwinian processes too. Rather than saying that one substance changed into another through a chemical reaction, he asked, why not say instead that one substance *evolved* into another? According to him, various eminent scientists had tried their best to prove that all present-day chemical elements were merely the 'products of evolution, or of historically originating combinations of, seven [earlier] primary elements, and that these last again are historical products of one single primitive element'. One day, Haeckel said, this original primary primitive element – God in His purest form, basically – might well be found, and this discovery 'would probably realise the alchemists' hope of being able to produce gold and silver artificially out of the other elements'. After all, every single thing in Creation, whether living or dead, was simply made up of 'different combinations of a varying number of atoms of one single original element', so why should lead not become gold by manipulating the aspect of God which lived within it? Furthermore, said Haeckel, for the true Monist, energy and matter were also merely differing aspects of the same ultimate thing – his beloved impersonal God. This notion led Haeckel to the remarkable claim that the First Law of Thermodynamics should henceforth be made to 'stand as the first article of our Monistic religion'. This law, he said:

> ... shows that the energy of the universe is a constant unchangeable magnitude; if any energy whatever seems to vanish or to come anew into play, this is only due to the transformation of one form of energy into another. In the same way Lavoisier's law of the 'conservation of matter' shows us that the material of the cosmos is of a constant unchangeable magnitude; if any body seems to vanish (as, for example, by burning) or to come anew into being (as, for example, by crystallisation), this also is simply due to change of form or of combination [of matter]. Both these great laws ... may be brought under one philosophical conception as the law of the conservation of substance; for, according to our Monistic conception, energy and matter are inseparable, being only different inalienable manifestations of one single universal being-substance.

Or, put in simpler terms – 'all is one', as we said before. This was not a message, though, which found an enthusiastic ear among the Catholic Church. However, said Haeckel, just so long as the Pope dropped his silly picture of God as a 'gaseous vertebrate', then there was no reason why the new religion of science could not join together with the old religion of Christianity in fruitful union. After all, as he showed an appreciative public in his bestselling 1899 book *Art-Forms in Nature*, which was filled with beautiful images of the patterns and symmetries to be found in the natural world when placed beneath a microscope, science could satisfy mankind's need for the numinous and the spiritual every bit as well as a good Mass could. The new Monistic science, he said, when taught in schools through his books, would inevitably 'unfold to the rising youth' of Europe:

> ... not only the wonderful truths of the evolution of the cosmos, but also the inexhaustible treasures of beauty lying everywhere hidden therein. Whether we marvel at the majesty of the lofty mountains or the magic world of the sea, whether with the telescope we explore the infinitely great wonders of the starry heaven, or with the microscope the yet more surprising wonders of a life infinitely small, everywhere does Divine Nature open up to us an inexhaustible fountain of aesthetic enjoyment ... now, at last, it is given to the mightily advancing human mind to have its eyes opened; it is given to it to show that a true knowledge of Nature affords full satisfaction and inexhaustible nourishment not only for its searching understanding, but also for its yearning spirit.

As well as access to beauty, though, Monism would also help mankind improve itself morally. 'Beauty is Truth, Truth Beauty', as Keats once wrote, and, said Haeckel, through such lovely picture books as his own, Monism would soon help the masses to realise the essential divinity of the world and themselves, and thus become better, more moral people. He was 'convinced', he said, sounding not a million miles away from a modern-day liberal humanist, that the 'rational morality of Monistic religion' was 'in no way contrary to the good and truly valuable elements of the Christian ethic, but is destined in conjunction with these to promote the true progress of humanity in the future'. In Monism, he said, both 'the ethical demands of the soul are satisfied, as well as the logical necessities of the understanding'. It is hard to see how Haeckel squared this belief in morals and progress with his professed anti-Semitism – his unfortunate ideas about each race having its own collective ancestral soul later fed into Nazism – but he seemed to think these doctrines compatible with peace and love.[27] Basically, then, Haeckel was attempting to create for himself a kind of rationalistic religion, in which science itself became a kind of depersonalised God with a new, more 'logical' Holy Trinity at its heart:

> Monistic investigation of Nature, as knowledge of the true, Monistic ethic as training for the good, Monistic aesthetic as pursuit of the beautiful – these are the three great departments of our Monism; by

> the harmonious and consistent cultivation of these we effect at last the truly beatific union of religion and science, so painfully longed after by so many today. The True, the Beautiful and the Good, these are the three august Divine Ones before which we bow the knee in adoration; in the unforced combination and mutual supplementing of these we gain the pure idea of God. To this 'triune' [threefold] Divine Ideal shall the coming twentieth century build its altars.

Bizarre though they may sound, these ideas were not as fringe at the time as we may now suppose. Haeckel was not some lone outsider, but a respected mainstream figure, the Vice-President of the University of Jena, whose books sold every bit as well as those of Richard Dawkins, preaching a more materialist creed, do today, a fact which led their author to declare himself 'convinced' that 'at least nine-tenths of the men of science now living' subscribed to the ideas expressed in them.[28]

Joined-Up Thinking

Whatever the truth of this claim, August Strindberg certainly held Haeckel's tenets dear. He declared himself a Monist, and began producing experimental artworks of the type we would now call 'Expressionist', in which blobs of paint coalesced with one another to create blurred pictures of storm-tossed landscapes. Unkind critics at the time suggested that the vagueness of the forms on his canvases meant that Strindberg simply couldn't paint, but he preferred to claim that the fact the sea, mountains and ground all merged into one another fuzzily was simply an innovative visual tribute to his latest hero Haeckel and his notion that 'all is one'. Inspired anew by Haeckel's theories, rather than trying to go against the new Monist creed of the First Law of Thermodynamics, as in his youthful experiments with perpetual motion, Strindberg now set out to prove instead that, in a living, God-permeated universe where atoms could feel love and hate, mind and matter were simply differing aspects of the same underlying thing.[29]

His childhood love of geometry and mania for measuring could now be re-explained as a nascent stirring of the religious impulse within him; if hidden laws of shape and number really did underlie the world, then might this not be just another way of saying that the visible world was not really as it seemed, and held a secret unity to it, just like Haeckel was saying? He certainly hoped so. There was something about the mainstream science of his day which repelled Strindberg. Partly, it was the trend towards increasing specialisation, and the inevitable fragmenting of knowledge down into different scientific disciplines, something he said had brought about a 'Babylonian confusion, where things are so far gone that a mineralogist can't understand a zoologist'.[30] It was no wonder that Strindberg drifted towards Monism; after all, if plants and animals really did exist in a continuum with inorganic matter, and had evolved from out of it as Haeckel said, then did this not hold out hope that one fine future day the mineralogist and the zoologist would finally be able to speak to one another again?

A century or so beforehand, Goethe had expressed a similar kind of hope when he spoke to the poet Friedrich Schiller (1759–1805) about his idea that, as he put it, Nature itself was 'God's living garment', an 'active and living' thing, which was constantly 'struggling from the whole into parts' through a process of evolution. Schiller, though, was not impressed by the notion. 'That's not an empirical experience,' he sneered. 'It's just an *idea*.'[31] In other words, Schiller was saying, Goethe's proposal was not really provable or falsifiable, and thus was not really science at all, merely a pleasant piece of poetic intuition. It was the same with Strindberg's pseudoscientific view about the world; it was primarily based upon an emotional need rather than any actual hard evidence as such. As Strindberg himself admitted about his approach to science, 'although I cannot formulate it distinctly, a kind of religion has been forming in me. It is rather a condition of the soul than a view of things based on dogmatic instruction; a chaos of sensations which condense themselves more or less into thoughts.'[32]

Here we have the notion of science as religion again, which in Strindberg ultimately transmuted into a bizarre form of laboratory-based occultism. Like Haeckel, his own brand of mystical pseudoscience was directly opposed to the prevailing scientific spirit of the day, the emerging creed of scient*ism*, in which all phenomena were being treated as if potentially explicable upon materialist terms. In order to counteract the increasing growth of rampant rationalism, Strindberg started meeting with a coterie of likeminded peers at a Berlin tavern named The Black Piglet, where all forms of irrationalism began to be worshipped instead, leading to an atmosphere which centred around the avowedly subjective, emotional and impulse-led spheres of sex, drugs, alcohol, poetry, music, art and occultism. Some of the other members of this little coterie had begun charting a new course between science and the occult prior to Strindberg's arrival, most notably a Polish Satanist named Stanislaw Przybyszewski (1868–1927) who, in an earlier stage of his career, had been a noted psychologist who had penned a detailed and well-respected study of the microscopic structure of the human brain. By October 1892, however, when Strindberg arrived in Berlin and Przybyszewski went to the station to meet him, he had become so alienated from his former way of life that he claimed to be able to see the 'ghost' of Strindberg's infant placenta following him off the train, a vision which he interpreted as meaning that he was too attached to women, and that one day the female sex would be his downfall.[33]

Carbon Copies

Whether troubled by a ghostly placenta or not, Strindberg was certainly feeling haunted by something at around this time, and that 'something' was materialist science. In 1894 he penned a book called *Antibarbarus* (*Against Barbarism*), in which he attempted to explain what he called 'the psychology of sulphur' in Monist terms, trying to show how, as a quasi-living thing, it had evolved from earlier elements like carbon. This programme of discovery was not merely theoretical, though. Parallel to writing his treatise, Strindberg began performing experiments in alchemy – a practice to which Haeckel had specifically lent credence, remember – trying to prove beyond doubt that trace elements of carbon could be found in sulphur, much as he thought relics of mankind's ape past could still be found in the allegedly similar hand structure of a gorilla and an elderly sailor.

Alchemy, we tend to think, was something that only took place many centuries ago, in that more superstitious and credulous time prior to the Enlightenment, not during the 1890s. In fact, this is quite wrong. Far from being a lone visionary, Strindberg was one of many people at the time caught up in the impossible quest to create artificial gold; according to one estimate in the French newspaper *La Paix* for 19 January 1883, there were no fewer than 50,000 alchemists living in Paris alone during that one single year.[34] While this number is undoubtedly exaggerated, it is clear that there were large numbers of people engaged in the art in Paris at the time. The question is: why?

The ultimate dream of alchemy was to transform base metals into gold; by proving all the elements had once been a unified whole, as Haeckel said, the Swedish poet-scientist aimed to demonstrate this was indeed possible. Strindberg, modest as ever, reckoned it would take him a fortnight at most to unpick the secrets of Creation, an estimate which proved slightly optimistic, to say the least. Nonetheless, in the end Strindberg finally succeeded in proving (at least to his own satisfaction) that sulphur did indeed contain small amounts of carbon. This finding may come as a surprise to most modern chemists, but then most modern chemists, unlike Strindberg, tend not to smoke large cigars while performing their experiments, carelessly dropping their carbon-laced ashes into laboratory vessels as they do so ...[35]

Strindberg, however, didn't care that his experiments could never be replicated by non-smokers. As one contemporary review of *Antibarbarus* put it, Strindberg was not really a scientist but 'purely and simply a poet' for whom 'the boundaries between fact, hypothesis and fantasy have been obliterated'.[36] Despairing at the way the world was rapidly being disenchanted and robbed of all its mysteries by late Victorian science's ever-accumulating advances, it seems obvious that Strindberg simply *wanted* to believe in Goethe's idea of Nature as 'God's living garment', and engaged in a kind of willing suspension of disbelief in his alchemy lab which allowed him to do so. The great playwright needed mystery in his life, and began decrying what he called 'the bankruptcy of science', a bankruptcy which Strindberg set out mercilessly to expose in his studies. In 1895, he wrote a somehow rather endearing capsule summary of his peculiar occult-scientific career, in which, besides claiming to have discovered argon, he made it comically clear just how much he felt the need for the inexplicable, and also just how much of an all-time genius he considered himself to be:

> Dedicated from my childhood to the natural sciences ... I had discovered how unsatisfactory the scientific method is ... The weakness of the system showed itself in the gradual degeneration of science; it had marked off a boundary-line over which one was not to step. 'We,' it said, 'have solved all problems; the world has no more riddles.' This presumptuous lie had annoyed me already in 1880, and during the following fifteen years I occupied myself with a complete revision of the natural sciences. In 1884 I doubted the supposed composition of the atmosphere. The nitrogen of the air is not identical with the nitrogen obtained by analysis of a nitrogenous body. In 1891 I visited the Scientific Institute in Lund in order to compare the spectrum analyses of these two sorts of nitrogen whose difference I had discovered. Do I need to describe the reception which the learned scientists gave me? Now in this year, 1895, the [re-]discovery of argon has confirmed my former hypotheses and given a fresh impulse to my investigations ... It is not science which is bankrupt, only the antiquated, degenerate science [of materialism].[37]

Frog-Men

How, specifically, was materialist science disenchanting the world during Strindberg's day, though? These (as it later transpired, inaccurate) 1894 words of the Nobel Prize-winning physicist Albert Michelson (1852–1931) might give a good clue:

> The more important fundamental laws and facts of physical science have all been discovered, and these are now so firmly established that the possibility of their ever being supplanted in consequence of new discoveries is exceedingly remote ... Our future discoveries must be looked for in the sixth place of decimals.[38]

This was not science as a heroic quest in search of unknown truth, but science as a grey realm of dull plodding and boringly minor refinement. 'We have solved all problems; the world has no more riddles,' Strindberg had his hated materialist scientists as saying, thus robbing the universe of all sense of mystery. When applied merely to the physical world, this kind of attitude seemed bad enough, but even more disturbing was the increasing sense that these same people were now setting their sights on disenchanting the hitherto sacred and immaterial realm of mind and spirit, too. In 1874, that arch-materialist T. H. Huxley gave a speech to the British Association for the Advancement of Science, detailing some rather gruesome experiments he had recently been performing upon frogs. Following the lead of the German physiologist Friedrich Göltz (1834–1902), who had reported his own findings about lobotomised amphibians in an 1869 paper, Huxley had removed 'the anterior division' of one particular frog's brain, or at least 'so much of it as lies in front of the optic lobes', thus rendering the poor creature to all intents unconscious, and yet still alive. In this condition, reported Huxley, the frog continued to respond automatically to external stimuli even though, presumably, it could have had no actual awareness of doing so. As he explained, if the leucotomy is performed 'quickly and skilfully', then

> the frog may be kept in a state of full bodily vigour for months ... but it will sit unmoved. It sees nothing; it hears nothing. It will starve sooner than feed itself, although food put into its mouth is swallowed. On irritation, it jumps or walks; if thrown into the water it swims ... To the ordinary influences of light, the frog, deprived of its cerebral hemispheres, appears to be blind. Nevertheless, if the animal be put upon a table, with a book at some little distance between it and the light, and the skin of the hinder part of the body is then irritated [with acid], it will jump forward, avoiding the book by passing to the right or left of it. Therefore, although the frog appears to have no sensation of light, visible objects act through its brain upon the motor mechanism of the body.[39]

The frog was doing all this automatically, not intentionally; as Huxley said, it was 'not acting from purpose, has no consciousness, and is a mere insensible machine'.[40] Thinking of the animal in this way, though, raised another question. Given that the unconscious frog could operate successfully in such a manner, did frogs really *need* to have any kind of consciousness in order to live? Huxley argued that non-lobotomised frogs' sense of self was merely the result of physical and other stimuli acting upon their body, brain and central nervous system. Rather than consciously deciding to hop away from pain, for instance, the ordinary non-lobotomised amphibian automatically fled from it, with it just *seeming* to the frog as if it had decided to escape, when in fact it had no choice but to do so. As Huxley summed this theory up:

> The consciousness of brutes would appear to be related to the mechanism of their body simply as a collateral product of its working, and to be as completely without any power of modifying that working as the steam-whistle which accompanies the work of a locomotive engine is without influence upon its machinery. Their volition [power of choice], if they have any, is an emotion *indicative* of physical changes, not a *cause* of such changes.[41]

Huxley's speech was (and remains) the most famous articulation of an idea now known as 'epiphenomenalism', a theory which, though it may take some time to establish, was to have great import for the thinking of Ernst Haeckel, August Strindberg and Nikola Tesla alike. To say that consciousness is an epiphenomenon is basically to say that our sense of self is a kind of illusion, a secondary phenomenon arising from the physical workings of the brain and body, which gives rise to the erroneous sense that we are in control of our own actions. Huxley was not the first person to describe animals and humans as being a form of automata, or living robots – the French physician Julien de Lamettrie (1709–51) had said much the same in his 1748 book *Man the Machine* – but it was the way in which Huxley went about spreading his idea which was to prove so objectionable to many. Maybe, he said, human beings, like lobotomised frogs, were but 'conscious automata', whose emotions, feelings and desires were 'simply the *symbols* in consciousness of the change which takes place *automatically* in the organism'.[42]

If this really was true, however, then it would prove awkward for the traditional notion of free will; if our conscious selves merely observed any actions which our bodies and brains performed, and did not cause them, then did that not mean we had no choice but to perform any given action, and that we could never have acted otherwise? Actually, Huxley claimed that his theory did not abolish the idea of free will, though his arguing on this point is confusing. He says, for instance, that 'if a greyhound chases a hare, he is a free agent, because his action is in entire accordance with his strong desire to catch the hare'[43] – but surely according to his own prior logic the greyhound's 'desire' to catch his quarry was merely 'an emotion *indicative* of physical changes, not a *cause* of such changes'? Most debatably of all, towards the end of his speech, Huxley went on to apply his hubristic notion of being able to 'predict the state of the fauna of Great Britain in 1869' from the initial state of all the particles in the universe at the point of its birth to the human mind as well. We are, he said, all ultimately 'parts of the great series of causes and effects which, in unbroken continuity, composes that which is, and has been, and shall be – the sum of existence'.[44] Or, in other words, what you had for breakfast this morning, or watched on TV last night, was in a sense already determined from the beginning of time, or at least from the moment of your birth.

Rise of the Robots

Perhaps unsurprisingly, epiphenomenalism is an idea which has historically found few friends, due to it seeming to go against all common sense and practical personal experience.[45] Those who have found themselves its champions, particularly in the present day, have often appeared to be motivated largely by a simple *desire* for it to be true (although, on their own terms, presumably they don't feel that they actually desire to do anything). Many such people appear to be militant atheists, eager to destroy the notion of mankind having a soul by first denying him access to a self. Paul Churchland (b. 1942), for example, a materialist philosopher, has gone so far as to label most people's foolish belief that they exist and are capable of having thoughts and desires as a kind of 'folk psychology', a mere superstition which will eventually be exploded by advances in neurological science.[46]

It is not as if such persons have no evidence at all for their position. For many years, it has been known that physical injuries to the brain can lead to sudden and dramatic alterations in personality and behaviour. The most celebrated case was that of Phineas Gage (1823–60), an American railroad worker who got a metal rod blown through his frontal lobe in 1848 while trying to dynamite a boulder. Formerly a studious, sober and upright citizen, the partial lobotomy caused by the rod had the extraordinary effect of transforming him into an unpleasant lout who took to gambling, swearing and using prostitutes.[47] T. H. Huxley gave an example of another man who had undergone an involuntary partial leucotomy, an unnamed twenty-seven-year-old French soldier who had 'been reduced to a condition more or less comparable to that of a frog' after taking a bullet to the head during battle. The soldier in question subsequently became subject to fits in which he lost all consciousness of his actions and behaved like a living automaton, just as Huxley's frogs had done. The way the soldier moved around while in this state makes him sound like one of those little robotic vacuum cleaners which 'learn' to adapt to their environment: 'If he is in a new place, or if obstacles are intentionally placed in his way, he stumbles gently against them, stops, and then, feeling over the objects with his hands, passes on one side of them,' explained the scientist. Furthermore, said Huxley, 'he offers no resistance to any change of direction which may be impressed upon him; or to the forcible acceleration or retardation of his movements'.[48]

In the case of Gage, then, it seems as if the man's personality was a direct result of the structure of his brain; alter it, and the personality alters alongside it too. With the French soldier, meanwhile, the idea of the human being as simply a kind of living automaton who responds automatically to external stimuli comes to the fore, implying that we could all to some extent still continue to function without certain aspects of our minds being present. Logically, then, if consciousness was merely a function of physical brain activity, and physical brain activity produced automatic responses, perhaps it might be possible to create a living automaton which didn't actually have a real brain at all, merely an artificial one? That was the opinion of Charles Bonnet (1720–93), a Swiss naturalist and philosopher, who wrote in his 1754 *Essai de Psychologie* of his belief that God could, if he so desired, 'create an automaton which should exactly imitate all the external and internal actions of man':

> In the automaton which we are considering, everything would be precisely determined. Everything would occur according to the rules of the most admirable mechanism; one state would succeed another state, one operation would lead to another operation, according to invariable laws; motion would become alternately cause and effect, effect and cause; reaction would answer to action, and reproduction to production ... The sense of the automaton, set in motion by the objects presented to it, would communicate their motion to the brain, the chief motor apparatus of the machine. This would put in action the muscles of the hands and feet, in virtue of their secret connection with the senses ... Words being only the motions impressed on the organ of hearing and that of voice, the diversity of these movements, and their combination ... would represent judgements, reasoning, and all the operations of the mind ... A close correspondence between the senses ...[could be assured] by interposed springs ... Give the automaton a soul which contemplates its movements, which believes itself to be the author of them ... and you will on this hypothesis construct a man.[49]

The unspoken implication, of course, is that this is exactly what God in fact did do. I am inevitably reminded of Richard Dawkins and his description of us all as mere 'gigantic lumbering robots' devoted entirely to the propagation of our genes.

Where There's No Will, There's a Way

Modern-day epiphenomenalists also have the findings of another researcher to draw upon for support, namely those of Benjamin Libet (1916–2007), a San Francisco-based neuroscientist who performed a series of experiments during the 1970s which have since become legendary, as they appear at first glance to disprove the existence of free will. Libet and his colleague John Eccles (1903–97) were interested in the recent discovery of something called 'readiness potential', an electrical signal which fires off in the brain around a second prior to any voluntary action being performed; before you pull the trigger on a gun, for instance, that readiness potential signal primes your finger to move. The natural presumption of how this worked was that the person pulling the trigger first of all made the decision to shoot, then the readiness potential kicked in, causing the finger to twitch. To test this theory out, Libet wired volunteers up to electroencephalographs to measure their brain activity, asked them to stare at a clock and then flick a wrist whenever they felt like it, while also indicating the precise time shown by the clock when they became aware of feeling this wish. Predictably, the point at which the volunteers knew of their desire to perform the flick preceded the flick itself; but, more surprisingly, the readiness potential signal preceded the person's actual felt desire by around 350 milliseconds. Disturbingly, it appeared that the brain had already decided to move the volunteers' wrists before they were even aware they wanted to do so – thereby implying they had no say in the matter, their consciousness thus being merely an epiphenomenon, the whistle accompanying Huxley's steam engine. Libet didn't like this finding, and quickly contrived a get-out clause; the brief gap between awareness of our felt intent to act, and our actually acting, gave us a small window to veto the action which the readiness potential had already prepared our muscles for, he guessed.[50] Rupert Sheldrake has since wittily summed up this position as 'Instead of free will, we have free won't'.[51]

This explanation of Libet's does seem a little contrived, and many materialists (or the rule-governed biological machines which imitate them, at any rate) were overjoyed at his findings – certainly rather more overjoyed than he was. One Libet lover today who quite happily admits that he doesn't believe in free will is the British neuroscientist Patrick Haggard. Haggard works at the Institute for Cognitive Neuroscience in London, where he has access to machinery which produces something called 'transcranial magnetic stimulation'. This involves using magnetic coils to stimulate certain parts of the brain in order to make a person perform a certain physical action, like a marionette being controlled by a puppet master. By allowing an assistant to target a specific part of his brain, Professor Haggard is happy to demonstrate to journalists how his hands, fingers, arms or legs can be moved around without him being able to do anything about it. 'I'm just a machine,' he declared during one typical 2010 interview, 'and she [his assistant] is operating me.' Instead of having free will, Haggard argues, all our actions are merely 'the last output stage' of a living biological computer. This computer seems to be programmed so as to be able to make some pretty remarkable decisions; according to Haggard, even the choice of films he sees at the cinema is really no choice at all, seeing as his ultimate decision 'must be determined somewhere in my brain' already, which must be annoying when his brain chemistry chooses a boring one. 'As a neuroscientist, you've got to be a determinist,' he argues. According to Haggard, 'There are physical laws, which the electrical and chemical events in the brain obey. Under identical circumstances, you couldn't have done otherwise; there's no 'I' which can say "I want to do otherwise".'[52] So, if you reject Haggard's findings as unlikely, simplistic or dogmatic in nature, then it's really not your fault – your brain biology has unavoidably pre-programmed you to do so.

Typical of the exaggerated claims made for transcranial stimulation was a study released in 2015, in which researchers pronounced that zapping a person's brain made them feel 32.8 per cent less likely to believe in God, and 28.5 per cent more positive towards immigrants. Reports summarised the scientists' findings as showing that 'some beliefs had a biological footing and that this could be altered using magnetism', perhaps in order to reduce the number of 'zealous acts' (what a euphemism!) linked to certain extreme religious and political ideologies. Or, in other words, both religious belief and right-wing attitudes are nothing more than malfunctions in the brain, which can potentially one day be 'cured' through the intervention of science. But what about the liberal humanist assumptions which themselves clearly underlie the trial itself? Can they be treated too, or are sufferers from those particular mental malfunctions simply beyond all hope? In fact, closer examination of the study reveals a murkier picture. There is no real evidence that the alleged results of the magnetic stimulation were anything other than temporary, the sample size of test subjects was very small, and the way the results were obtained seemed subjective and open to criticism; obviously, you can't 'measure' a person's attitudes towards immigrants empirically, so participants were simply asked to read an essay by a foreigner being rude about their own country and then give an opinion about the issue. What if some of them just didn't like to give an honest answer? And how can someone possibly be 32.8 per cent less religious than another? The day when people can be cured of sympathising with ISIS by electricity still seems quite some way off if you ask me.[53]

This was the kind of attitude which so irritated people like Strindberg when it first began to be properly expressed back in the late nineteenth century; experiments like Haggard's are actually not an entirely new thing. It is beyond dispute that his mind-control machine works; stimulate a certain part of my brain with an electromagnetic current, and you can indeed make my little finger twitch. Haggard's general finding is obviously a valid one, even if the wider implications he and others have then drawn from it are highly debatable. Even more debatable, however, were the conclusions drawn by a number of Victorian proponents of a dubious and now largely forgotten fad called 'phreno-mesmerism'.

Mesmeric Personalities

Phrenology is now remembered as the very epitome of quackery. Its founder was a Viennese physician, Franz Josef Gall (1758–1828), who had set out to map the brain by splitting it up into twenty-seven separate physical parts, each of which, he said, was responsible for a certain specific aspect of human personality; the 'organ of combativeness', for instance, was located in the lower rear of the brain, along with the 'organ of acquisitiveness'. A person in whose brain these organs were large would naturally be greedy, arrogant and argumentative in nature, whereas someone in whose brain these organs were small would be meek and generous. According to Gall, the average brain type not only of individuals but also of each race and nation differed, thus accounting for diversity in national character.[54]

Gall's theory was an elaborate one but, we now know, complete nonsense. However, that did not prevent two American doctors named R. H. Collyer (1814–91) and Joseph Rhodes Buchanan (1814–99) from developing a technique called phreno-mesmerism, in which a person was placed into a mesmeric or hypnotic trance and then had various points on their heads prodded by the mesmerist. Apparently, this had the effect of setting off the precise phrenological organ in the brain which was supposed to lie beneath the head area being poked, thus allowing the mesmerist to control their patient in a way which would have made Patrick Haggard green with envy.[55] Dr John Elliotson (1791–1868), one-time senior physician at London's University College Hospital, and the man who did more than any other to popularise mesmerism in England, provided a typical account of the results he obtained when employing the method on a female patient:

> On placing the point of a finger on the right organ of attachment, she strongly squeezed my fingers ...[which I had] placed in her right hand, and fancied I was her favourite sister; on removing it to the

> organ of self-esteem, she let go my fingers in her right hand, repelled my hand, mistook me for a person she disliked, and talked in the haughtiest manner ... The finger upon [the organ of] benevolence silenced her instantly, and made her amiable ... I could thus alter her mood, and her conception of my person at pleasure, and play upon her head as upon a piano ... Oh, that Gall could have lived to see this day – these astounding proofs of the truth of phrenology![56]

Except, of course, phrenology was *not* true, and we have to presume that persons like Elliotson's unnamed patient were simply acting under the influence of the hypnotic suggestions being given to them by their manipulators.

Other eminent Victorians were also busily trying to reduce human beings down to the status of fleshy robots during the nineteenth century, most notably the great neurologist Jean-Martin Charcot (1829–93), director of Paris' Salpêtrière Hospital, and a man whose researches August Strindberg followed closely. Known as 'the Napoleon of the Neuroses', he spent the latter decades of the 1800s looking into the pressing issue of hysteria, a mental disease then thought to be approaching epidemic proportions. Traditionally, hysteria has been viewed as an inherently female disease; the word itself means something like 'of the womb', seeing as this was where the illness was once thought to originate. In the texts of the influential classical physicians Hippocrates (460 BC–370 BC) and Galen, the womb was viewed as a kind of 'animal' which wandered around inside women, from the skull to the big toe, in search of moisture, something which supposedly made the female body intrinsically weak and unstable. If the womb got trapped inside a woman's chest, for instance, it could suffocate her.[57] Even into the nineteenth century, despite the fall from favour of the old Galenic and Hippocratic models of medicine, the idea of hysteria being an aberrant expression of female physiology lingered on, especially in Charcot's Salpêtrière. One of the first doctors to accurately associate different areas of the brain and nervous system with the occurrence of different neurological symptoms, Charcot had the pet idea that the entire realm of what appeared to be the psychological would one day prove ultimately reducible down to an expression of cerebral physiology.[58] This was understandable, because he was the first man ever to have established that the tremors associated with multiple sclerosis were caused by certain lesions which showed up on the spinal cord during a dead sufferer's later dissection. As such, when he came to turn his mind towards hysteria, Charcot initially expected to find that it would also turn out to have some similarly organic basis, perhaps in the form of a hitherto unnoticed type of lesion of the brain. However, autopsies upon dead hysterics showed evidence of no such thing. Therefore, unable to relinquish his purely material conception of mind, Charcot turned his attention instead towards the old view of hysteria having its origins within the specifically female organs of the human body – with truly bizarre results.[59]

Pressing All the Right Buttons

Charcot tried to explain hysteria by inventing the concept of so-called 'hysterogenic zones' on the female body; points of alleged sensitivity which, when stimulated, would either produce hysterical symptoms or cause them to stop. Needless to say, these zones centred with suspicious predictability on and around a woman's breasts and genitalia, something which gave rise to much innuendo at the time. In particular, Charcot thought the ovaries were the most significant of what might be termed his 'magic buttons'. Despite repeatedly claiming that hysteria should not be thought of as being a disease of the female reproductive system (he claimed to have demonstrated that the behaviour of male hysterics could be controlled through a timely squeeze on the testicles from his expert hand)[60], in practice Charcot appears to have been unable to escape from this time-honoured notion of the illness, even inventing a bizarre device, known as the 'ovary compressor', to combat hysterical attacks. This alarming-looking item was strapped to a patient's abdomen and worked rather like a wearable vice. It was equipped with a blunt knob on a long screw which could be turned around to tighten it, enabling more and more pressure to be applied to the woman's crotch with each twist. If this didn't work, it was always an option for doctors at the Salpêtrière to lay a female hysteric down on the floor and then hit her in the groin with as much force as was possible until the desired effect was finally produced. Remarkably, many patients actively *requested* this treatment and claimed it brought them some relief! Maybe it did; but of what kind?[61]

Just like John Elliotson with his phrenological mesmerism, or Patrick Haggard with his transcranial magnetic stimulation, Charcot claimed to be able to scientifically operate his harem of female robots like a medically qualified puppet master. When one of Charcot's patients, for instance, a frightened eighteen-year-old girl named Marie Whitman, was admitted to the Salpêtrière in 1877 suffering from bouts of paralysis, fainting and convulsions, she was quickly examined to work out the extent of her clearly faulty hysterogenic zones. It turned out she had two particularly sensitive ones on the outer edge of each breast. Pressing one of these magic breast-buttons at a time had the effect of making Marie perform an involuntary act of swallowing, it was found. Placing 'simultaneous energetic pressure' on both breasts at once with your hands, though, 'immediately produce[d] a convulsive crisis' in the fair maiden, deduced Charcot's dedicated assistants.[62] I'm sure it would be the same for most girls.

It was not only Marie's breasts which proved curiously responsive to such close attention, but also her groin, with the young lady apparently finding Charcot's ovary compressor so well adapted to her personal needs that she sometimes wore it for days on end, while groups of male doctors eagerly experimented with screwing her ever tighter with its hard metal knob. So reliably predictable were Marie's responses to having her ovaries compressed like this that one of her supervising physicians, Dr Paul Richer (1849–1933), specifically compared her to 'one of those music boxes that play several different tunes, but always in the same order'. If Richer wanted to see her perform the second stage of her symptoms, then he should squeeze her ovaries twice; if he wanted to observe the third stage, she needed three twists of the strap-on. Some of these displays were even conducted before a live theatre audience![63] So obsessed by their materialist conception of mind did Charcot and his men become that, eventually, he and Dr Richer began retrospectively re-diagnosing various people from throughout history who had claimed to be possessed by Satan as actually suffering from hysteria, leading to the publication of a book, *The Demonically Possessed in Art*. Here, the good doctors examined various old paintings and drawings, carefully assessing the way that exorcising priests had placed their hands and fingers on certain parts of the demoniacs' bodies, something which revealed that, quite accidentally, the holy men were actually dispelling the 'demons' (i.e. medical symptoms) by pushing down on their victims' hidden hysterogenic zones.[64]

Nowadays, this all seems deeply suspect. Nevertheless, Charcot's contemporary reputation was immense, and it was not until after his death that his hysteria-related experiments came to be largely discredited. So what was going on? While no such thing as hysterogenic zones actually exist, it has since been proposed that hysteria is what is termed an 'iatrogenic illness', one which is unconsciously moulded by a doctor and his patient during their interactions – the phrase means 'brought forth by a healer' in Greek. The women who came to the Salpêtrière were genuinely ill, exhibiting on arrival symptoms like paralysis, fainting, hallucinations, and pseudo-epileptic fitting. Many of them were profoundly damaged, coming from appalling backgrounds involving deprivation, physical abuse and rape. They were not simply frauds. However, living in an age when the traditional symptoms of female hysteria were well known, and being admitted into a hospital which was filled with other women exhibiting those very same symptoms, and well-respected and powerful doctors who expected to see them likewise exhibited in their new cases, a place where diagrams, drawings and photographs of the various stages of hysteria filled the walls and even the ceilings, the female patients soon conformed to the signs of the disease which were expected of them. The whole thing was basically just an act, albeit frequently an unconscious one, seeing as many of Charcot's patients were actually in hypnotic trances when they had their buttons played. One recent historian of Charcot's doings, Asti Hustvedt, has wisely said that while hysteria wasn't actively *manufactured* at the Salpêtrière, as some critics have claimed, 'it was most definitely cultivated'.[65]

Some people, though, wished to cultivate an image of the human mind which was not quite so robotic in nature. When Sigmund Freud (1856–1939) – who had briefly studied under Charcot – and his followers rose to prominence during the 1890s, for instance, they reclaimed hysteria as being a disease of the psyche rather than of the body. As hysteria's victims shifted from the slums to the suburbs under the influence of Freud (who treated mostly middle-class patients), so too the disease's theoretical origins shifted from lesions and mammaries to leisure and memories. The human mind was in the process of being reclaimed once more as an immaterial thing, not a purely material one. Charcot's patients, the French physician theorised, suffered from genetic degeneration handed down from parent to child – a commonly held view at the time. The poor-quality biological inheritance of the poor was, ultimately, what made their invisible lesions, defective hysterogenic zones and other such hysteria-producing bodily flaws appear, Charcot thought.[66] Freud's middle-class patients did not want to hear this unpleasant message being applied to them, though, so instead in psychoanalysis biology was miraculously translated into upbringing. You could still blame your parents for your woes, but for the things which they had thoughtlessly done to you as a child, not for the faulty genes they had so carelessly contributed to your body.

Mind over Matter

Other people also wanted to reclaim the human mind – and the human soul – from the materialists and the epiphenomenalists. Stanislaw Przybyszewski's sudden transformation from viewing the brain beneath a microscope to viewing the ghost of August Strindberg's placenta floating behind him at a Berlin railway station in 1892 could stand as a good symbol for some wider intellectual currents of the time. The late nineteenth century was one of the high-water marks of the worldwide Victorian spiritualist craze, a time when the brain was being reimagined by some as the seat of hitherto unimaginable supernatural powers. Curiously, some of the people who tried to recast the brain as a potentially paranormal organ, like Przybyszewski himself, came from backgrounds which you might have thought would have given them good reason to consider it a purely material thing. John Elliotson, for instance, the eminent physician whom we met earlier playing a girl 'like a piano' through phreno-mesmerism, also became convinced of the reality of clairvoyance, noting approvingly that one alleged psychic with whom he sat would wrinkle her forehead in just the spot that was meant to be occupied by the phrenological 'organ of eventuality' [anticipation] while conjuring up images of distant scenes.[67]

Far from banishing the ghost from the machine, people like Charcot had accidentally conjured it back up, at least for those who wished to see it. During many of the experiments which took place at the Salpêtrière, Charcot's patients apparently displayed uncanny powers, such as heightened senses (one patient could hear conversations taking place hundreds of metres away behind closed doors), telepathy and clairvoyance. One of Charcot's colleagues, Dr Jules Luys, became well known for some weird experiments in what he called 'medication at a distance'. Hypnotising hysterics before holding up sealed test tubes filled with various substances before them, Luys found that, even though his patients did not consume their concealed contents, they still produced the appropriate medical response; the emetic syrup ipecac automatically induced vomiting, for instance, while alcohol made the women act drunk.[68] Charcot himself proved able to produce stigmata-like wounds on some of his hypnotised patients; by merely suggesting to one trance-bound hysteric that some hot wax had just dripped onto her arm, for instance, he was able to induce a sense of extreme pain, followed by her skin reddening and blistering into a large burn.[69]

These may seem like strange feats for a materialist to perform, but they were part of an entire scientific programme, now largely forgotten, of trying to reclaim the supernatural as in fact being wholly natural. Charcot always denied that the wonders his hysterics performed were in any sense paranormal – but by this, he meant not that they were the results of delusion or fraud, simply that they were not performed through the agency of spirits or demons. Instead, he would have said that certain extraordinary but little-understood properties of the human brain were involved in such cases. A number of other scientific thinkers at the time would have agreed – including, as we shall see, Nikola Tesla and August Strindberg.

Body and Soul

Cases like those of Phineas Gage and T. H. Huxley's zombified frogs could not simply be denied, however, and it was becoming increasingly obvious that the personality (or soul, if you preferred) of any living being had at least something to do with the structure of its brain. But was the soul reducible down *solely* to an epiphenomenon, or only partly? One way of arguing that there was more to life than pure neurology was adopted by Ernst Haeckel. Though he still often used the word 'soul', Haeckel found himself unable to declare in light of recent medical findings that human consciousness was anything other than a biological process; personality, he said, 'is a mechanical work of the ganglion-cells'. We can, he said, 'as little think of our individual soul as separated from our brain, as we can conceive ... the circulation of our blood apart from the action of the heart'. Our current high human levels of consciousness were merely the result of millions upon millions of years of evolution, he admitted. Haeckel speaks of a 'cell-soul' being present even in the most primitive of single-celled organisms. This he defines as being 'a sum of sensations, perceptions and volitions', which would enable the organism to move, eat, reproduce and so forth. Our human soul, he says, differs from the simple mind of such an amoeba 'only in degree'. This all meant that there would be no survival of the individual soul beyond death. The human personality would die with the brain which produces it, said Haeckel, quoting his hero Goethe to the effect that 'matter can neither exist nor act apart from spirit, neither can spirit apart from matter'.

This sounds like pure materialism, but it is not. Haeckel used this line of reasoning to argue that, whilst what people like T. H. Huxley were saying was true, it rather missed the point somewhat. Seeing as, in Haeckel's Monistic view, all was one, the brain itself, whilst simply a kind of biological machine from which consciousness arose as an epiphenomenon, was nonetheless still made up of living atoms which were but a tiny part of the all-pervading God who embodied His own universe. In Monism, he said, 'an immaterial living spirit' – that is, a soul without a brain – 'is just as unthinkable as a dead, spiritless material' – that is, a brain without a soul. While personal immortality was just a childish myth, nonetheless the idea of *impersonal* immortality was an unquestionable truth for him. When a person died, said Haeckel, the immortal atoms which had once made him up remained in existence, and recombined to create some other expression of God instead. Once again, Haeckel summed up this doctrine in terms of the First Law of Thermodynamics:

> Immortality in a scientific sense is conservation of substance, therefore the same as conservation of energy as defined by physics, or conservation of matter as defined by chemistry. The cosmos as a whole is immortal. It is just as inconceivable that any of the atoms of our brain or of the energies of our spirit should vanish out of the world as that any other particle of matter or energy could do so. At our death there disappears only the individual form in which the nerve-substance was fashioned, and the personal "soul"

> which represented the work performed by this. The complicated chemical combinations of that nervous mass pass over into other combinations by decomposition, and the kinetic energy produced by them is transformed into other forms of motion.[70]

So, we are both immortal and mortal at once, depending upon your perspective. The brain, for Haeckel, might well be the cause of *individual* personality and soul, just as Huxley said, but it was also simultaneously a physical medium through which the impersonal and immortal soul of God could choose temporarily to manifest. Thus, to say that mind is an epiphenomenon was, to Haeckel, to say that each man is in fact a partial manifestation of God, the means through which the universe comes to know itself. While expressed in scientific terms, this idea of Haeckel's isn't really science; it is more a quasi-religious belief, which can never be wholly proved or disproved. You either accept it or you don't. If you don't accept it, you are downgrading man's place in the cosmos. If you do accept it, you are ennobling him once more. August Strindberg chose the latter option.

What a Nutcase

Strindberg was not religious in any conventional sense. Though his opinions shifted, during the 1890s he rejected many aspects of Christianity, its authorities and teachings. God Himself, though, was a different matter. Strindberg believed in God; but at this point it was the impersonal scientific God of Haeckel, a God who lived in matter. The poet-scientist knew that God lived in matter as a result of personal experience, because he had once seen Him while looking through a microscope. During 1895, while pursuing alchemy in-depth, Strindberg underwent the most severe mental crisis of his life, an experience he later immortalised in his truly strange 1897 memoir, *Inferno*. Here, he recorded the unexpected result of his attempts to monitor plant growth through a magnifying lens:

> For four days I had let a nut germinate … It looked like a diminutive human brain. One may imagine my surprise when I saw on the glass-side of the microscope two tiny hands, white as alabaster, folded as if in prayer [growing out of the brain]. Was it a vision, an hallucination? Oh, no! It was a crushing reality which made me shudder. The little hands were stretched out towards me, immovable, as if adjuring me. I could count the five fingers, the thumb shorter than the others – real women's or child's hands.[71]

At this point a friend came to call, who agreed that he could see the vision too. The hands and the brain were really there, beneath the lens – and yet, at the same time, they were not. In fact, the 'hands' were the first two sprouting leaves of the infant walnut tree, which simply resembled hands. The 'brain' was just the nut itself, its ridges and patterns looking somewhat brain-like under the microscope. There was nothing supernatural there at all. The real mind which Strindberg was seeing in matter here was his own; obsessed with finding Haeckel's God ensouled within His own Creation, Strindberg had become unhinged. Everywhere he looked, he saw weird simulacra – things which accidentally resemble other things – and interpreted the resemblances as being real. Burning coal during his alchemical experiments, Strindberg found that the charred lumps left in the grate resembled a variety of demonic forms; a 'human trunk with twisted limbs' to which was attached the head of a cockerel, for example. Placing these items on his table, a painter he knew came around one day and took them for real sculptures, innocently asking who had made them.[72]

Together with a sculptor friend, Strindberg then claims one night to see the figure of Zeus reclining in his bed; it turns out it is only the shadow cast by his rumpled bedsheets. Nonetheless, the sculptor says that this is a perfectly valid way for an artist to get his inspiration. The man in question, says Strindberg, 'is a seer', who sees the figures of Christ and Orpheus inherent within a block of stone before he carves them. Strindberg, impressed, interprets his friend's ideas as being 'a new-discovered art of Nature' called 'naturalistic clairvoyance' which heralds nothing less than 'a new-born harmony of matter and spirit'. Orpheus was able to make stones dance through his music in the old Greek myth; by putting his mind into matter, and eliminating the distance between the two, sees Strindberg, so can the sculptor, turning dead marble into something living. The artist shows Strindberg a pencil-sketch outline of a Madonna. Then, he tells him he has copied its graceful outline from some plants he saw floating on a Swiss lake. It is much more beautiful than it ever could have been had he simply invented its shape himself.[73] Good Monist as he is, Strindberg begins now to see the artistic applications of Haeckel's scientific philosophy. If man and matter are simply aspects of one another, then maybe eliminating the distance between them completely will enable him to create a great work of art himself? The trouble is, such an attempt will end up tipping him over the edge from extreme eccentricity into genuine madness …

Hell's Kitchen

One morning in 1896, Charlotte Futterer, the owner of a small Paris café, entered the kitchen of her establishment only to be confronted by a very strange sight. There in front of her stood one of her regular customers, caught red-handed in the act of performing some bizarre ritual. The man, who was clad only in his shirt and a pair of underpants, had arranged all of Madame Futterer's saucepans in a big circle on the floor and was dancing around them, chanting. When Futterer asked her semi-naked patron what he thought he was doing, he answered calmly that he was chasing away evil spirits. Madame Futterer was less surprised by this peculiar sight than you might expect her to have been, however, for she knew full well that this customer of hers was a bit of an odd one. During hot weather, when she left her windows open, she had often seen him climbing in through them instead of using the door, which he said was being closely guarded by invisible demons. On one particularly memorable occasion, this same man had caused a huge explosion in the kitchen, ruining all her food. When Madame Futterer went in to see what was going on, he had an explanation ready to hand. He was sorry he had destroyed her kitchen, he said, but he had been trying to magically create gold in one of her saucepans, an experiment which had gone horribly wrong. This man, pretty obviously, was August Strindberg, who by now had become a complete and utter lunatic.[74]

As his descent into alchemical obsession continued, Strindberg became totally unable to distinguish between reality and fantasy. Coincidences proliferated, some banal, others unbelievable. Books began to open for him at the most appropriate passages. At one point, he finds one lying open on his table at a certain page where a splinter points towards the written phrase 'Is God not Science itself?'[75] Maybe this was a hint. Passing by a second-hand bookstall one day, he comes across a text by the French chemist Mathieu Orfila (1787–1853), which he opens at random. The page he alights on contains a passage denying that sulphur is a pure element, just what he himself is trying to prove. Strindberg takes this as a sign of providence – on that same walk he sees his initials, 'A. S.', painted on a window above a silver cloud and below a rainbow, which he interprets as a sign of covenant between himself and the Science-God. Maybe his quest is blessed? Strindberg quickly performs more experiments, claiming to find proof that sulphur is composed of a holy trinity of carbon, oxygen and hydrogen. Wandering through a graveyard a few days after this momentous discovery, he then comes across 'a monument of classical beauty' dedicated to ... the chemist Orfila, whom Strindberg henceforth deems to be his 'friend and protector', a real guardian angel.[76]

On a walk in the country, Strindberg later spies a tiny Roman knight in grey armour lying on the ground. Getting closer, he sees it is just a piece of scrap metal. It seems to point, however, towards a wall, where he sees the letters 'F. S.' written on the bricks with coal.

He interprets these as being the chemical symbols for iron and sulphur, and thinks that the universe has sent him a message about which elements to burn in order to create gold.[77] Another time, stopping to examine a public monument, he finds two oval pieces of cardboard on the ground, one with the number 207 printed on it, the other with the number 28. These, he said, were the atomic masses of lead and silicon. He later performs experiments with these very substances in his laboratory, and claims to have successfully created gold.[78] Soon, he is boasting about this feat to actual chemists. One of them believes him, and arranges a meeting to which Strindberg does not turn up, sending instead a telegram mysteriously claiming that 'the time has not yet come' for him to reveal his occult secret to the world. It is becoming increasingly clear that he is a fantasist.[79]

If fantasist he was, however, then Strindberg was not alone. Other people, too, were busily trying to eliminate the distance between matter and spirit by exploring the issue of coincidence at around this time, such as the eminent French meteorologist and astronomer Camille Flammarion (1842–1925). The author of such learned tomes as *Thunder and Lightning* and *The Atmosphere*, Flammarion also penned books with rather more esoteric titles like *The Unknown*, *Death and Its Mystery* and *Haunted Houses*, the readership for each being by no means mutually exclusive. Flammarion noted down several extraordinary coincidences he underwent himself, which were apparently genuine. There was the time, for instance, that some pages of a book he was writing were carried off by the wind only to drop down in another street at the feet of his surprised publisher, who gratefully retrieved them – these pages consisting of a chapter about the incredible powers of the wind![80] Flammarion was particularly interested in coincidences surrounding the moment of death, such as clocks stopping at the exact minute a person expires:

> Can we try to interpret these 'coincidences' as symbolic? What is a clock or a watch? It is an instrument for measuring *time*. Now time is the essential element of life, and leads to death. In the universal psychic force which governs all things there is an unknown intelligent principle associated with all events, both great and small ... Would not the stoppage of an instrument of measuring time correspond to the stoppage of a life? And would it not thus have a sense, a significance, instead of being a causal effect of an unknown cause? ... Chance, placed at the service of the calculus of probabilities, does not explain these coincidences.[81]

'Everywhere in Nature, in the directive force of terrestrial life, in the signs of instinct in plants and animals, in the general spirit of things, in humanity, in the cosmic universe, everywhere there is a psychic element,' proposed Flammarion.[82] He seemed, like Haeckel, to conceive of the universe as being somewhat coterminous with mind, saying that 'it is not matter which governs the universe, it is the dynamic and psychic element'.[83] Strindberg would have agreed.

A Sick Mind

Whether you believe any of this or not, these were genuine intellectual currents at the time among many men of science, philosophy and literature. In a Monist world, where matter lives, coincidence would be expected to occur almost as a matter of course. 'The spirits had become naturalistic like the times and were no longer content with visions,' Strindberg said, with extraordinary coincidences now fulfilling the exact same function instead.[84] Some of the supposed 'messages' sent to Strindberg by the living universe, though, were clearly of his own imagining. Seeing his path barred by a large dog one day, Strindberg interprets it as being a Hell-Hound, but it is merely an ordinary Great Dane. Finding his drink has been spilled in a bar, he claims Satan did it. Given a restaurant table located near a toilet, he finds an occult conspiracy in the fact. Noticing a scrap of paper with the word 'vulture' written on it, he is convinced he is going to die, an idea reinforced by a skull with 'two glowing eyes' formed out of borax left over from one of his experiments at the bottom of a saucepan. At the height of his derangement, Strindberg claims to encounter some pansies with human faces which try and talk to him: 'They shake their heads as though they wished to warn me of a danger, and one of them, with a child's face and large eyes, signals to me: "Go away!"'[85]

By now Strindberg was extremely ill, both mentally and physically. He had been trying to make gold and burning sulphur in search of its hidden carbon so long and so furiously that his flesh had begun to peel off: 'the cracked skin of my hands becomes worse, the fissures gape and become full of coal-dust; blood oozes out and the pains become so intolerable that I can undertake nothing more'. Eventually, he ended up being admitted to hospital with blood poisoning. Here, Strindberg began to blame 'unknown powers which had persecuted me for years' for his illness, speculating that evil forces were trying to punish him for having deserted his wife and child in favour of a career in alchemy. Or maybe his true sin was really a greater one? Like Goethe's Faust, the philosopher and proto-scientist who sells his soul to the Devil in return for knowledge, Strindberg had been trying to seek out the hidden truths of Creation with his experiments, and perhaps these secrets were meant to remain hidden? Possibly these two sins, against his family and against the world, were in fact related. 'Obliged to choose between love and knowledge, I had decided to strive for the highest knowledge,' Strindberg admitted, Faust-like, thus condemning him to 'the solemn and terrible silence of the desert in which I defiantly challenge the unknown.'[86] His real sin, he says, was hubris, 'the one sin which the gods do not forgive'. Painting himself as 'an imitator of Orpheus', Strindberg had set out 'to reanimate Nature, which had been done to death by the scientists', but now Nature was *too* alive, and the Monist gods were fighting back.[87] Or were they?

Eventually, the 'unknown powers' which persecuted Strindberg began to feel less like Mephistophelean demons, and more like the curses of people he had wronged. Desiring to see his daughter again, Strindberg tries to engage in a practice known as *envoûtement*, a hypothetical form of psychic curse which involved sending out beams of electrical thought-waves through the ether in order to either communicate with people telepathically or do them harm. In 1896, Strindberg had written an essay on this very subject, *The Irradiation and the Extension of the Soul*, in which he had tried to explain the alleged powers of Charcot's patients at the Salpêtrière in pseudoscientific terms as involving the electrical transmission of thoughts and feelings.[88] If Dr Charcot could make burns appear on a hysteric's arm through electrical willpower alone, then what else might be possible, Strindberg wanted to know? One day, he generously lends a man a coat; but, later, the man takes it off. Why could this be? Might it be because he has now warmed up? Or might it be because Strindberg's 'nervous [i.e. electrical] fluid has become stored up in it and through its opposite polarity subjugated him' into submission? 'Have I become a wizard without knowing it?' he asked, quite seriously.[89] Then there was the fact that various prominent astronomers with whom Strindberg disagreed kept on dying after he had published his own competing theories about the solar-system; he wasn't accidentally killing them off with electrical hatred, was he?[90] Feeling lonely, and in search of 'some catastrophe which unites two hearts', he decides to manipulate his estranged wife into a reconciliation by staring at a picture of his younger daughter Kerstin and sending out electrical curses so that she will fall ill, thus leading his wife to summon him to the child's sickbed. The *envoûtement*, however, goes wrong and it is his eldest daughter, Karin, who actually gets sick soon after.[91]

Presumably this was just coincidence – but, as we have just seen, 'coincidence' no longer existed to Strindberg, and the experience got him thinking. If he could do harm to distant people through his thought-waves, then might not other people try and do the same to him, as revenge for his many sins? When he heard that Stanislaw Przybyszewski had been arrested and accused (wrongly, as it turned out) of gassing his mistress and children to death, Strindberg really started to get scared. In fact the woman had committed suicide, taking her kids along with her, but he was not to know this. Strindberg had fallen out with Przybyszewski some time earlier, and when he began suffering auditory hallucinations of a Schubert piano-piece that Przybyszewski had constantly played during their earlier period of drunken friendship, he started to fear the worst.[92] This apparent murderer had knowledge of both the occult and the physiology of the brain, so ought really to be a psychic mastermind. Strindberg soon finds 'proof' of this idea when he finds two twigs lying in a nearby park, one of which resembles a 'P', the other a 'Y', the first and last letters of his enemy's name (provided you deliberately misspell it, that is!).[93] Hidden electrical machines, possibly operated by Przybyszewski himself, eventually begin to assault him with their invisible influences after dark, he says. They become so bad that, one night, Strindberg has no option but to jump out of a window to escape them in his nightshirt, landing in a mass of thorn bushes.[94] Haeckel's Monism, which says that mind and matter are both simply different aspects of the same thing, has taken on a sinister turn, allowing the brain of an evil genius to attack the mad playwright from miles and miles away.

Hell on Earth

However, there was an even more disturbing possibility than this. If there was no such thing as life after death, as Haeckel had also taught, and God was inherent in the world all around us, then might this not mean that each individual person was, while still alive, actually living in either Heaven or in Hell? Locked up in hospital among syphilitics with rotting noses and missing eyes after his own skin had started to peel and putrefy, Strindberg began to feel as if he was in actual purgatory. He took to reading the Book of Job and identifying himself with that particular individual who had also been horribly tested by God.[95] After his release, he sought out physical signs that he was really living not on Earth, but in a kind of Monist Hell – the *Inferno* of his memoir's title. Looking into a zinc bath in which he makes experiments one day, Strindberg is shocked to find that the evaporation of iron salts has left a pattern strongly resembling a country landscape. He sees 'hills covered with forests of firs, lying between them plains covered with fruit trees and cornfields', the presence of a river, and 'the ruins of a stately castle' sitting on top of some hills 'with precipices of stratified formation'. The alchemist interprets this as some kind of message – but what does it mean?[96] Visiting his estranged wife and children in the Austrian countryside, he soon finds out. Staying with his mother-in-law's sister in the tiny hamlet of Saxen, Strindberg finds the place strangely familiar – because, he realises, it is the very same location he had recently seen depicted in his zinc bath! The hills, the medieval castle, the fir trees, all were the same.[97] What was going on?

Strindberg found the answer in the writings of yet another scientist turned mystic, Emmanuel Swedenborg (1688–1772). One of Sweden's greatest-ever men of science, Swedenborg had been a noted inventor whose creations included the first mercurial air-pump, and had served as Assessor at Sweden's Royal Board of Mines from 1716 to 1747. He also established Sweden's first-ever scientific journal, *Daedalus Hyperboreus*, and was an expert on metallurgy and astronomy. At the age of fifty-six, however, Swedenborg began undergoing trances in which he claimed to speak to angels and demons, and historical figures like St Paul, Moses, Plato, Cicero and even Jesus Christ Himself. Conversing with these wise men of history, said Swedenborg, allowed him to devise yet another of those by now familiar-sounding philosophies in which religion and science were merged, with God infusing the observable world of matter with his divine spirit. To study Nature was thus to study God, he said. In addition, Swedenborg also claimed to have travelled to other planets, where he had met moon-men who spoke through their stomachs, and in his book *Conjugal Love* gave explicit descriptions of angelic sex-sessions he professed to have seen taking place in Heaven.[98]

Swedenborg's main idea was that there existed some kind of mystical correspondence between the earthly world of man and matter, and a higher, spiritual reality. In a God-suffused universe, in which matter was somehow alive, might not man and the Earth contain within them higher spiritual realms like Heaven and Hell? Rather than being a physical place, a big fiery pit in the ground, perhaps Hell was really a kind of psychological state which men could endure while still alive, and which could transform the physical landscape around them into a sort of inferno? Strindberg certainly began to feel this way. Opening one of his new guru's books, he came across a description of Hell which sounded bizarrely similar to the scenery around Saxen where he was staying. So close was the match that Strindberg started wondering whether Swedenborg might not actually have visited the hamlet.[99] Gradually, however, Strindberg came to realise that the demons tormenting him were really psychological in nature. Swedenborg, he said, had 'pointed out the only way to salvation; to seek out the demons in their dens within myself, and there to slay them by repentance'.[100] At last, Strindberg admitted that it was his sin of abandoning his family in search of scientific discovery which had led him down into the real-life Hell of mental illness. The 'powers' tormenting him were nothing more than his own hubristic desire for scientific knowledge:

> The fire of Hell is the wish to rise in the world; the powers awaken this wish and allow the damned souls to get all they want. But as

> soon as the goal is reached, and the wish is fulfilled, everything is seen to be worthless and the victory is null and void. Oh, vanity of vanities![101]

Winding down his Faustian scientific quest, Strindberg soon returned to sanity. He renounced occultism and alchemy, went back to literature, and got better. Strindberg's Monistic nightmare, where the boundaries between self and the world had not so much been abolished as utterly obliterated, was finally coming to an end. At last, he was redeemed!

All of which makes for quite a story – *Inferno* is a truly great book, and Strindberg had high (but sadly misplaced) hopes it would win him the Nobel Prize for Literature.[102] Transforming himself into a real-life Faust showed, if nothing else, a real commitment towards his art, and did indeed transform the base material of Strindberg's life into a kind of gold. It is not, though, the kind of utilitarian outcome we are now taught to expect from science. Strindberg achieved absolutely nothing of any practical scientific use at all (other than proving you should never smoke in the laboratory, maybe). And yet this doesn't in any way matter. He treated science quite openly as a myth, and made the dead universe feel temporarily like a living thing again. Science as a whole does not have only the one single meaning inherent within it. Individuals can imbue it with new meanings and use it for their own ends – in this case, to write a classic of world literature. So long as the real, hum-drum, day-to-day reality of science goes on alongside such unusual (some may say insane) projects, making real, objectively true findings which can be applied practically, then does this really matter? It depends upon the nature of the myth you wish to use science to create, I suppose ...

Paranoid Android

The myth which Nikola Tesla used science to help create for himself was both similar and yet opposite to the myth Strindberg used it to construct. Whereas Strindberg tried to re-enchant matter as an emanation of spirit, Tesla did his best to do the opposite by disenchanting spirit as an emanation of matter. Both, though, used the language of science to try and argue that each were really just differing aspects of the exact same thing.

If Strindberg was strange, then Tesla was even stranger. He subscribed to the late Victorian myth of epiphenomenalism so strongly that he saw fit to describe himself as a mere 'meat-machine' – but if machine he really was, then he seemed to be constantly upon the verge of severe malfunction. Interestingly, the American Artificial Intelligence pioneer Marvin Minksy (b. 1927) also later called himself a 'meat-machine', by which he meant that the human body was merely a mess of flesh and blood which acted as a 'teleoperator for the brain'. Ultimately, Minksy thought that 'our brains themselves are machines', which led him to set out on his quest to create robotic and computer-based AI – that is to say, 'machines that manufacture thoughts'.[103] We tend to think of such attitudes as being quintessentially modern or futuristic in nature, but this is not necessarily so. Even during the late 1800s, Tesla was a pioneer of a field of science he termed 'telautomatics' – what we would call 'remote-control', albeit with a bit of primitive AI thrown in, too. He developed one of the first-ever remote-control toys, a boat which he demonstrated to an astonished world in 1898 at Madison Square Garden. So new was this concept at the time that some of those present thought it contained a tiny hidden monkey steering it, or that Tesla was moving the craft telepathically.[104]

There was no Captain Monkey, of course, but there may have been a little bit of telepathy involved in the feat – or, at least, telepathy as Tesla appeared to define it. The boat was actually controlled by radio waves, but Tesla conceived of it as being an extension of his own body. The receiver on the boat he deemed an artificial 'ear', which responded to the directions its operator beamed out to it from his radio control set.[105] It was an ingenious solution, but one based upon the fact that Tesla thought himself to be a robot – and the reason why Tesla thought himself to be a robot was because he underwent a number of very bizarre experiences of a variety it is almost tempting to label as being 'psychic' in nature. From childhood onwards he had been plagued by a series of incredibly weird visual phenomena; even into adulthood, whenever he had a great new idea, or encountered a dangerous situation, he said that 'flashes of light' would appear before his eyes, with 'all the air' around him being 'filled with living flame'. When he closed his eyes in bed, meanwhile, he would see strange coloured lights appear inside his eyelids, viewing 'innumerable scintillating flakes of green', followed by 'a billowy sea of clouds, seemingly trying to mould themselves into living shapes'.[106]

There was definitely something wrong with Tesla's brain wiring. During a spell of stress in his early manhood, he claimed to have developed super-powered bat senses, being able to 'see' objects in the dark by sensing their presence within twelve feet of him through 'a peculiar creepy sensation' on his forehead. Worse, Tesla's hearing became incredibly sensitive; he said he could hear the ticking of a watch three rooms away, and that a fly landing on a table would cause a loud 'thud' in his ear. He could even feel the vibrations of distant trains echoing throughout his body. 'The ground under my feet trembled continuously,' he recalled. 'I had to support my bed on rubber cushions to get any rest at all.'[107] Tesla was obviously suffering from a bout of mental illness here, but the experience left a lasting mark, as it convinced the young inventor that the whole world was connected to him, and that human beings were governed by the exact same laws as the universe was:

> Though we may never be able to comprehend human life, we know certainly that it is a movement, of whatever nature it may be ... Hence, wherever there is life, there is a mass moved by a force ... All life manifestation, then, even in its most intricate form, as exemplified in man ... is only a movement, to which the same general laws of movement which govern throughout the physical universe must be applicable.[108]

Given this, Tesla penned peculiar articles with titles such as *The Problem of Increasing Human Energy*, in which he wrote things like 'we may conceive of human energy being measured by half the human mass multiplied with the square of a velocity which we are not yet able to compute' and came up with complex equations designed to explain our species' behaviour scientifically.[109] Even the thoughts which passed through Tesla's head were best conceived of as being physics-governed acts of movement, it appeared. Ever since his earliest infancy, Tesla had been seeing hallucinations of an exceedingly annoying nature. According to him, whenever a word was spoken to him, he would instantly see a solid-looking, three-dimensional apparition of the object floating in front of his eyes, interfering with his vision; say the word 'shoe', and a big shoe would appear there immediately before him, as if by magic.[110] While this was generally a nuisance, on occasion it could prove useful to the young Tesla – there was the time, for instance, that he was saved from drowning when he suddenly witnessed 'a familiar diagram illustrating the hydraulic principle that the pressure of a fluid in motion is proportionate to the area exposed' hovering before him, something which allowed him to work out a mechanically valid means of escape.[111] Also quite handy was the way that, while at school, a phantom blackboard would materialise in front of him, displaying the answers during maths lessons, and leading his teachers to acclaim him as a true *wunderkind*.[112] How to account for all this oddness? To Tesla, the answer was obvious; thoughts were physical things which obeyed physical laws, and he – like all other people – was nothing more than a humanoid robot. Realising that the things he saw before his eyes were invariably determined by things he had been thinking or talking about beforehand, and that the things he thought about were, in their turn, precipitated by events and objects he saw in the world around him, he concluded that his mind was 'automatic'. As such, Tesla became a committed epiphenomenalist, stating, 'I have, by every thought and every act of mine, demonstrated, and do so daily, to my absolute satisfaction, that I am an automaton endowed with power of movement, which merely responds to external stimuli beating upon my sense organs, and thinks and acts and moves accordingly.'[113] This Charles Bonnet-like realisation was the basis for Tesla's later invention of his telautomatic boat:

> With these experiences it was only natural that, long ago, I conceived the idea of constructing an automaton [the boat] which would mechanically represent me, and which would respond, as I do myself, but of course in a much more primitive manner, to external influences.[114]

Well, that's all very neat and tidy isn't it? Except that, when it came down to it, Nikola Tesla, like all human beings, turned out to be rather more complex in his mode of operation than a mere remote-controlled toy boat could ever hope to be ...

Mental Radio

Tesla was always most adamant that he did *not* believe in the paranormal. Indeed, he openly scoffed at the idea – making it all the more surprising that he repeatedly boasted of his alleged psychic experiences to those he thought would be willing to listen. One person to whom he liked to speak about such matters was his friend John J. O'Neill (1889–1953), a Pulitzer Prize-winning science journalist with the *New York Herald Tribune* who, in 1944, published the first full biography of the mad genius. To him, Tesla often explained his theory that, as O'Neill had it, 'our experiences, which we call life, are a complex mixture of the responses of our component atoms to the external forces of our environment'.[115] However, he also told O'Neill of an occasion during 1892 when he apparently had a prophetic vision of his mother's death, when he had seen 'a cloud carrying angelic figures of marvellous beauty', one of whom was his mother, floating across his hotel room and singing 'an indescribably sweet song of many voices'. At first Tesla thought this was a ghostly visitation, but later he rationalised it away as just being yet another hallucination. He had known his mother was ill beforehand, the image of the angels in a cloud was one he eventually recognised as having seen in a painting some time earlier, and there was a church near to the hotel whence the heavenly singing may well have emerged.[116]

O'Neill, though – an avid and rather gullible fan of the paranormal – refused to accept such prosaic explanations. According to him, Tesla was simply trying to reconcile his direct personal experience of psychic phenomena with his materialist outlook of epiphenomenalism.[117] O'Neill located the problem in the fact that, during Tesla's day, all alleged psychic events were explained by most in terms of the intervention of spirits; if you had a premonition of the imminent death of a family member, then a ghost had passed it on to you. Tesla, though, thought all talk of ghosts was pure rot. To him, people had no souls to return from beyond the grave, seeing as their personalities and thoughts were merely the consequence of the interaction of matter and energy in the brain; and yet, he still appeared to have occasional glimpses of a world 'beyond the veil' himself. O'Neill spoke to one of Tesla's nephews, who told him of an occasion when he had warned several friends not to get onto a certain train – a train which had then promptly crashed. According to the nephew:

> [Tesla] explained this [premonition] in a mechanical way, saying he was a sensitive receiver that registers any disturbance. He declared that each man is like an automaton which reacts to certain impressions.[118]

Or, in other words, he was rather like his telautomatic remote-controlled boat; a kind of robot, responding to external stimuli which governed his every action. Just as the boat was controlled by invisible waves from his radio set, so Tesla's actions here, he said, were governed by the extreme sensitivity of his senses to some unknown – but wholly natural, and not at all ghostly – force inherent in the universe which was telling him all about the future train crash. Who knew what this force might have been? Tesla, perhaps with his toy boat in mind, speculated that maybe some kind of electrical radio waves were emitted by certain people's brains as the inevitable after-result of their thought processes, allowing psychic communication between people to occur. 'Suppose I make up my mind to murder you,' Tesla once threatened a journalist who was interviewing him. 'In an instant you would know it.'[119] Furthermore, seeing as these waves would be physical things, perhaps it might even be possible for others to physically feel them being broadcast? O'Neill gives a strange and credulous account of a dinner of Tesla's he had once attended, in which the host, annoyed by some trivial slight, supposedly sent out malicious brain waves of 'highly attenuated energy-bearing fluid' against his fellow diners, causing them to feel a bit hot on the various parts of their bodies where these psychic rays then fell.[120]

Like believers in *envoûtement*, Tesla tried to locate all of what had previously been known as 'the supernatural' in the mysterious powers of electricity and the new hidden forces like x-rays which were beginning to be uncovered throughout his own lifetime. He was not the only prominent man of science to have tried doing such a thing. According to the coincidence-loving French meteorologist Camille

Flammarion, for instance, 'The universe is a dynamism. And it seems that everything is electrical. World soul, animal electricity, magnetic fluid ... are all diverse names for this same principal of movement. The psychic and physical worlds are associated.'[121]

Tesla, equally as unwilling to abandon the notion of psychic phenomena as he was reluctant to drop his adherence to epiphenomenalism, found in such arguments a convenient get-out clause. The strangest result of Tesla's attempts to transform the immaterial world into something more material, however, came in his bizarre and implausible theories about the human eye.

The Story of the Eye

It is a great shame for medical science that Nikola Tesla apparently never booked himself in for an appointment with an optician – if he had done, then surely our knowledge of human optics would have been advanced by several hundred years. According to Tesla, he had a highly remarkable pair of eyes. By his own account, he had been born with brown eyes, but later saw these fade into a grey-blue colour because of the 'excessive use of his brain' he had made while inventing things.[122] This was because, as he explained in his insane 1893 paper 'The Action of the Eye', whenever he had a good idea, his eyes literally lit up and glowed with a 'distinct and sometimes painful sensation of luminosity' which was visible to others 'even in broad day-light'.[123] Tesla even alleged that it was possible for men of scientific genius to see in pitch-blackness by using their special light-emitting eyes as torches, something which he warned his readers not to try themselves unless by some miracle they happened to be as clever as he was: 'Probably only a few men could satisfactorily repeat [this feat], for it is very likely that the luminosity of the eyes is associated with uncommon activity of the brain and great imaginative power. It is fluorescence of brain action, as it were.'[124]

Because of the incredible 'fluorescence' of Tesla's own brain actions, he even ended up devising a plan – sadly never enacted – for building a machine which would be capable of broadcasting his thoughts so that they could be seen by others by 'analys[ing] the condition of the retina when disturbed by thought'.[125] After all, if thoughts were physical things, as Tesla proposed, then it should prove possible somehow to record them. This all turned out to be a complete delusion, naturally, but Tesla's own powers of visualisation were so extreme it is actually quite easy to see how he came to believe in this fantasy. According to him, the reason why he came up with so many inventions was because, rather than working them all out with intricate designs on paper, he was able simply to think of the devices he wanted and literally build them up, component by component, in front of his very eyes; he could see his creations floating there as solid 3D objects before him, constructed precisely to scale and with their parts moving around and working exactly as they would in real life. He could then leave them running at the back of his mind for days on end, conjure them back up, and inspect them for visible signs of wear and tear! Later, once he had perfected these devices mentally, he would build them physically in his workshop, and find they functioned precisely as he had foreseen – his thoughts were, literally, being made solid.[126]

Tesla's most famous account of seeing one of these 'mental machines', as he called them, in action had all the force of a genuine, full-blown mystical vision. The invention in question was none other than his amazing A/C motor, Tesla's greatest gift to humanity, which was revealed to him fully formed during one beautiful afternoon in February 1882, when he was walking through a Budapest park with a friend named Szigeti, admiring the sunset and reciting aloud from memory certain verses penned by his literary hero, Goethe. Staring at the dying red blaze of the sun and descanting Goethe's lovely words, Tesla suddenly stopped stock-still in a rigid pose, and went into a trance; then, just as abruptly, he snapped out of it and shouted the words 'Watch me! Watch me reverse it!' Szigeti was confused. 'Are you ill?' he asked. No more than usual, it seemed. 'Don't you see it?' shouted Tesla. 'See how smoothly it is running!' He then picked up a branch and drew a diagram of his motor, complete in every detail, in the dust on the pavement, for Szigeti to share in the revelation he had just received. Tesla had made the most significant scientific discovery of his life – and yet the discovery had been made through means which are surely better described as being poetic or religious in nature than technical or logical. 'In an instant the truth was revealed,' Tesla later wrote. He made it sound as if he just met God.[127]

Poetic Licence

It has often been said of Tesla that he was every bit as much of a poet or mystic as he was a scientist, and indeed his recall of verse in several languages was prodigious, a skill he apparently inherited from his mother. He once even helped co-author an English-language edition of various Serbian epics.[128] Nowadays, though, we just don't expect science and poetry to mix. The classic example of their supposed incompatibility is the 'helpful' note allegedly sent to the Poet Laureate Alfred, Lord Tennyson (1809–92) by the Victorian computer pioneer Charles Babbage (1791–1871) pointing out a statistical inaccuracy in his lines 'Every moment dies a man/Every moment one is born':

> I need hardly point out to you that this calculation would tend to keep the sum total of the world's population in a state of perpetual

equipoise, whereas it is a well-known fact that the said sum total is constantly on the increase. I would therefore take the liberty of suggesting that in the next edition of your excellent poem the erroneous calculation to which I refer should be corrected as follows: 'Every minute dies a man/And one-and-a-sixteenth is born.' (I may add that the exact figures are 1.167, but something must, of course, be conceded to the laws of metre.)[129]

Many people who have tried to argue their 'science' through the medium of verse have been, quite frankly, nothing but lunatics. Otis T. Carr (1904–82), for example, was a UFO-loving fantasist and alleged friend of Nikola Tesla who claimed to have produced a perpetual motion machine which he said was 'powered entirely by the sun's immutable pressure-energy known as gravity'. Lest you mistakenly consider this statement to be nothing but meaningless nonsense, Carr graciously explained his theory of sun-derived perpetual motion in the following haunting stanzas:

All energy is atomic;
The light from the sun,
The pull of gravity,
The fission of uranium,
The fusion of hydrogen,
All that moves,
Or grows, or glows,
Everywhere on Earth
And throughout the Universe.
It is all atomic energy.
Its ultronic presence is everywhere ...
So, how to borrow of this energy
From Nature has been quite
A problem.[130]

But, thanks to Otis, it was a problem no more! The wisdom that the amazing Mr Carr was able to express through his muse was simply inexhaustible. Elsewhere in his oeuvre, for instance, we will find the following invaluable information:

While you may safely blast a dog
Or a man away into space, you will
Never safely blast him back again.[131]

These lines, of course, were written at least a decade prior to Neil Armstrong (1930–2012) landing on the moon – and then blasting himself safely back home to Earth again. To be fair to Carr, though, he didn't take his dog up there with him, so that part of the prediction still stands.

Contrary to such unmitigated nonsense, however, the connection between poetry and science is in fact rather an old one. Far from beginning with Strindberg, the poet-scientist was already alive in the ancient world, where the Roman poet Lucretius' (*c.* 99 BC–*c.* 55 BC) *De Rerum Natura* (*On the Nature of Things*) was a versified explanation of early atomic theory.[132] One of the most interesting of the poet-scientists was Erasmus Darwin, whom we have met briefly several times already. A real polymath, Erasmus was the first man to fully work out what went on during photosynthesis, and apparently once sketched the design for a basic internal-combustion engine, though it was never actually built. He was also somewhat eccentric, notoriously becoming so rotund in the belly that he had a big semi-circle cut into his dining table so he could still sit down at it to make himself even rounder. He was also, like his grandson Charles, an evolutionary theorist, but unlike his more celebrated descendant chose to express these ideas in poetry, not prose, with epics like 1789's *The Loves of the Plants* finding a surprising degree of popularity amongst the public.[133] Erasmus' proto-evolutionary philosophy was well-expressed in his 1803 work *The Temple of Nature*:

> While Nature sinks in Time's destructive Storms,
> The wrecks of Death are but a change of forms;
> Emerging matter from the grave returns,
> Feels new desires, with new sensations burns.[134]

With thinking like that, perhaps Ernst Haeckel was a reader?

The Great Goethe

Undoubtedly the greatest poet-scientist of all was Nikola Tesla's literary hero, Johann Wolfgang von Goethe, the 'German Shakespeare'. Far from simply writing about men like Faust, Goethe had also dabbled in alchemy himself during his youth, and devoted almost as much time to science as to literature. He was even responsible for making one genuinely significant scientific discovery, finding and describing the intermaxillary bone in the human jaw in 1784. The interpretation he placed upon this breakthrough was telling. All vertebrate animals have these bones, but it was thought that humans did not, something which was used to argue that mankind occupied a special place in God's Creation, standing apart from the animal world. However, Goethe showed that the intermaxillary bone was actually present in the human jaw, but fused into the surrounding bones so closely that it was hard to detect. Goethe's conclusion was that every creature in existence was 'only a tone, a shade of one great harmony'.[135] It is easy to see why Haeckel and Strindberg so liked Goethe. He pursued a path of evolutionary Monism, proposing that there was one single primal animal, the *Urtier*, from which all others had later sprung. Arguing that animals' vertebrae, which all looked different and yet were clearly all versions of the same basic thing, were the basis for all subsequent evolutionary development, Goethe became convinced that metamorphoses of the vertebrae had led to the later development of skulls and spinal columns. The body structure of every animal, Goethe said, was merely an expression of transformations of an initial primal substance of bone.[136]

More famously, Goethe also sought an *Urpflanze*, an ideal plant from which all others had later developed. He spent time in Sicily, searching for a sample of this initial perfect plant, but later seemed to modify his notion into one merely of 'plant-ness', an idea of development from one primal plant substance which later led to the appearance of each separate species. Goethe found what he sought in the leaf, from which basic structure he argued all other parts of the plant had ultimately developed, with petals being simply colourful leaves, for instance. He then used these ideas to argue that every living thing in Creation participated within one greater whole, which was God; a plant was an aspect of God in plant form, an animal an aspect of God in animal form. Goethe gave the process of plant development seven stages, a number deliberately chosen for its spiritual implications. This was less pure science than a projection of religious desires onto the external world; the *Urpflanze* represented for Goethe a kind of holy union of opposites, with the plant containing both male and female sex organs, growing up from seed and later producing seeds of their own from which other variations of the plant then grew. Like Haeckel's later conception of God, the ideal *Urpflanze* was thus in a constant state of 'becoming' in the physical world through each physical example of its own manifestation.[137]

While he grew to discard any literal belief in alchemy after the early experiments of his youth, Goethe nonetheless kept hold of the more symbolic aspects of the discipline and continued to obsessively project them out onto the world around him. He became interested in geology, seeing in granite a representation of the fabled Philosopher's Stone. Like the Stone, granite was red in colour, and generally consisted of three separate substances, quartz, feldspar and mica, which together made up more than its individual components, which Goethe saw as paralleling the similar phenomenon of the Holy Trinity. He also viewed granite as analogous to the *Urpflanze*, arguing that it was a 'primal rock' from which all other minerals and metals were later formed; granite 'steps out of itself', he said, being the 'seed' from which all other rocks grew, just as all things in Creation grew from out of God. Just as alchemists spoke of transmuting substances from a fluid state to a fixed, he spoke of rocks themselves forming through a sudden transformation from liquid to solid form; the illustration he gave was of how a glass of water, chilled to near freezing point, would suddenly solidify into hard ice if shaken. He interpreted the veins and ribbons found in rock types like marble as being a permanent record of this process.[138]

You can certainly see what it was Tesla saw in Goethe that he so liked; both men, after all, had a view of Nature as being 'God's living garment'. When he had been gifted the vision of the A/C motor during that glorious red sunset in 1882, it appears that Tesla had also been gifted what John J. O'Neill termed 'an illumination that revealed to him the whole cosmos, in its infinite variations and its myriad forms of manifestations, as a symphony of alternating currents'. Many mystics down the years have been granted a similar insight. Few, however, have then planned, as Tesla did, to create an 'electrical harmonium', which would enable him to 'understand the motif of the cosmic symphony of the electrical vibrations that pervaded the entire universe'.[139] This was an interesting idea. After all, if man and the universe were both simply expressions of some underlying oneness, if thoughts were reducible down to electrical wave patterns and so were the laws of the cosmos itself, then this meant that both man and the universe were one, weren't they? Tesla may never have specifically said so, but it seems at some level he was really just yet another Monist.

All Spaced Out

As with August Strindberg, it appears the attempt to eliminate the distance between spirit and matter sent Tesla completely mad by the end. There can be no other scientist in history who, due to a combination of eccentricity, spirituality and errant genius, has provided such an inspiration to the lunatic fringe. Following Tesla's death in 1943, dozens of books have appeared on the market making claims about him which are so outrageous that they almost make the man himself sound normal. If you believe the various rumours which have since sprung up (and you really shouldn't), then Tesla was, among other things, the reincarnation of an ancient Atlantean engineer, the secret inventor of teleportation and time-travel devices, the man behind that hoary old optical illusion known as 'The Face On Mars', and an immortal, non-human being,[140] Perhaps most absurdly, he is persistently rumoured to have buried the hatchet with his old rival Marconi over the issue of who had really invented radio, and helped him to establish a secret hi-tech city hidden away somewhere in the Venezuelan jungle from where the two men regularly made trips to the moon in self-built flying saucers.[141]

Partly, these mad rumours are Tesla's own fault. For one thing, he managed to convince himself that he came from a family of super-centenarians (even though he didn't), and was certain that, simply by drinking whiskey, he would live to be 150. By the end of his life, so keen was Tesla to live forever that he was surviving purely off milk and vegetarian dishes, and was so hyper-fussy about his hygiene that if a fly landed on the same table from which he was eating, he would order his entire meal be binned and a new one provided.[142] However, because of a combination of his unerring ability to conjure up hallucinations of any future inventions which he hadn't quite perfected yet, and his belief that he would live to well over 100, he had lots of potential ideas stored away up in his skull which he had never bothered to write down.[143] When it became known that he had died, this provided a convenient gap for fantasists to exploit, by claiming to have uncovered details about his future blueprints for things like spaceships and death rays. One of the most extreme such kooks was Otis T. Carr, whose god-awful poetry we encountered a short while back. Carr was a dubious fixture on the 1950s American UFO scene, who ended up being sent to prison for fraud in 1961 after somehow managing to con naive investors into believing in his ability to create a flying saucer which would fly him to the moon and back within a matter of hours. He had announced that he would take off in his new spaceship from a fairground in Oklahoma one day in April 1959, but suspiciously felt 'unwell' on the day of proposed blast-off, and stayed at home in bed instead.[144]

In 1955, Carr had founded a fake company called OTC Enterprises, which he claimed was devoted to applying the secret inventions of Nikola Tesla towards such areas as space travel and perpetual motion. His fake flying saucer was therefore meant to be an extension of the things Tesla had (supposedly) been working on during his own lifetime. If you were a true believer in the reality of UFOs at this time, these ideas may have appeared relatively plausible, seeing as Tesla had made genuine attempts to communicate with Martians at various points during his life, and even professed to have received radio signals from the Red Planet, which was not then definitely known to be unpopulated.[145] Another man who exploited these facts was Arthur H. Matthews (1893–?), a Canadian who, like Carr, claimed to have been a good friend of Tesla's, so much so that the great man had let him in on a few of his most closely guarded secrets – in particular, the design of a special radio set to be used for communicating with Earth-orbiting spaceships. If Tesla really did create this device, though, then it had one big design flaw: for some inexplicable reason, it only worked inside Matthews' own house in Quebec, which was remote and not easily visited. However, in many ways it turned out that this secluded location was also quite advantageous. It was so private that spaceships could land there on a regular basis without being seen by anybody else, Matthews said, which was rather useful as the aliens who piloted them were curiously shy.[146]

The Dove from Above

It seems incredible now that anybody could ever fall for such palpable drivel, but at least one woman at the time did. She was Margaret Storm, a New York-based Theosophist who in 1957 did something extraordinary: she recast Tesla as a new Christ, and set up her own miniature religion in his honour. In that year, Storm published an obscure book called *Return of the Dove* – a work so cranky it is actually printed entirely in green ink.[147] Seeing as Storm modestly describes her New Gospel as being 'a swift historical review covering the past nineteen million years',[148] her precise teachings are far too complex to describe fully here, so what follows is a simplification, albeit probably a merciful one. Essentially, Storm's creed centres on the idea of the reincarnated ghost of Sir Francis Bacon plotting to bring about a modern-day Great Instauration. Bacon had tried to return us all back to Eden, Storm said, by causing the new Jesus, Nikola Tesla, to be born upon Earth rather than on Venus, as he should have been, in order to save mankind from the curses of work, sickness and war by enabling us to master electricity through the use of his A/C motor. Such Tesla inventions would ultimately lead to 'unlimited abundance for all … down to the last dog and cat', she said.[149] Storm claimed to have proof that Bacon, now dubbed 'The Lord of Civilisation', was wandering across America in reincarnated form, displaying special self-cleaning tablecloths to housewives as a taste of the scientific wonders which were yet to come.[150]

However, while Tesla was certainly a scientific utopian, it was becoming increasingly obvious as the 1950s wore on and Cold War nuclear annihilation loomed that science hadn't saved the world after all. Why could this be? Storm provided the millions of people across the globe asking this same question with a convenient answer in the form of a conspiracy theory. Unable to accept that scientific advances had not yet led to humanity's salvation, Storm declared that the military-industrial complex had all ganged up against Tesla and suppressed his best inventions, such as perpetual motion and free wireless electricity for all, so the New Messiah's full work had not yet been completed. As such, once Tesla had died, his ghost (which was not really a ghost at all, but something called an Evolved Ascended Master … it's complicated) continued to contact certain favoured disciples here on our Earth-plane, giving them details about how to build spherical spaceships, matter transformers, perpetual motion motors, anti-war machines and various other amazing new devices. Decrying death as an 'ugly delusion' and a mere 'habit',[151] Storm describes how following the spiritual example of Tesla will one day enable us all to become immortal, transforming us into 'evolving gods'[152] who will achieve eternal youth before eventually shedding our physical bodies completely and learning to live our disembodied lives in perfect spiritual peace and harmony with each other and the universe. Once we have achieved all this, we will be back in Eden – because we used to have all these superpowers 19 million years ago anyway, but then lost them due to some appalling moral catastrophe.[153] Sir Francis Bacon's ultimate plan, apparently, is not only to take us back to Eden, but to abolish all recollection of us ever having not been there in the first place by wiping the cosmic memory banks clean of all such unpleasantness.[154]

As an added bonus, Storm also explains why it is that sparrows don't have to pay death taxes,[155] and details infallible methods for creating magical Tesla-derived 'tubes of light' with the power of your mind, which can then be placed around your children to prevent them from being shot, or erected around your pet dog to stop it from being raped.[156] Basically, then, Storm was a Rosicrucian several centuries too late – albeit a Rosicrucian with no apparent understanding of any actual science whatsoever. Her initial impulse to develop this loony creed seemed to have been some information which had supposedly been imparted to Arthur H. Matthews by orbiting spacemen over his special Tesla radio set. According to Matthews:

> Nikola Tesla was not an Earth-man. The space-people have stated that a male child was born on board a spaceship which was on a flight from Venus to the Earth in July, 1856. The little boy was called Nikola. The ship landed at midnight, between July 9 and 10, in a remote mountain province in what is now Yugoslavia. There, according to arrangements, the child was placed in the care of a good man and his wife, the Rev. Milutin and Djouka Tesla.[157]

With this information in mind, Storm began combing through John J. O'Neill's biography of Tesla, searching for 'evidence' that he was not truly of this world. By obsessively reinterpreting his various eccentric habits and foibles, she was easily able to do so. Why did his eyes glow? Because he was highly evolved and had a golden extra-terrestrial brain.[158] Why did he have no financial sense? Because there was no cash on Venus, and so he 'had not had any previous training in handling money'.[159] Why did he think drinking whiskey would make him live to 150? Because he was using his advanced alien powers to 'transmute' the alcohol into 'pure energy' with special health-giving qualities.[160] (Tesla's predecessor Jesus had very similar powers, allowing him to transmute water into wine.[161]) Why did he have no interest in sex? Because he was born on a UFO sexlessly by aliens crossing beams of electricity together, thus probably accounting for his distaste for physical union.[162] Why did he think he could fly as a child? Because he *could* fly, because he was an alien – he was always levitating around New York, but had since mastered the art of invisibility too, so nobody had ever seen him in flight ... except, perhaps, for a certain pigeon.[163]

Ah yes, that '*certain pigeon*'. Storm's book may well have been full to bursting with bizarre claims, but none were quite so bizarre as the one made by Tesla to John J. O'Neill sometime before his death in 1943. Up until this point, it had been believed that Tesla had never fallen in love – but he had. Just not with a woman. Apparently, there was a special pigeon in his life, one to whom no other pigeon could ever compare, for a variety of genuinely unique reasons. Tesla's own account of this pigeon's incomparable dove-like loveliness begs to be cited in full:

> I have been feeding pigeons, thousands of them, for years ... But there was one pigeon, a beautiful bird, pure white with light grey tips on its wings; that one was different. It was a female. I would know that pigeon anywhere. No matter where I was, that pigeon would find me; when I wanted her I had only to wish and call her and she would come flying to me. She understood me and I understood her. I loved that pigeon ... Yes, I loved that pigeon, I loved her as a man loves a woman, and she loved me. When she was ill I knew and understood; she came to my room and I stayed beside her for days. I nursed her back to health. That pigeon was the joy of my life. If she needed me, nothing else mattered. As long as I had her, there was a purpose in my life. Then one night as I was lying on my bed in the dark, solving problems as usual, she flew in through the open window and stood on my desk. I knew she wanted me. She wanted to tell me something important ... I knew she wanted to tell me she was dying. And then, as I got her message, there came a light from her eyes – powerful beams of light ... It was a real light, a powerful, dazzling, blinding light, more intense than I had ever produced by the most powerful lamps in my laboratory. When that pigeon died, something went out of my life ... I knew my life's work was finished.[164]

Tesla really did say this. It was not Margaret Storm's fantasy, although she did later build part of her religion upon this alleged encounter, saying that such magical birds were things called 'Twin-Rays' which

Francis Bacon called down from Heaven to act as special carrier pigeons to take messages between himself and his disciples. According to her, the bird was a mere 'undercover pigeon', a dove-like angelic messenger which took the mundane form it did so as not to attract undue attention to itself.[165] Only Tesla, with his magic, visionary eyes, could see its real nature.

Being Human

From today's perspective, we might be reluctant to give credence to the views of a man who claims to have fallen in love with a laser-emitting holy pigeon, but there is no doubt that Nikola Tesla was a scientist of the first order. Maybe having a highly developed fantasy life is actually a rather good quality for a scientist to have? After all, science is, in its highest sense, a truly creative act. When we think of serried ranks of technicians in white coats performing mundane (but necessary) tasks and measurements in sterile labs, it can often seem quite dull and plodding, but to a Newton or an Einstein, a Darwin or a Lavoisier, the dawning of the Great New Idea which overthrows so many old ones within their minds must have been an exhilarating thing. By definition, the person who comes up with a theory as revolutionary as those of general relativity or evolution by natural selection must be in possession of a highly developed imagination. With their incredible ideas, such people create the world anew, just as great artists and writers also create their own new worlds in their novels, plays and paintings. No wonder there was such an unconscious affinity between the lives and thinking of both August Strindberg and Nikola Tesla.

The idea of epiphenomenalism, and of us being naught but Richard Dawkins-style 'biological machines', seeks to reduce us all down to the status of mere robots – the telautomata spoken of by Tesla and previously proposed by Charles Bonnet. When Tesla explained this same notion to John J. O'Neill one day, the journalist was having none of it, telling his friend, 'I don't believe a word of your theory … and, thank God, I am convinced you don't believe a word of it either. The strongest proof I have that your theory is totally inadequate is that Tesla exists. Under your theory we could not have a Tesla!'[166]

How true. Someone like Patrick Haggard might profess that we are all measly biological telautomata who are reducible down to nothing but a mechanical process of rules-governed interaction between abstract chemical and neurological impulses, but what kind of transcranial magnetic stimulation, precisely, can make a grown man fall in love with a pigeon? Or make a playwright see a walnut saying a prayer? Or cause a deluded Theosophist to worship a dead scientist as the new Jesus? Our modern-day secular Western liberal humanist conception of ourselves as inherently rational actors, which now seeks to reduce us down to the point where we are not really even actors at all, but mere automatic stimulus-response mechanisms, seems to falter in the light of such profoundly illogical life stories. When it comes to the workings of the human mind, perhaps the most rational approach, rather wonderfully, is to adopt a completely irrational one.

The Birds, but Not the Bees: Spontaneous Generation, Alchemy and the Living Universe

The universe is in effect a cranks' charter, in that it will obligingly reflect back to the theorist any ideas projected onto it.

John Michell[1]

We are constantly being told these days that the world's bees are in danger of dying out. All across the globe, from America to Austria, these tiny striped insects are suffering from a mysterious malady known as 'colony collapse disorder'; beekeepers everywhere are opening up previously healthy hives only to find that most of the bees they had left living there happily only days beforehand have suddenly all died off or else simply disappeared to pastures new.[2] Considering the central part such creatures play in the pollination of our planet's crops, this could be bad news indeed – as Albert Einstein himself (supposedly) once put it, 'If the bee disappeared off the face of the Earth, man would have no more than four years left to live.'[3] Things have now got so bad that some orchards in China have begun employing men with pollen-coated sticks to go around pretending to be giant bees, jabbing at apple blossoms to ensure the future survival of their harvests.[4]

The reasons for this alarming decline are, as yet, unknown. Some point the finger at parasites or pesticides, others blame global warming and yet others claim the iniquities of modern industrial bee-farming methods may be to blame. As far as I know, however, nobody has so far made the suggestion that the planet's bees are dying out on account of their lamentable failure to have sex with one another. Indeed, if anyone *did* venture to make such a suggestion, it would be safe to presume that he would be roundly ignored. If that same person then said that the solution would be for someone to kill a few lions and then leave their corpses to dry out in the sun for a bit until bees began to magically pour out of their mouths, then he would probably be carted off in a straightjacket. After all, only a complete madman could ever bring themselves to believe in such an absurd fairy tale ... right?

The Spirit of the Beehive

Not at all. The idea that bees were celibate, and born spontaneously from the mouths of dead animals, was once very widely believed, right across Europe, North Africa and the Middle East – and not only by lunatics. The great Aristotle himself, in his *De Generatione Animalum*, a book discussing the reproductive processes of animals, gave a whole section over to the problem of how it was that bees sired their young, finally admitting that 'none of them has ever been seen copulating', and going on to say that the whole question was something of a mystery. If bees didn't have sex, then they must surely have come from somewhere – but where? The sun, maybe? Or did they grow on trees like grapes? Possibly they sprang from olives? Some Greeks thought they were born from flowers, an idea whose origins are surely obvious, but the notion of them germinating from dead animals seems to have been the most popular theory.[5] So suffused did Western and Middle Eastern culture become with the idea that a reference to it even appears on tins of Lyle's Golden Syrup!

If you have ever bought a can of this saccharine substance, then you may have found yourself being rather puzzled by the logo inked onto its label; namely, the corpse of a dead lion, surrounded by a swarm of what appear to be carrion flies. What can the marketing men be trying to say with it, exactly? In fact, the entire thing is an allegory. Those bluebottles are actually bees, and the choice of motto to go along with the image gives us a clue as to its import: 'Out of the strong came forth sweetness.' When this design was originally chosen, the average consumer was obviously deemed by Messrs Tate and Lyle to be rather more Bible-literate than most shoppers are now, for these words are taken from Judges 14, which tells the once well-known tale of 'Samson and the Lion'.

In this story, the Old Testament's most famous strongman is wandering about the land of Israel as usual until suddenly, somewhere near the vineyards of Timnath, he is set upon by a young lion. Never fear, however, for, just like Pop-Eye eating his spinach, 'the spirit of the LORD came mightily upon him', and Samson 'rent [the lion] as he would have rent a kid', tearing it limb from limb with his bare hands. After dismembering the lion in this way, Samson spies a comely young maiden who 'pleased [him] well' and claims her as his wife. A feast is then declared among her kinsmen, at which Samson proposes a game. He tells his hosts that, if they can solve his riddle, he will give them gifts of linen. If they fail, however, they must give him the same in return. His hosts agree, and Samson asks them what he is talking about in the words 'Out of the eater came forth meat, and out of the strong came forth sweetness.' Of course, the answer is 'a dead lion'. Samson had returned to the slain beast and found 'meat' (the word once simply meaning 'food' in general) aplenty in the lion's corpse, in the form of sweet, sweet honey, which had been produced by a veritable 'swarm' of bees which was now living in there, nesting between the dead animal's bones. This being the Old Testament, the rest of the story then descends down into an ultraviolent orgy of wife-swapping and mass murder which we need not detail here, but the basic point of Samson's riddle is obvious: bees are born from the bodies of dead lions. They had to be – the Bible itself specifically said so.[6]

Passport to Bugonia

This was all useful information for the pre-modern beekeeper. The only problem being – where was it possible for the average countryside apiculturist to get hold of a dead lion, precisely? An ancient Roman farming manual called the *Geoponica* had the answer. Just as a poor man who couldn't afford a horse could always buy a far cheaper donkey to help him out around the farm, a beekeeper without any spare lions to hand could simply use an ox as a substitute bee host instead, via a process known as *bugonia*. The recommended method for transforming an ox into a beehive was as follows. First, you had to get some rags and shove them up every orifice (yes, *every* orifice) in the ox's body, so that its soul would be unable to escape from its corporeal frame once you had killed it. Then, you had to take the ox indoors and brutally beat it to death with some blunt instrument without actually drawing any blood; any cuts to the animal's flesh would allow its spirit to flee, and that would be no good to anyone. Once you had bashed the bovine beast into oblivion, you then had to leave its body to rot for thirty days lying on a heap of thyme, sealing up all the windows and doors in its temporary tomb with mud as you exited. Now the ox's soul, trapped inside its body and prevented from flying away up to Cattle Heaven, would have no option but to transform itself into a swarm of bees instead. The end result? A lovely new store of bee-born honey – albeit one that would probably have registered a faint tinge of beef.[7]

How could anybody ever have thought that such a thing would actually work? The question is surprisingly easily answered. People thought that leaving the corpses of large animals around to rot would result in bees taking up their homes in them because sometimes bees *do* live in them. Or, at least, they might *seem* to do so. Actually, the idea is a case of mistaken identity. It took until 1894 for the solution to this particular problem to be found, by the Russian entomologist C. R. Osten-Sacken (1828–1906), who proposed that the swarms sometimes seen emerging from dead animals were in fact not bees at all, but drone flies. These misleading creatures, known to science as *Eristalis tenax*, look just like honeybees in almost all respects: shape, size and colour. If you look closely, you might notice they have two wings as opposed to a honeybee's four, but people don't tend to get up near to clouds of insects to take a closer look as a general rule. Suggestively, drone flies do lay their eggs in decaying flesh, so it would simply be a matter of time before some were observed emerging from within dead lions and oxen, thus giving rise to the whole myth. Furthermore, sealing up a dead ox within a darkened room would provide excellent conditions for drone fly larvae to thrive, clinching the matter in the eyes of many.[8]

As Osten-Sacken himself admitted, however, drone flies do not produce any honey; they might look like a bee, sound like a bee and act like a bee, but whatever it is that they produce from within their abdomens is most emphatically not sweet, and entirely unfit for human consumption. So what were the writers of tomes like the *Geoponica* actually eating, then? The answer, presumably, is 'nothing' – they were lying. There were certainly some charlatans around who took satisfaction in claiming to have proved the old theories right, though. In the Rosicrucian Samuel Hartlib's bizarre 1655 book *The Reformed Common-Wealth of Bees*, for instance – an insane attempt to claim that England's bees supported the government of Oliver Cromwell – the tale is told of an old Cornish countryman named 'Mr Carew', who claimed to have successfully caused bees to grow out of some calves he had slaughtered and then buried underground. According to Hartlib's book, beekeepers up and down the land rejoiced at this news and set out to try and emulate the mysterious Mr Carew – but, if they succeeded, then they don't appear to have ever told anyone about the fact.[9]

Osten-Sacken's basic assumption was as follows. People in the past were not idiots, and would have tried experiments aiming to replicate these wonders; sometimes drone flies would have hatched, and they thought these were bees. They would have made no honey for the farmers to scoop from the rotting animal hides, but the manifestation of the 'bees' seemed like proof of the basic theory which, when backed up by the testimony of the Holy Bible, they might have considered effectively proven before their own eyes. The absence of the honey could then be attributed to other factors – the local climate, say, or some deficiency in the dead animal's preparation – and chalked up to experience. They could then go on to tell their friends and neighbours anecdotes about 'that time I nearly made honey-bees appear out of a dead calf', and spread the myth even further. In such a way, legends (though not any actual bees, sadly) are born.[10]

Sexual Bee-Haviour

But why was it that, for over a millennium, people chose to believe that bees were entirely celibate? The answer is best expressed by asking another question: have *you* ever seen any bees hard at it? If so, then you are a lucky fellow indeed. It isn't until 1859 that we have a definite account of an unambiguous act of bee-related passion being first witnessed by the eyes of man, that man being one Reverend Millette of Whitemarsh, Pennsylvania, who doubtless got an entire fire-and-brimstone sermon out of the shocking event.[11]

In actual fact, the whole idea of bee-based chastity is complete nonsense. Not only do bees have sex, they do it in open-air group sessions in which the male stabs his penis into the queen's private parts so enthusiastically that it actually falls off and he dies, the penis remaining dangling from the queen bee's abdomen like an arrow from its bullseye while she simply shrugs and prepares herself nonchalantly to take her next victim inside her royal passage. There are three kinds of bees living in a beehive. The queen bee is the only true female, being the only one capable of laying eggs. The worker bees – who form the vast majority of bee-kind – are essentially genderless, being only *potentially* female, if fed on so-called 'royal jelly' by their fellow citizens of the hive. The drones, meanwhile, few in number, are the only males in the colony, and it is these who act to fertilise the queen so that she can lay the next generation of workers. A few times during her life, the queen bee will fly out of the hive, followed by a small trail of her chosen suitors. These drones, aflame with mindless desire, chase after the queen and compete to land their stiffened sceptres within her crown jewels. Even if they succeed, the moment of ecstasy is short-lived; just as some bees' stings can disembowel them once discharged, so their penises have been manufactured in much the same way by cruel Mother Nature. The drones' testicles, meanwhile, are – comparative to their tiny frames – huge, taking up as much space hidden away inside their bodies as their stomachs do. Returning back to the hive so filled up with semen from these bloated organs that her entire abdomen is distended, the queen quickly sets about laying thousands upon thousands of eggs, thus ensuring that Nature's weird but necessary cycle can begin again anew.[12]

So, far from being walking adverts for abstinence, bees instead have profoundly deviant sex lives. The trouble is that during her mating flight the queen can store up enough drone sperm to last her for as long as five or six years, making these flights of fornication rare sights indeed – but, eventually, scientists would slowly begin building up an incontrovertible body of evidence that such seemingly unbelievable sessions of insect intercourse did indeed occur. One of the key figures in this quest was a Dutchman, Jan Swammerdam (1637–80), who had gained a taste for natural history during his youth after becoming fascinated by his father's large collection of strangely shaped vegetables. Training as a doctor, Swammerdam's medical career never really took off, seeing as he preferred conducting autopsies on dead insects rather than examining live humans. Beginning work at six in the morning, Swammerdam toiled unpaid until late at night, inventing his own tiny tools to cut up insects beneath that other wonderful Dutch invention, the microscope. In particular, he became obsessed with bees, which he grew to think were greater even than God. Enchanted, Swammerdam set out to prove that drones had penises, which it turned out they did. The penniless scholar spent five whole years cutting them up and then drawing disturbingly detailed pictures of the minuscule organs, after which he suffered a nervous breakdown. Bees' penises, said Swammerdam, were 'wonderfully small and delicate ... a thing very worthy of contemplation', with an 'admirable structure' and 'full of exquisite art'. However, their actual function was obscure. If they didn't have sex, why did drones need them? Swammerdam's suggestion was that male bees sprayed microscopic sperm particles through the air, which then floated into the queen's waiting ovaries, but it turned out he was wrong.[13]

The man who finally solved the problem was a rich Swiss gentleman named François Huber (1750–1831), the greatest beekeeper of all time. Even though he was blind, Huber still managed to prove that bees had sex outside the hive. He forced his compliant wife to read all available literature about bees then pass on her findings to him, before training his faithful servant, François Burnens, to endure twenty-four-hour foodless stints on beehive observation duty, getting stung half to death in the name of science. While Burnens never actually saw any bees mating, he did manage to spot the queen flying back to her hive one day covered in semen, after which she quickly swelled in abdomen and began laying eggs. Further investigations showed that the queen was jabbed full of snapped penises like a horrible sexual pincushion. At last, it was proved that bees were not celibate, and all these people could stop their slightly worrying obsession with the matter.[14] But why, precisely, *were* they so obsessed with it? Perhaps it is because at stake in such investigations was nothing less than the very nature of life on Earth itself.

Early theories of precisely how it was that any life forms managed to reproduce at all were in themselves extremely confused. For example, there was the bizarre idea that men and women might secretly be exactly the same sex, but with their genitalia turned inside-out. Boys, it was proposed, were the result of their mothers' wombs being hotter than those which held baby girls, leading to men's organs growing more fully and protruding out of their body; vaginas were actually the same as penises, but 'folded' inside the groin, alongside hidden female testicles. Even into the Renaissance, doctors used the same words to refer to both the penis and the vagina, and the testicles and the ovaries. Thus, during sex it was presumed that the woman also ejaculated, the mixture of male and female sperm producing the resultant baby. It wasn't until 1672 that the Dutch anatomist Regnier de Graaf (1641–73) published a text conclusively proving that women had no balls.[15] Another old belief held that there was a secret duct hidden away somewhere inside the male body connecting his brain to his penis; when a man climaxed inside a woman, he shot out a fluid portion of his own mind to give the foetus the gift of consciousness.[16] Even odder, some said that all life had been created simultaneously by God at the same time as He had created the Earth. By looking inside seeds under microscopes, it had been discovered that tiny parts of the adult plant were already present within. From this it was concluded that, with humans, Eve's children had already been there within her eggs before they were even born. However, inside her female offspring lay other even smaller eggs containing their own future children, and so on forever. This doctrine was termed 'pre-formationism', and some scientists thought these tiny future beings lived within sperm, not eggs; through lenses, researchers claimed to see tiny donkeys in donkey semen, and miniature horses floating in horse semen.[17] With beliefs like these freely flourishing within educated minds, is it any wonder that so many people thought it entirely probable that bees might spawn from dead livestock?

The Generation Game

There was also a wider belief during the past in something called 'spontaneous generation', or 'abiogenesis'. Many creatures were supposed to be born from other animals, not just bees. Wasps, for instance, might come from dead donkeys, or scorpions hatch from a rotten crab. It is easy to see how the belief that maggots were spawned from putrid meat arose; likewise the notion that storing cheese in a darkened room could make mice appear. Less plausible was the thought that snakes could germinate from things which looked a bit sinuous, like a dead man's spinal column laid out to resemble an 's'.[18] Another idea held that horse hairs placed in water would transform into baby eels, a myth which could be accounted for by the presence of water lice and various tiny worm-like creatures lurking in pond water.

Seeing these animals appear in the same water as the horse hairs, to which they could easily cling, it would be only natural to think the legend had been directly proven.[19] Erasmus Darwin seemed to believe that similar creatures could be generated by leaving a pasta-like paste formed from flour and water exposed to the air until it became sour, at which point 'a great abundance' of tiny, eel-like bacteria called *Vibrio anguillarum* would appear in the substance, the motions of their flagella being 'rapid and strong'.[20] A further fable was that reptiles and amphibians could be generated by the action of the sun's rays upon wet mud. In *Anthony and Cleopatra*, Shakespeare has Lepidus speak of how

> Your serpent of Egypt is bred now of your mud
> by the operation of your sun; so is your crocodile.[21]

Of course, this wasn't actually how crocodiles were born; they just laid their eggs beneath layers of Nile-mud so it looked as if when the baby reptiles hatched out they were emerging fully formed from the sludge. Many people still thought that mice and reptiles had sex, but that abiogenesis was an alternative way for them to reproduce. The resultant babies might look the same as their normal cousins, some authorities said, but there was one big difference: animals created spontaneously from rotten mud, flesh and cheese were sterile.[22]

It took a long time for these beliefs to be disproved. The first big step came from Italy's Francesco Redi (1626–97), who in 1668 published *Experiments on the Generation of Insects*, proving that, when meat was placed to rot in a series of sealed and unsealed containers, only the exposed flesh developed maggots. The conclusion was obvious – maggots hatched from eggs laid by flies. *Bugonia*, too, was a mere myth: seeing as 'bees are very dainty animals', said Redi, they show no taste for rotting flesh whatsoever. He tried to hatch various insects – ants, wasps, flies – from a variety of meats – lamb, dog, buffalo, even tiger – but always his findings were the same. However, Redi was a man of his time, and still believed that human intestinal worms grew via abiogenesis, rather than through the accidental ingestion of their eggs in food. Maybe he just didn't want to think about what he might be eating.[23] With the invention of the microscope in Holland around 1608–10, the belief in spontaneous generation should have simply died out – but it didn't. On 9 October 1676 Antoni van Leeuwenhoek (1632–1723), a Dutch lens maker, wrote a letter to England's Royal Society informing them about the weird world of microscopic creatures he had found lurking in samples of rain and canal water, hitherto unknown beings he termed 'infusoria' and 'animalcules'. At first he wasn't believed, but gradually it became obvious Leeuwenhoek was telling the truth. Greedily examining any substance that came to hand beneath his microscope, he found that it was almost always full of infusoria; even when he had diarrhoea, Leeuwenhoek smeared some brown stuff on a slide to see what miniature monsters lurked within.[24]

Looking at cheese beneath a lens showed no tiny mice being formed in embryo, then; but it did show swarms of tiny animalcules. How did these new creatures reproduce? Was it via abiogenesis? Leeuwenhoek was keenly against this idea, but others disagreed. In 1748, John Needham (1713–81), an Irish priest, boiled up a sealed flask filled with gravy to kill off any lurking infusoria, then four days later blobbed the gravy under his microscope. It teemed with life; spontaneous generation, he concluded, was real. Others merely concluded that Needham had botched his experiment. Over the next 150 years, variations of the gravy-boiling test were performed repeatedly, providing wildly contradictory results.[25] It wasn't until as late as 1877 that the Englishman John Tyndall (1820–93) finally demonstrated once and for all that no such processes actually occurred via a series of impeccably thorough investigations involving heated test tubes locked within a sealed box filled with filtered, sterilised air.[26] At last, it was safe to swallow a slice of cheese without also accidentally ingesting a microscopic rodent as an unwanted extra. Unfortunately, scientists instead discovered that cheese was filled with armies of tiny mites and bacteria, thus putting people off eating it for a different reason; the first film to experience calls to be banned in Britain was a 1903 silent documentary, *The Cheese Mites*, in which a man patiently examines some Stilton with a magnifying glass. The resultant footage was deemed too disturbing for public consumption by outraged representatives of the British cheese industry.[27]

Going Crosse-Eyed

Spontaneous generation is essentially a myth – but that hasn't stopped various scientists down the years claiming to have seen the phenomenon at work with their own eyes, both before and after Tyndall had conclusively disproved the idea. Most notorious was Andrew Crosse (1784–1855), an English gentleman scientist and MP who was running electrical current through some volcanic rock one day in 1836, when he accidentally created life – or so it seemed. These were the days when galvanism was all the rage, and scientists were busily electrocuting everything from corpses to vegetables, just to see what would happen. By running a current through his rock, Crosse hoped to create crystals via a process of electrolysis, causing the rock to separate into its various components in crystalline form. Special potassium silicate-based fluid dripped down onto the electrified rock, forming not only Crosse's desired crystals, but something rather less expected, too – nipples.

According to Crosse, these nipples – 'a few small white excrescences' – grew from the charged iron-ore stone at the point where fluid dribbled onto it. After eighteen days, the nipples enlarged and sprouted long hairs. By the twenty-sixth day, these filaments had begun to resemble insects, a suspicion confirmed some days later when they detached themselves and commenced walking around. Viewed beneath a lens, each insect looked like 'a microscopic porcupine', and they swarmed over the rock until eventually there were enough to cover the whole table. What could they be? Crosse put some in an envelope, sent them off to an expert in Paris, and awaited a reply. It turned out they were mites, of the genus *Acarus*, but of a type hitherto unknown. Once word got out, the British press christened the mites *Acarus galvanicus*, or 'electric mites', and smeared Crosse as a new Frankenstein. Cheap horror stories with titles like *The Electric Vampire* and *Death of a Professor* began to be churned out, in which hordes of Crosse's mites ate people's bones from within, leaving them formless piles of flesh, and giant, saucer-eyed weevils sucked their creator's blood dry in revenge. Worse, local farmers blamed Crosse for causing the failure of their crops by angering God, and an exorcism was carried out on nearby hills to rid the area of evil. Crosse suffered threats of violence, and demonstrators gathered outside his house. The pitchfork-wielding villagers need not have bothered, however. Crosse was not working in fully controlled conditions, and the electric insects have since plausibly been identified as ordinary domestic cheese mites which just happened to have laid their eggs on Crosse's rock – as Crosse himself actually half-suspected. The only thing which really grew from out of nothing here was the subsequent media-led hysteria.[28]

Crosse was not the first or last to commit such errors. In 1872 another Englishman, Henry Charlton Bastian (1837–1915), published his epic *The Beginning of Life* in which he purported to have observed the spontaneous generation of microbes through a lens – findings which, others soon discovered, were probably down to Bastian not taking the correct precautions to prevent airborne germs from contaminating his samples.[29] In 1906, meanwhile, John Butler Burke's (1873–?) book *The Origin of Life* featured startling claims of new half-crystalline life forms called 'radiobes' being conjured up from radium samples in his lab at Cambridge. According to him, when he dropped this newly discovered wonder material into test tubes of sterilised bouillon, a breed of bizarre and tiny creatures emerged from the freshly radioactive soup. Over the course of a few days, Burke could see these new animalcules progressing through a number of distinct developmental stages; firstly they manifested as tiny specks which then split into two little dots, which then adopted the shape of either dumbbells or frogspawn. From this point on, they apparently turned into organic crystals, before finally ending up adopting the guise of microscopic biscuits – but biscuits which were alive! According to Burke, these animals were not simply bacteria, and nor were they just normal crystals, as they had nuclei and so 'must be organisms'. They were pretty strange organisms, though, as they dissolved in water, vanished like miniature vampires when exposed to sunlight, melted away at temperatures of 35°C, and disintegrated of their own accord after a fortnight or so in any case. Christening them 'growing atoms', a stage prior to that of the development of living biological cells, Burke released micro-photographs of his mini-beasts to the world – but, as critics pointed out, some of these 'photographs' appeared in fact to be drawings which Burke had made with his own hand. Even Burke's apparently genuine photos showed the radiobes rather indistinctly, and required a fair amount of squinting and subjective interpretation before you could see the miniature blobs they showed as being little crystal animals rather than as, say, random patterns of molecules or small air bubbles. When other experimenters tried and failed to reproduce his findings, interest in Burke's claims died off every bit as quickly as his radiobes did once tipped out from their tubes of irradiated soup and exposed to the harsh glare of the sun.[30]

Another believer in the reality of spontaneous generation was the French anatomist Felix Dujardin (1801–60), who in 1835 claimed to have discovered a substance later named 'protoplasm' by crushing microscopic animals and mixing their remains together to create a kind of thin, pulpy jelly he said was alive; from out of protoplasm the first primitive life forms spontaneously spawned, and eventually evolved into all the higher animals, said Dujardin. T. H. Huxley believed in the stuff, and in the 1890s identified it with a kind of organic slime present on the sea floor from which minuscule sea creatures then grew. Protoplasm, he said, was 'the one kind of matter' from which the organs of all living things were ultimately made, 'the clay of the potter' which Nature had since spun into a million wondrous shapes. No doubt Huxley was delighted we all came from sea slime, not from God. The only problem was that this briny protoplasm of his didn't really exist; it turned out to be an artificial substance produced by a chemical reaction between saltwater and the alcohol Huxley had used to preserve his bottled sea specimens.[31]

Protoplasm did not die with Huxley, however, and in 1934 the living jelly made a very strange comeback indeed with the publication of a pamphlet, *The Reincarnation of Animal and Plant Life from Protoplasm Isolated from the Mineral Kingdom*, by an amateur English biochemist named Morley Martin (?–1937). Martin had his own private laboratory, and in 1927 took some rock from the Azoic era, an old-fashioned term for Earth's earliest geological period, and heated it into ash. Then, by adding various chemicals, he professed to have made prehistoric protoplasm. By further subjecting this protoplasm to x-ray bombardment, Morley claimed to have caused a race of tiny prehistoric fish to rise from the dead – he said there were 15,000 in one batch of protoplasm alone. There was a problem, though. During their early stages of life, these microscopic fish fed happily off the protoplasm, but as they grew larger, began to eat each other or starve. Therefore, Martin developed a special serum to nourish the mini-beasts, but, frustratingly, never revealed what it was. His basic theory was that the fish were not really dead, merely lying in a state of suspended animation in the Azoic rock, and that he had found a way to resurrect them by reincarnating their life force into newly formed protoplasmic bodies. Martin didn't just revive fish, though; he also created crustaceans. First of all, he saw 'globules' appear in the protoplasm, which became vertebrae in a spinal column, before claws and a head appeared in front of Martin's very eyes.[32] Here is his own description of the reincarnation process:

> The liquid protoplasm split up into innumerable globules ... which took the form of an animal which was obviously *progressively* forming. I photographed it every few minutes, which proved progressive formation of its body ... Its separate globules ... elaborated a body, head, eyes; with legs obviously deficient in number, as some were at first quite rudimentary and were withdrawn again into its body as though it were conserving material for more important purposes. Other rudimentary legs appeared and were also withdrawn in like manner; but those which had not been malformed progressed in formation. After a few hours, all its globular condition ceased to be, the globules having been elaborated into a well-formed animal body; whereupon it *slowly commenced to walk*, apparently trying to get away from the focused light [of the microscope] ... It was being born again into a new body ... within the field of the microscope.[33]

The level of truth of these claims can very easily be gauged by considering the fact that there was no life on Earth during the Azoic era to be resurrected in the first place; its very name comes from the Greek *a-zoön*, meaning 'no animals'. Morley Martin was either lying or seeing things which simply were not there. Perhaps it was a little bit of both.

Even more unhinged was the work of the American homeopath Charles Wentworth Littlefield (1859–1945). When patients came to him suffering from cuts, Littlefield used to recite a short prayer to help them heal, and wondered how the process worked. The real answer was that most cuts heal by themselves anyway, but Littlefield disagreed. Taking a sample of the organic body salts which help clotting occur, Littlefield prayed over it one day while idly thinking about a chicken. Then, he examined the sample beneath a microscope. To his surprise, he found the salts now resembled the very same chicken about which he had just been thinking! Presuming he was telepathic, Littlefield began contacting the ghosts of famous dead scientists for advice, and in 1919 published an unnecessarily massive book, *The Beginning and Way of Life*, in which he revealed that, by concentrating really hard upon small piles of salt, he had managed to make their crystals resemble America's national mascot, Uncle Sam. Sometimes, the strange shapes came about by chance; noticing that one salt sample had transformed into a picture of 'a woman carrying a dog under her arm on a windy day', Littlefield was initially puzzled. Then, he remembered. He had accidentally thought of a woman carrying a dog under her arm on a windy day some time previously, and the salts had reacted accordingly. Equally as odd, when Littlefield just left the salts alone, they spontaneously developed into tiny animals like crabs, apes, fishes and even miniature humans. Most were not alive, except for a race of microscopic octopuses, which for some reason were. This, said Littlefield, was how life on Earth began – by tiny octopuses generating themselves spontaneously from piles of psychic salt. In the years since, very few persons have found themselves agreeing with him upon this point.[34]

What was going on? How were serious people (and a fair few nutcases) managing to see living things beneath a microscope which were not really there? In much the same way, I would suggest, as August Strindberg once saw a tiny brain living inside a walnut; namely, by accidentally engaging in an obscure form of alchemy.

Science Is Golden

What exactly is alchemy? The cartoon answer would be to say that it was the quest to transform base metals like lead into gold through quasi-chemical, quasi-magical means within a laboratory, a kind of get-rich-quick scheme for quacks and charlatans. However, there were other impossible dreams involved in the art too; the dream of achieving immortality, for instance, or of creating artificial life. Seeing as alchemists did carry out genuine chemical experiments in their laboratories, and invented many useful instruments and techniques, the standard narrative has it that modern chemistry 'grew' out of alchemy, like a flower from dung. This is not entirely untrue. Certainly, the list of genuine discoveries and inventions made by alchemists is a long and distinguished one.

A partial catalogue might include Johann Friedrich Boetticher (1682–1719) being the first European to successfully make porcelain, hitherto a closely guarded secret of the Chinese; Ramon Llull (*c.* 1235–*c.* 1315) being the first to prepare bicarbonate of potassium; and Johann Rudolph Glauber (1604–70) discovering sodium sulphate. In a purely practical sense, European alchemists also contributed greatly to various industries of the day, helping invent and improve processes of dyeing, tanning and ceramics, and the manufacture of substances like gunpowder, alcohol, cosmetics and glass. The first systematic knowledge of the chemical elements undoubtedly came from alchemists, some of whose texts contain what are sometimes described as the first rudimentary Periodic Tables. Al-Rāzi (*c.* 854–*c.* 925), an Islamic alchemist, for instance, tried to classify all mineral substances into one of six categories according to their fundamental properties – salts, for example, were things that dissolved in water, whereas spirits were flammable. The language of early chemistry also inherited the lexicon of alchemy, with alchemical terms like 'butter', 'flower', 'vitriol' and 'spirit' being used to describe different types of chemical substance well into the 1700s. Even the word 'gas' originated with an alchemist, namely the Belgian Johann Baptiste von Helmont (1580–1644), who coined it from the Greek word for 'chaos' – and who also once claimed to have successfully transmuted mercury into gold with the aid of a mysterious powder obtained from a stranger.[35]

However, to say that alchemy was simply an early, primitive form of chemistry now rendered obsolete by modern knowledge is only part of the story. The trouble with the subject is that it is so complex and self-contradictory that almost anything you say about it is both true and untrue simultaneously. For instance, if I say that alchemy was primarily a mystical enterprise, someone will point to several specific alchemists who set out only to perform practical chemical experiments to do with things like dyeing processes. Then again, if I say that alchemy was primarily just early chemistry, someone else will point out several other alchemists who pursued their work purely in pursuit of spiritual revelation. It is generally accepted today that the 'gold' sought by many alchemists was essentially metaphorical in nature – simplistically, let's say it was a code word for 'enlightenment' – but this was not *universally* true. The fact that almost all alchemical texts are written in an extremely baffling manner, with symbol layered upon symbol in a way utterly incomprehensible to the uninitiated, only makes the situation even more confusing.

The response of the eminent Swiss psychologist C. G. Jung (1875–1961) to the topic when he first began researching it in the 1920s was typical: 'Good Lord, what nonsense! This stuff is impossible to understand.'[36] As Jung eventually came to see, however, alchemy is not *meaningless*, as such. The problem is rather the opposite; it has *far too many* meanings, each of them equally valid, all at once. Perhaps this was why the thousands of different alchemical texts so regularly contradicted one another. Jung came to see these contradictions as inevitable, because each individual alchemist's engagement with the so-called 'Great Work' was just that – an individual project. There was no one single authoritative definition of the Philosopher's Stone which each man sought because it was a different thing for each who sought it. In simple terms, this famously ruby-red Stone (or *lapis philosophorum*) was the secret ingredient needed to make lead become gold, a process known as 'transmutation'. A truly miraculous material, the Stone could also, depending upon which version of the myth you preferred, cure all ills, or be made into an elixir of life to render you immortal. No alchemist ever openly comes out and says what the Philosopher's Stone is in simple terms, though. Imagine some old text says sulphur is the key substance in the Stone. This would sound clear enough – except that this 'sulphur' will not really be ordinary sulphur at all, but some other symbolic form of the element going by the same name. Drink real sulphur and you will die, not achieve immortality. Alchemists never describe with any clarity what the Stone is, but instead describe around it, in a way which often hovers upon the very edge of meaninglessness. According to one famous account, given in the anonymous 1526 treatise *Gloria Mundi*, the substance worked up by an alchemist to become the *lapis philosophorum* is

> familiar to all men, both young and old, is found in the country, in the village, in the town, in all things created by God; yet it is despised by all. Rich and poor handle it every day. It is cast into the streets by servant-maids. Children play with it. Yet no one prizes it, though, next to the human soul, it is the most beautiful and the most precious thing upon Earth and has the power to pull down kings and princes. Nevertheless, it is esteemed the vilest and meanest of earthly things.[37]

Because of such wilful obscurities, this substance which becomes the Stone (known as the *prima materia*) has been sought in a number of common but valueless substances – blood, urine and dung have all been boiled in glass vessels in the vain hope of making gold grow from them. This *prima materia* has been called air, earth, fire, water, sulphur, iron, lead, sky, cloud, sea, moon, sun and dozens more besides; it is proverbially said to have a thousand names, none of them literal. Even more confusing than the nature of the *prima materia* is the nature of the fire with which it is to be cooked. As well as using ordinary furnaces in his lab, the alchemist also had to make use of some kind of 'secret fire' at the same time, a fire which, strangely, did not burn. Some even said that this fire was really a water – but one which did not wet the hands.[38] So what was it?

Finding Yourself

As we saw with Strindberg, one of the main effects of engaging in alchemy over a long period of time was to eliminate the space which existed in the alchemist between spirit and matter. If 'all is one' as the Monists taught, then the successful alchemist ought to be able to see his own psyche being reflected back to him in the inanimate chemical world of his surroundings; Strindberg seeing figures emblematic of his own fears in crumpled bedsheets, for instance. Alchemists, through obsessive contemplation of their work, might well have ended up having visions of strange dramas taking place within their furnaces and test tubes, an experience which could lead to great spiritual satisfaction. If so, then it would make perfect sense that each successful alchemist would achieve this result differently. No man's epiphany is identical with that of his neighbour, and so neither would they share the same Philosopher's Stone. Maybe one alchemist found enlightenment burning mercury with tin; maybe another did it by boiling sulphur with antimony. Neither was correct, neither was wrong. Both were merely individuals, seeking their own individual revelation.[39]

Jung's own alchemical epiphany, for example, was clearly coloured by his own pre-existing obsessions and theories. According to his own personal philosophy, developed over long years as a practising psychologist, each individual person had as an unconscious goal for their psyche something he called 'individuation', an ideal union of all possible opposing features and characteristics within the self – a perfect self-realisation of the conscious and unconscious minds which would be equivalent to the perfect union between matter and spirit. The unconscious, Jung said, was constantly striving to turn over and around aspects of itself until it finally achieved such a goal (which it actually never would, as no person is truly perfect). A person's overall psyche, then, was constantly being formed and reformed by a never-ending process of transference of contents between the unconscious mind and conscious ego with the unstated aim of achieving this unachievable goal. Not all alchemists may have realised it, but this, in psychological terms, was what they were really trying to achieve, Jung thought. Rather than a pursuit of physical chemistry, the Great Work was really more a pursuit of the mind.[40]

In terms of the actual laboratory techniques performed by the alchemist, this mental or mystical process was, said Jung, itself directly mirrored by a procedure known as 'reflux distillation', whereby a liquid is heated within a vessel and evaporates into a gas. This gas rises and condenses on the walls of the glass, forming liquid droplets. These beads then drip down back into the original body of liquid, only to be heated and evaporated again, in an endless cycle, often symbolised by the figure of the Ouroboros, or serpent eating its own tail, a traditional emblem of eternity. The tangible liquid and intangible gas are thus seen to be at some underlying level mere aspects of one another, not two wholly separate things, a realisation which, through repeated close observation, the alchemist is supposed to apply metaphorically to his own self and the world of inanimate matter surrounding him.[41]

Firing the Imagination

According to such interpretations, it would therefore seem that the 'secret fire' which had to be exploited by the alchemists during their operations was really that of the imagination – certainly, that is the opinion of Patrick Harpur, one of the best modern writers on the topic. The Latin term *imaginatio* recurs repeatedly within the texts of the alchemists, where it is accorded the utmost importance in terms of the successful operation of the Great Work. In the German alchemist Martin Ruland's (1569–1611) text *Lexicon Alchemiae*, for instance, this *imaginatio* is defined as being 'the star in man, the celestial or supercelestial body' – that is to say, that aspect of the macrocosm, the heavenly world of stars, planets and the God who created them, which resides within the fleshly microcosm, or 'little world' of man.[42] For Ruland, by making use of this star of imagination living within his own material body, the alchemist can somehow manage to perceive the animating spirit of God which lies hidden within other material bodies like metals and minerals, too. Another alchemist, the Belgian Gerhard Dorn (1530–84), agreed:

> There is in natural things a certain truth which cannot be seen with the outward eye, but is perceived by the mind alone, and of this the Philosophers [alchemists] have had experience, and have ascertained that its virtue is such as to work miracles ... As faith works miracles in man, so this power [the imagination] ... brings them about in matter.[43]

Alchemists, then, would see things – no real surprise, as many of the substances they were experimenting with were toxic and could potentially cause hallucinations. See *what*, though? Whatever it is your psyche expects to see, I suppose. An interesting example of one of the strange things alchemists sometimes claimed to observe was the salamander, a 'spirit of the fire' shaped like a large, foot-long lizard which twisted and turned within the furnace, immune to all flames. It would be easy to view salamanders as simply some kind of personification of the dancing furnace flames themselves. One obscure sub-type of salamander, for instance, was the pyrallis, an alleged inhabitant of the copper-smelting forges of Cyprus. As small as a fly, with four wings, a dragon's head and a four-legged body made from bronze, the pyrallis could supposedly be observed swarming among the sparks emitted during smelting. If these mini-dragons should ever fly out from their homes within the fiery furnaces, however, then they would instantly die and shrivel in the cooler external air.[44]

It seems obvious that the pyrallis swarms simply *were* the sparks in the Cyprian forges, seen through poetic or visionary eyes. Quite what their larger salamander cousins were is a matter for more debate, although there is no doubt that people in past ages did claim to actually see them. The most famous such account comes from the celebrated Italian Renaissance goldsmith Benvenuto Cellini (1500–71), who in his *Autobiography* told of a strange childhood experience when his father Giovanni had excitedly called him down to the basement. Breathlessly, Giovanni pointed out something strange within the cellar's fireplace – a lizard, with star-like markings running down its side, dancing within the hottest part of the fire, apparently unharmed. Suddenly, Benvenuto received a smack on the head from his father, making the five-year-old break into tears. Giovanni's explanation was that he was only punching Cellini to make him better remember the occasion, for the sighting of a real, live salamander was an event so rare he did not wish him to ever forget it. Feeling guilty, he then kissed young Benvenuto and gave him some coins, no doubt as a further *aide-memoir* of the miracle. Demonstrative chaps, these Italians.[45]

One possible explanation for this bizarre episode is that Cellini's salamander was really there, in a literal sense – because a type of animal broadly fitting its description does actually exist. Real-life salamanders are not lizards, but a type of long-tailed amphibian which, to the untrained eye, would look very much like one. Distributed almost worldwide, they like to hide out within piles of damp wood, something which has sometimes been used to account for the legend. If you toss some moist logs on to a fire then they will initially burn quite slowly, giving any concealed salamander time to escape from the flames, thus appearing as if it were immune to the heat.[46] Doubtless some accounts of salamanders really are explicable in such terms – but not, surely, all of them. The high temperatures at which alchemists often cooked their metals would simply have burned such amphibians to a cinder, and their fires were not all log-based anyway. Moreover, salamanders in real life do not have stars running down their bodies like Cellini's did – but pictures of salamanders in some old alchemical texts do.[47] The starry salamanders of the alchemists were not really there, in any literal physical sense – but they were, nonetheless, still seen. Maybe it was the star within man that sometimes allowed him to see the salamander within the fire – and the stars within the salamander?

Unrealistic Projections

Alchemists called their application of the Philosopher's Stone towards another metal in order to transmute it 'projection', a word which also seems appropriate to describe the way they would 'project' aspects of their own psyches out into the world of dead chemical matter. For Jung, this meant that the weird pictures and descriptions of things like dragons, salamanders, green lions, black ravens, suns, rainbows, stars and moons with which the alchemists filled their manuscripts could sometimes have been descriptions of actual hallucinations they had undergone while performing their experiments.[48] As a possible illustration, Jung cites an extract from a German text of 1762, in which the author gives instructions for how any alchemist can have a full-blown vision of the solar system within the confines of his own laboratory. First of all, the author says, you should take seven pieces of metal corresponding to the then known seven planets and drop them into your heated crucible 'after the order in which they stand in the heavens'. This being done, the alchemist should then 'make all the windows fast in the chamber' so that 'it may be quite dark within', and allow the metals to melt into one another, before finally dropping seven fragments of the Philosopher's Stone into the mix. At this point, says the author:

> ... forthwith a flame of fire will come out of the crucible and spread itself over the whole chamber (fear no harm), and will light up the whole chamber more brightly than sun and moon, and over your heads you shall behold the whole firmament as it is in the starry heavens above, and the planets shall hold to their appointed courses as in the sky. Let it cease of itself, in a quarter of an hour everything will be in its own place.[49]

This certainly sounds like a good step-by-step guide to inducing hallucinations. The alchemist stands in a darkened, sealed chamber, free of all sensory stimuli other than the glowing and molten metals which lie within his heated crucible, inhaling various potentially noxious substances and fixing his entire attention upon an act which, by the time he was writing, had acquired all kinds of occult implications and trappings to it. Eventually, a flame of 'secret fire' emerges, and he gets his vision. It is a process not a million miles away from a fortune teller concentrating so intently upon a crystal ball that, eventually, they end up seeing events playing themselves out upon the polished surface which are not really there. The 'psychic' may claim these to be visions of the future, much as the alchemist may claim his own visions to be accounts of genuine physical processes, but an alternative interpretation would be to say that all these sights – while genuinely seen – are really the projections of the visionary's own unconscious.

Daydream Believers

Most people would dismiss such reveries as mere daydreaming. However, the secret of the Philosopher's Stone was often said to come to alchemists in their dreams, and several alchemical texts adopt the structural device of framing the stages of the Great Work as the contents of an inspired dream.[50] Jung himself maintained that his initial interest in the subject was foreshadowed in a dream,[51] and was interested to note that certain modern-day scientists, meditating upon the hidden secrets of matter themselves, could sometimes draw inspiration from the dream world. The most famous such tale is undoubtedly that of the German chemist August Kekulé (1829–96), and his strange discovery of the chemical structure of benzene.

In 1865, Kekulé published the paper which made his name, in which he solved the then pressing problem of how precisely a molecule of benzene, a type of liquid hydrocarbon, was structured. He theorised that it was a kind of 'ring', made up of six alternating double and single bonds of carbon; picture a hexagon, each of whose corners represents an atom of carbon attached to a hydrogen atom, with the actual sides of the hexagon being the links between them. Three of these sides would be drawn with a single line, and three with a double line, in an alternating pattern, representing the single and double bonds between the atoms. This hexagon, taken as a whole, was the benzene ring, a discovery which proved hugely important to the field of organic chemistry, giving chemists a theoretical underpinning which allowed them to create new carbon-based chemical compounds which had not previously existed before in Nature – something which had equally as profound effects upon industry as upon science. Kekulé's work led ultimately to the creation of an industrial process called polymerisation, from which we get various materials essential for the modern world, such as Bakelite, polyester, nylon and any number of plastics.

Naturally, Kekulé's fellow chemists wished to celebrate him for this achievement, and in 1890 a so-called 'Benzolfest' (sounds fun) was arranged by the German Chemical Society, where scientists and industrialists gathered to commemorate the great man. Here, Kekulé gave a now-famous speech in which he claimed to have had the idea of the benzene ring while half-dozing in front of his fireplace one day in 1862. The daydream, he said, was of the Ouroboros, or snake eating its own tail, that symbol of infinity so beloved of the alchemists. Just as the Ouroboros represented a closed ring without end, so too did the benzene molecule, he realised upon waking – or, at least, so he told his audience. Recent research has cast doubt upon whether or not Kekulé really had the dream, or whether the whole anecdote was really meant as a joke. After all, in the same speech, Kekulé also claimed to have made his other great discovery, that carbon atoms had a propensity to link with one another and form long chains, following a daydream; supposedly, he enjoyed a vision aboard a horse-drawn London omnibus one day in 1855, in which the traffic became dancing carbon atoms and molecules, with the larger vehicles dragging the smaller ones along behind them in great chains. According to the American chemistry professor John H. Wotiz, the most prominent sceptic as regards Kekulé's claims, all this talk of inspiration from dreams was merely a cunning strategy upon behalf of the German chemist to avoid having to give any lesser-known foreign rivals (various of whom, Wotiz says, independently discovered the benzene ring before Kekulé) any due credit.[52]

Whether it truly happened or not, Kekulé's alleged snake dream has certainly had a long afterlife. As well as being repeatedly trotted out by pop psychologists who tell their needy clients about the need to 'listen to your dreams', it was picked up on by Jung, who interpreted it as showing a variation upon the so-called 'Chemical Wedding' of the alchemists in which two opposing concepts (spirit and matter), represented by the opposing figures of king and queen, married with one another, thus creating a kind of miraculous unity made up of both apparently incompatible poles. For Jung, the Chemical Wedding and the Ouroboros were the same thing in different guises, namely symbols of the ultimate unity of all things, and Kekulé's great insight about the snake-like benzene ring, which allowed new substances to be created from out of old ones, thus stood revealed as some kind of modern-day scientific equivalent of the Philosopher's Stone.[53]

The Serpent's Tale

That, at least, is the positive reading of Kekulé's dream. An interesting negative take upon it was given by the American novelist Thomas Pynchon in his 1973 book *Gravity's Rainbow*, a picaresque account of the Nazis' ballistic rocket programme, in which he has the Ouroboros act as a symbol not of eternity, wholeness, or the never-ending wheel of Nature, but a purposeful *breaking* of that closed yet virtuous cycle. Providing as it did the basis for the initial development of the whole scientific-industrial complex, Kekulé's snake-like benzene ring becomes for Pynchon a modern manifestation of the serpent from the Garden of Eden, whispering into mankind's ears the temptation to play God, to create new forms of matter which were not present on Earth before. With the advent of modern industry that Kekulé's discovery unleashed upon the world, a fresh era in history comes around, where a new cycle, based upon profit and plundering the planet for raw materials, replaces the old one, leading only to disaster.[54]

In his book Pynchon has board members of IG Farben, the German chemical giant which manufactured Zyklon B gas for use in Nazi death camps, and which would never have been able to rise but for Kekulé's revelation, attending séances in search of other such pieces of useful inspiration from the unseen world. One spiritual entity contacted via such means speaks in cryptic terms of modern industrial processes really being some kind of horrific parody of alchemy pursued for wholly materialist means. Just as, in the ghost's words, there is a place where such different-seeming materials as coal and steel meet (in coke, a super-heated and condensed form of coal which is an essential ingredient in the steelmaking process), so there is a place where modern industrial materials meet with the matter of organic life – namely, in coal itself, the powerhouse of the entire industrial revolution.[55] Coal, of course, is ultimately formed through the pressure of immense geological forces upon organic matter. Plants, animals and sea creatures die, are compressed beneath huge masses of rock and, as the aeons pass, become transformed into coal – and thence, millions of years later, into the coking coal used to help manufacture the steel which makes up Nazi tanks and missiles. Viewed from a certain perspective, these lethal weapons are actually made partly from dead prehistoric plants and animals, the life force transformed into a death force, spirit become matter, alchemy reversed.

As the ghost says, modern industrial chemistry is no longer the movement from death to rebirth traditionally represented by the Ouroboros of reflux distillation, but 'from death to death-transfigured'. The profit-hungry, gold-seeking directors of IG Farben, says the spectre, must, like the modern-day reverse-alchemists they are, look 'into the hearts of certain molecules' and find lurking there not the living spirit in matter like the old alchemists, but something dead and insensate, waiting to be exploited for their own ends. It is, after all, the hidden chemical structure of the matter needed to create any given commercial product which ultimately 'dictate[s] temperatures, pressures, rates of flow, costs [and] profits', says the ghost.[56] Looked at in this jaded way, Kekulé's snake was not really the Ouroboros, then, but Satan in disguise.

Theatre of the Absurd

Such ideas, while dealing with science, are clearly not science in themselves; *Gravity's Rainbow* is a work of fiction, not a textbook (although, unusually for a novel, it does feature the odd equation ...). In Jungian terms, ideas like Pynchon's are projections of the psyche out onto the modern world of industrial chemistry. The structure of the benzene ring is a simple, indisputable fact. The meanings which have then been projected outwards on to it are not. For Pynchon, it acquired a negative meaning; but others could equally portray Kekulé's serpent as having whispered positive things into mankind's ear. After all, many of the inventions and products brought about by the advent of industrial chemistry and polymerisation have greatly improved the lives of millions. That apple of knowledge nudged our way by Kekulé's serpent one sleepy afternoon back in 1862 needn't necessarily have been a poison one.

In his novel, Pynchon turned the apparently objective facts of science into a subjective story, and so did many alchemists in their own books. Rather than an actual science, perhaps alchemy is better thought of as a kind of endlessly confusing, Pynchon-like dream narrative – specifically, a dream narrative which the alchemist experiences while still half-awake, like Kekulé during his own states of half-conscious reverie. If it is a narrative, however, then it is one which apparently has two storylines playing out simultaneously, one of which is chemical, the other psychological. One of the most famous of all alchemical texts, for instance, published by the antiquary and founder of Oxford's Ashmolean Museum Elias Ashmole (1617–92), in 1652, was called the *Theatrum Chemicum*, or 'Chemical Theatre', and it bore an appropriate name indeed. As Patrick Harpur has pointed out, 'Alchemical recipes [often] read like plays or psychodramas' whose ingredients act as dramatis personae, with things like sulphur and mercury being given names like Sol and Luna (Sun and Moon) or Rex and Regina (King and Queen), and taking on the role of characters who enjoy (or endure) such adventures within the alchemist's glass as being eaten by a green lion or torn to pieces by eagles and ravens.[57]

But can apparently inanimate things like light and oxygen also become 'characters' within a more modern scientist's lab? The case of the great English chemist Humphry Davy (1778–1829) suggests so. Davy, now most famous for having invented the miner's lamp in 1815, was an experimenter and theorist of prodigious talent, but also at heart a great Romantic. In 1798, he put forward some of his earliest scientific ideas in a series of essays published as *On Heat, Light and the Combinations of Light*. Lit up himself with youthful enthusiasm, Davy claimed that light fuelled all chemical reactions, including combustion, and that human bodies 'fed' upon it as it penetrated through our skin and into our bloodstreams, stimulating our nervous systems and allowing us to see, feel and think. Therefore, said Davy, light was the prime moving force behind both chemical reactions in the physical world, and the imaginative powers of man – he should have called it a 'secret fire'. Trying to build his idea into a grand 'Theory of Everything', Davy even proposed that our sun and all the other stars in our galaxy had been deliberately placed there by God as gigantic cosmic light-producing batteries, allowing not only the hidden laws of chemistry and combustion, but also life and thought themselves, to be beamed out across the universe. It was a nice idea but, predictably, one which he later recanted ...[58]

Wavy Davy

The greatest alchemists of the Renaissance would surely have considered Davy to be one of their own had they known about some of his achievements, especially his ability to make apparently fundamental substances 'transmute' into new ones. By performing electrolysis on various alkaline earths like soda and potash, he could transform them into their parent elements, in this case sodium and potassium, both of which were previously unheard of. Davy managed to successfully isolate no fewer than five other new elements in this way, too – strontium, barium, magnesium, calcium and aluminium. The accepted definition of an element is that of a fundamental substance, consisting entirely of atoms of only one kind, which cannot be broken down any further. The great Lavoisier's classification of soda and potash as elements was considered definitive, so when Davy proved him wrong, showing that even these allegedly 'fundamental' substances could be broken down further, he caused a scientific sensation. Even today, no man can equal Davy for the number of new elements he introduced into what we now call the Periodic Table.[59]

Such achievements led Davy to conclude that the then prevailing scientific picture of the universe was somehow incomplete; below the level of the elements, there apparently existed some strange and unseen world where all things were really one. Davy liked to think of the cosmos in this way as he was at heart something of a Nature mystic, a Romantic like Keats or Wordsworth. His first passion had been poetry, and he enjoyed wandering through the wild but beautiful landscape of his Cornish youth imagining himself to be somehow a part of the landscape around him rather than any fully separate being, feelings he later put into verse. Indeed, Davy was closely acquainted with various prominent literary men, such as the poet Samuel Taylor Coleridge

(1772–1834), from whom he absorbed accounts of the writings of the new German Idealist philosophers of the day, men like Friedrich Schelling (1775–1854), Johann Fichte (1762–1814) and in particular Immanuel Kant (1724–1804).[60]

As understood by Davy, the ideas of these men acted as confirmation of his own ideas about the visible world of elements and matter which surrounds us being a mere trick of perspective. According to Kant, in his 1781 *Critique of Pure Reason*, the world as it truly is cannot be known to us. Things in themselves are unknowable; we can only know them through our own perception of them. Imagine you had sunglasses permanently welded on to your eyes; everything would appear dark, even though it wasn't. Our minds, Kant might have said, would constitute just such a pair of irremovable sunglasses. Humans see the world in a certain way because our sensory organs are quite simply *structured* to see it in such a way. In particular, he said, our minds perceive the world as something which exists within both time and space, categories which may appear fundamental to us, but which, to God, simply might not exist. The world men absorb through their minds and senses Kant termed the 'phenomenal' world, and the world as it *really* is, which could be perceived only by God, he termed the 'noumenal' world. Science, he said, seeing as it is practised by humans, can only ever deal with the phenomenal world, the world as it *appears* to be, and thus its conclusions must only ever be accepted as conditional and imperfect.[61]

Kant's early writings were more concerned with what we would now call science than philosophy *per se*, concerning as they did earthquakes, meteorology and cosmology. Pursuit of scientific knowledge led with Kant towards a more abiding interest in metaphysics. You could argue that it was much the same with Humphry Davy. Kant's disciple Friedrich Schelling developed his predecessor's theories into something he called *Naturalphilosophie*, which aimed to present a worldview in which chemical matter itself was revealed as a pure chimera. A lump of granite may seem solid, inert and (in human terms) permanent, but beneath the phenomenal appearance of the granite lay a truer, noumenal version of the rock, a dynamic world of flux and energy which mankind's sensory organs were simply incapable of perceiving. Davy seemed to agree. During his last days, the poet-chemist determined that the best way to get this message across was through verse, penning a series of poetic dialogues on his deathbed. One, *The Chemical Philosopher*, consists of a conversation between Davy and 'the Spirit of the Unknown', who reveals to him that the visible world is but a illusion, and chemistry a kind of microcosm of the mind of God, mastery over which will one day enable mankind to perform all manner of divine wonders.[62] The inescapable conclusion to draw is that Davy, pantheist to the end, thought the entire universe and everything in it was alive somehow, sharing within the hidden mind of God which underlay it all. Humphry Davy was not, literally speaking, an alchemist, but he certainly shared some aspects of the old alchemical worldview. For Davy, matter could never be allowed to be reduced down simply *to* matter; to do so would be, in essence, to kill off God by killing off His whole Creation.

A Mercurial Individual

As we have already read, however, Humphry Davy was not the only man to have seen living forces concealed within dead matter. I think that when people like Andrew Crosse and Morley Martin saw living things in lifeless rock, what they were really seeing was something called the 'Spirit Mercurius'. To discover what precisely Mercurius was, let us examine the alchemical researches of no less a figure than Sir Isaac Newton. Here, we can detect signs that the metals and inorganic substances he manipulated within his own laboratory were thought of by him as participating within a kind of 'vegetable' nature, with Newton describing them as trees which 'grew' into one another, mixing their varied qualities – 'The Tree of Diana' was an alchemical term used to describe the branch-like patterns which substances could sometimes spread out into during experiments. Some alchemists felt that metal contained the 'seed' of plants, and that plants thus contained particles of metals, thereby demonstrating yet again that all was really one.[63] In his undated text *Clavis* (*The Key*), Newton outlined his attempts to create something called 'philosophical mercury' by cooking ordinary mercury and molten gold jointly in a glass, hoping to make them impregnate one another:

> They [the mercury and gold] grow in these glasses in the form of a tree, and by a continual circulation the trees are dissolved again with the work into a new mercury. I have such a vessel in the fire with gold thus dissolved, where the gold was visibly not dissolved by a corrosive into atoms, but extrinsically and intrinsically into a new mercury as living and mobile as any mercury found in the world. For it makes gold begin to swell, to be swollen, and so putrefy, and to spring forth into sprouts and branches, changing colours daily, the appearances of which fascinate me every day.[64]

This new, 'philosophical mercury' seems alive not just because it grows towards golden perfection like a seed grows into a tree, but also because, as Newton says, it is 'living and mobile'. Mercury can indeed seem almost as if it is alive under certain circumstances – hence its old name quicksilver, where 'quick' meant 'alive'. Despite being a metal, it behaves like a liquid at or above room temperature; pour it into some channels, and it will move in a way perfectly explicable through the ordinary laws of fluid dynamics, but which *looks* as if it has a will of its own. It was natural, then, that the alchemists should come to recast mercury as Mercurius, a strange spirit which helped effect transmutations from one substance into another, and to animate the world of inanimate matter – allowing lead to 'grow' into gold, for instance. Mercurius is a confusing figure, who seems to embody all opposites and contradictory qualities within the same entity; he is both animate and inanimate, the matter in spirit and the spirit in matter, a liquid and a solid, a literal substance (actual mercury) and a metaphorical one (philosophical mercury). He is, in short, almost impossible to make any sense out of; and yet, at the same time, his position as the presiding spirit of the alchemists makes perfect sense when considered in a certain non-literal way.

Look, for instance, at the alchemists' conception of what matter was in the first place. The standard view was that all physical matter was composed of three basic symbolic substances called the *tria prima*. Basically, the alchemists had inherited the old idea of Empedocles that the material world was made up of four contrasting elements, namely earth, water, air and fire. This notion had later been modified by that other great Greek thinker Aristotle, who theorised that these four elements were really but aspects of one single underlying 'primary substance' called the *protylē*, which had within it the potential to take on the four different qualities of the elements (dryness, moistness, coldness and heat) in varying combinations and quantities. Where something was burning, for instance, the element of fire predominated in the *protylē*, and the element of earth predominated in a handful of soil or any other solid substance. The four elements, then, were more like *qualities* than things in themselves, differing aspects through which the underlying oneness of the *protylē* could manifest. Aristotle also put forward the idea of a so-called 'fifth element' called aether, which was permanent and unchangeable, and filled the seemingly empty vastness of space above the moon.[65]

It was these ideas which alchemists later remodelled into the *tria prima*, or three basic substances underlying all Creation. First in this trinity was salt, corresponding to earth, then sulphur, corresponding to fire, and finally mercury, which corresponded to both air and water simultaneously. None of these were *actually* salt, sulphur or mercury, of course, and each of the three then also had its own threefold nature, containing both of the other two within itself to some degree. Just as importantly, however, when these elements manifested in the world as some specific material substance – tin, say, or copper – then they were held together somehow by a mysterious fourth aether-like principle known as the 'azoth', a term derived from the Arabic word for 'mercury'. Furthermore, the Sanskrit word for alchemy, *rasayana*, means literally 'knowledge of mercury', suggesting that this strange liquid metal was considered central to the Great Work wherever it was practised. The Spirit Mercurius, it seemed, was thus in some sense the azoth itself, a kind of divine principle underlying the whole of Creation. Among European alchemists, the idea took root that mercury was the so-called 'first principle' of metal, meaning that all metals were, at some level, really mercury, a kind of 'father metal' that was in theory extractable from them all, and which would ultimately, given enough time, grow into gold – so, when Strindberg sought carbon in sulphur, what he was really looking for was a modern form of Mercurius. Interestingly, when placed in a solution of nitric acid, mercury undergoes a rather spectacular chemical reaction in which bright red crystals form and sink down to the bottom of the mixture, while a thick red vapour hangs over the surface of the solution. Might this have been the famous Rubedo, or final stage in the finding of the ruby-red Philosopher's Stone?[66]

If so, then the substances produced would not have brought an alchemist any riches; those red crystals were not rubies. When considered in conjunction with the crimson vapour which hung over them, however, they did embody the most famous dictum of the mythical founder of alchemy Hermes Trismegistus – namely 'what is below is like what is above, and what is above is like what is below'. 'Hermes Thrice-Great' was a legendary character, not a real one, a sage who is supposed to have introduced the alchemical arts into Egypt during ancient times, laying their rules and methods down in his 'Emerald Tablet', a kind of alchemists' charter. Hermes, though, was also the name of a Greek god, the wing-footed messenger of Olympus known to the Romans as ... Mercury. 'Hermes Trismegistus' is just another way of saying 'Mercury Thrice-Great', then, a being who is thrice-great, presumably, as he contains within him, and so holds together, the three *tria prima* of salt, sulphur and mercury. You could almost say he is a personification of the azoth. As such, the famous doctrine of 'as above, so below' was the most important of Hermes' teachings as it taught one thing above all – namely, the ultimate underlying unity of all matter both with other forms of matter (or of *protylē* with elements), and with the human psyche, of macrocosm with microcosm, of man and world.

Rising from the Ashes

Given this kind of thinking, it should come as no surprise that alchemy also had another main goal: creating life via a kind of deliberately willed spontaneous generation. Look, for example, at a process described in a 1723 German text called *Aurea Catena Homeri*, in which the alchemist is directed to collect some rainwater in a tumbler and let it stand untouched for several weeks. Eventually, sediment will begin to accumulate at the bottom, which must then be extracted. Within this sediment will be found gritty particles deemed to be the 'seeds' of future metals, and any organic-looking forms the 'seeds' of plants. Leave it alone for long enough and things like maggots will begin to be seen in there too, an entirely natural process which is not really spontaneous generation at all, but which the experimenter can fantasise might be. Inanimate matter would thus appear to have come alive, through the grace of Mercurius.[67]

Another form this quest took was called 'palingenesis' (Greek for something like 'rebirth'), which involved the idea of resurrecting a living thing from its very ashes. Most commonly, this feat was performed with plants, leading to the alternative term 'vegetable phoenix' being coined. If you burned a plant, mixed it with various chemical compounds and heated it, then its apparition was supposed to float up, before eventually falling back down into a pile of dust again at the bottom of the alchemist's vessel when the procedure was over. A variant procedure, 'icy palingenesis', involved mixing the ashes of a plant with water and leaving it out overnight during winter; in the morning, an icy pattern of its roots and leaves would appear on the frozen water, a temporary frosty ghost. Presumably this latter technique really could work, seeing as ice often has fronds and branch-like patterns waiting to be seen in it anyway. Concentrating too closely upon the smoke from a burning pile of plant ash might also play tricks on the mind, causing you to see random wisps as stems and stalks. Some alchemists got carried away, however, and started speculating that maybe graveyard ghosts were really the dust of the dead reassuming their old shapes under the influence of obscure atmospheric conditions. If so, then maybe men, as well as plants, could also be conjured from the dead? Sir Kenelm Digby (1603–65), one of those early Royal Society members who were half-alchemist, half-chemist, certainly claimed to be able to resurrect crayfish through palingenesis. He burned one to dust, then mixed its ashes with sand and water and placed it within a sealed container. A few days later a new infant crayfish appeared, presumably because some tiny eggs lay already hidden in the sand, a mistake similar to that made by Andrew Crosse centuries later.[68]

Possibly palingenesis was really just another poetic way of describing things like reflux distillation; metals 'died' as you boiled them, with their vapours producing crystals on the side of glasses which resembled flowers.[69] As usual, however, many people took this symbolic language literally. For Robert Boyle, for example, palingenesis worked as follows. Plants appeared to have been absolutely destroyed when burnt to ash, but really were not; their component atoms remained, as did a kind of 'seminal essence' or 'memory' of what they used to be. It was the same for animal and human bodies. Even if a man is eaten by wild beasts, his 'atoms are preserved in all their digestions, and kept capable of being reunited' once defecated out again, though he might smell a bit afterwards. According to Boyle, it was even possible that Jesus had resurrected himself via palingenesis![70] Weirder yet was a *bugonia*-like form of palingenesis detailed in a medieval text called *The Book of the Cow*, falsely attributed to Plato. Here, alchemists are instructed how to kill and then regenerate cows for purposes unknown. First, you kill the cow and behead it. Then you sew the head back on, and stitch all its orifices shut. For the next part of the procedure you will need to get your hands on a large dog penis from your nearest stockist, and beat the cow's body with it until all its bones are broken. After this, you should extract the pulp from inside the cow, mix it with herbs and leave it to rot into mulch, periodically adding a pinch of powdered bee. Then, you repeat the whole process in reverse, and are left with a living cow.[71] Presumably this is a metaphor for some cyclical laboratory process, but it also carries within its final stages an uncanny foreshadowing of the later work of Felix Dujardin and Morley Martin, grinding up tiny creatures into protoplasm then seeing them live again.

Hello, Little Man

The Book of the Cow also contains instructions for how to create a human being via abiogenesis. First of all, you have to get some human sperm – most alchemists could supply this themselves free of charge – and smear it across something called 'the stone of the sun', whose identity is debatable. Then, you find some poor cow and shove it up her vagina. The sun stone acts as a plug to stop the semen dribbling out, and the cow is locked away in a darkened room until, eventually, it gives birth to a baby humanoid. This is then taken away and bathed in special dust, allowing it to grow human skin. This foetus is kept inside a glass vessel for three days, fed on blood for a week, and then nourished with milk and rainwater for a year. At this point, it will emerge from its glass prison and call you 'daddy'.[72]

Such an artificial creature is known as a 'homunculus' – a 'little man'. The idea probably has its origins with the Islamic alchemists, in particular the great eighth-century practitioner Jābir ibn Hayyān (*c.* AD 721–*c.* AD 821), or 'Geber' as he was known in Europe, whose texts were thought by most to be essentially unreadable. Seeing as his writings combine accounts of genuine laboratory chemistry with far-fetched claims about creating a winged baby by pumping a woman full of bird sperm, or making snakes grow out of rotten hair, this fact is understandable. Indeed, it is from negative descriptions of his confusingly written works that we get the term 'gibberish' – originally 'Geber-ish'.[73] Jābir was born in Persia, the son of an apothecary, from whom he learned medicine. He didn't confine his studies to this field, however, and not only developed various useful chemical techniques, he also invented the word 'alkali', realising that such substances were inherently different from acids, and devised various items of chemical apparatus like the alembic, a vessel used to distil liquids. He may even have been the first person to create hydrochloric, sulphuric and nitric acids. Jābir also realised that by heating up minerals and ores to certain temperatures, you could make their constituent parts separate out from one another, and made systematic records of his observations. He even helped develop early industrial processes like tanning, glassblowing and dye manufacture, and devised practical techniques to prevent metals from rusting.[74] He was also one of the first to provide us with clear instructions for creating a homunculus, a procedure he called *takwin*. First, you got something called 'essence' – probably sperm – and left it inside a mould to rot. This mould was later placed inside a series of perforated spheres submerged within a bath of warm water, which were then rotated in such a way as to imitate the movement of the celestial bodies. In this fashion, the microcosm of human sperm was made to interact with the macrocosm of the heavens, thus forcing them to merge, infusing matter with spirit until a child was born.[75]

Undoubtedly the most famous account of giving Mercurius a fleshly body was provided by Paracelsus (1493–1541), a German-Swiss medical genius, chemist, astrologer, mystic, drunkard, notorious loudmouth and alchemist. In his book *De Rerum Natura*, Paracelsus instructs as follows:

> Let the sperm of a man by itself be putrefied in a gourd-glass [and] sealed up ... in horse-dung, for the space of forty days, until so long as it begin to be alive, move and stir ... After this time, it will be something like a man, yet transparent, and without a body. Now after this, if it be every day warily and prudently nourished and fed with ... man's blood, and be for the space of forty weeks kept in a constant, equal heat of horse-dung, it will become a true and living infant.[76]

Paracelsus actually claimed to have created a homunculus himself this way, which was twelve inches tall and in severe need of education. An alternative method, meanwhile, was to dig up the ground where a man had just been hanged. Seeing as people being hanged emit all kinds of bodily fluids, it was said that the dying man's sperm would fall on the soil below, and grow into a plant called a mandrake. Some mandrakes have roots which look a bit like dolls with arms and legs, and if you picked one, took it home and fed it on blood, milk and honey, it was supposed to become a homunculus. Or, then again, you could simply ejaculate inside a hen's egg, bury it in horse manure, and await developments.[77] The only trouble was that homunculi, being man-made, didn't have souls, so were free to sin. The Spanish alchemist Arnald de Vilanova (1235–1311) supposedly realised this while treating a big man-shaped pile of semen with drugs to make it live, so smashed the messy monster up with a hammer.[78] Some people, like the fifteenth-century Spanish bishop Alonso Tostado (1400–55), tried to argue that at least one homunculus had a soul, however, seeing as Jesus Christ, being the product of a virgin birth, must have also been a 'little man' who had spontaneously generated Himself within Mary's womb. Tostado speculated that Jesus had just suddenly formed out of Mary's retained menstrual blood, like micro-organisms within Dujardin's protoplasm. Remarkably, he was not charged with heresy.[79]

Test-Tube Babies

Did anyone ever really manage to make a homunculus? No. But perhaps some people really did think they had *seen* them, growing in their little jars within ill-lit laboratories. Or maybe the idea of the homunculus was not to be taken literally. Another description of homunculi, for instance, comes from an anonymous text of 1616, *The Chemical Wedding of Christian Rosenkreutz*, an allegorical tale in which the legendary Rosicrucian journeys through a magical land, undergoing trials which are really symbolic representations of various stages of the alchemical process. At one point, after witnessing the beheading of a king and queen, the hero palingenically resurrects a phoenix-like bird by placing its ashes into special moulds, whereupon 'there [then] appeared two beautiful bright and almost transparent little images, the like to which man's eye never saw, a male and a female, both of them only four inches long ... they were not hard, but limber and fleshy, as other human bodies, but they had no life'.[80]

By feeding these homunculi the blood of the phoenix, which itself had earlier been fed on the blood of the decapitated royals, Rosenkreutz restores the homunculi to life, they grow to full size, and transmute back into the kings and queens the alchemist met at the start of his journey, their heads now reattached.[81] Again, the tale is probably an allegory for reflux distillation, or something similar. The king and queen were personifications of chemical substances and so, presumably, were their transparent homunculoid forms. With this in mind, it has been suggested by some that the homunculus is itself really just yet another synonym for the Philosopher's Stone or the *prima materia* which goes to make it up. For the alchemists, metals must first 'die' in their furnaces and vessels before being reborn anew as the adept extracts that metal's 'spirit' and infuses it into a new one – by melting it and creating alloys, like Jābir did. This method directly paralleled the way that, with spontaneous generation, oxen must first die and rot before bees and maggots could then be reborn from them. Newton, we will remember, thought he could grow alchemical 'trees' in his lab, by heating and mixing mercury; the metals were alive, imbued with the Spirit Mercurius. To men of the past there was a graduated scale of life present in the world, a so-called 'Great Chain of Being', in which metals were more alive than rocks, plants more alive than metals, animals more alive than plants, and so forth. With each stage, a little more of the 'life force' – Mercurius – slipped through, causing the thing in question to be more animate and lively. Some instructions provided for creating homunculi are indistinguishable from those for creating the Philosopher's Stone; sometimes the stone is even referred to as an 'infant' a 'golden boy', or an 'embryo' being grown in a glass. The *prima materia*, then, was a kind of 'seed' which, through careful nurture, would grow into the bouncing baby Stone, as represented by the homunculus.[82] So, while no alchemist could truly generate artificial life any more than they could make gold, the wise alchemist could still experience a kind of miracle through contemplation of the homunculus nonetheless. According to Ronald Gray, a scholar who studied Goethe's adventures in alchemy:

> The seed of the metal, to which [the homunculus] so closely corresponded ... was ... in its most important aspect a symbol of divine life in Man, the germ of a desire for God. This was the seed which the religious alchemists desired to implant and cultivate ... Seen in this light, the homunculus corresponds not only to the seed of the metal and the central spark [of life], but also to the incarnation of God or Christ within the adept ... 'the man within' ... the possibility of a new man arising from the old Adam.[83]

In this view, the homunculus was yet another image of the alchemist's soul, seen externally, inhabiting dead matter. Some scientists are still conjuring up homunculi in this sense even today, although they might not be consciously aware of the fact. Take the example of computer scientists engaged in the attempt to create Artificial Intelligence (AI). The term AI was coined by Chris Langton (b. 1948), a one-time hacker who liked to run a software program on his PC called *Life*, which aimed to replicate lifelike processes in mathematical form on-screen. Sitting there one day while otherwise occupied, he 'suddenly felt a strong presence in the room' and looked up to see that 'an interesting configuration' had formed on his computer; he felt that he had actually been witness to the creation of an artificial life form, which had generated itself spontaneously inside his machine from inanimate silicon – in other words, a homunculus.[84]

The Crystal Gaze

Despite all these fantasies, we must recognise the fact that, sometime during the distant past, at least one genuine act of abiogenesis must have occurred. If life had not sprung from dead matter at some point, there would be no life at all. Scientists are still unsure as to how precisely this occurred, however. Darwin himself thought life might have begun in 'some warm little pond' filled with an extremely lucky chance mixture of chemicals.[85] In 1828, the German chemist Friedrich Wöhler (1800–82) managed to synthesise urea from purely inorganic chemical sources; previously, the only way to get it was from the urine of living things. This suggested that there was no fundamental difference between the chemistry of organic and inorganic material, thus implying that living things could, after all, be generated spontaneously from non-living matter under certain fortuitous conditions. You might not get full-blown bees, but single-celled primitive micro-organisms must come from somewhere.[86] In 1906, German chemist Walther Löb (1872–1916) demonstrated that mixing ammonia, carbon dioxide and carbon monoxide in heated and electrically charged water would create amino acids, so-called 'prebiotic chemicals' which were the earliest building blocks of life. Seeing as he presumed all these substances were present during Earth's earliest days, this was supposed to be a simulation of the famous 'primordial soup' from which it was guessed life must have emerged.[87] The most well-known such experiment was performed in 1953 by the American chemists Stanley Miller (1930–2007) and Harold C. Urey (1893–1981), who put ammonia, methane, hydrogen and water in a sealed flask, electrocuted it to simulate prehistoric lightning storms, and found about 2 per cent had become amino acids. However, it later turned out that these experiments were fundamentally flawed, as these chemicals were not necessarily present in Earth's earliest atmosphere after all.[88]

It's a long step from amino acids to electric cheese mites or homunculi, but nonetheless the parallels between these experiments and those of people like Andrew Crosse are clear. Spontaneous generation was making a comeback, rising from its ashes like a vegetable-phoenix! The clearest example of a scientist projecting his own desires – or soul, if you prefer – into inanimate matter was the German chemist Otto Bütschli (1848–1920), who in 1890 announced he had created artificial protoplasm by introducing potassium carbonate into samples of various oils. Looking at this alleged protoplasm beneath a microscope, he claimed he could see it swim around in an amoeba-like fashion when placed on water. Bütschli's eyes, however, deceived him. The goo swam around because oil, placed on water, will create surface tension, not because it was alive and ready to spawn amoebas.[89] Another instance of an accidental alchemist was the German geologist Otto Hahn, who published a book in 1880 filled with lovely pictures of microscopic fossilised shells, sponges and corals he claimed to have found inside meteorites, thus suggesting life actually came from outer space. It later turned out, however, that these 'fossils' were really inorganic crystals on to which Hahn had projected qualities of life. The only thing really inhabiting the meteorites was the Spirit Mercurius – and he disappeared as soon as Hahn averted his gaze.[90]

The notion of crystals being somehow alive is an interesting one, and has been proposed several times down the years. In 1917, Ernst Haeckel published his *Crystal Souls: Studies on Inorganic Life*, exploring the wonderful way that crystals form into complex shapes and patterns in an apparently spontaneous manner. For Haeckel, crystals had a bit more life to them than rock, and a bit less life than single-celled organisms; he thought that protoplasm was really a kind of liquid crystal. Really, then, he was a believer in the Great Chain of Being.[91] Of course, crystals aren't actually alive, even though when examined microscopically they demonstrate extremely beautiful geometrical lattice-like structures so complex they can look at first sight as if they must be more than mere minerals. Crystals also have the power to reproduce from 'seeds' placed within a supersaturated chemical solution, though as far as I know not even Haeckel claimed to have seen any having sex.

A more recent attempt to imbue crystals with lifelike properties came from Rupert Sheldrake, who we may remember from earlier daring to suggest that Nature has a memory. One of Sheldrake's best illustrations of this idea involved the fact that, when chemists synthesise new chemicals, they frequently find it hard to make them crystallise properly. Then, after crystallisation has successfully occurred the once, it becomes ever easier for it to occur again and again, in laboratories all over the world. Sheldrake proposed that maybe these crystals built up a 'habit' of forming, a habit which was reinforced with each instance of subsequent formation; or, in other words, the crystals were somehow 'learning' how to reproduce themselves from memory. Even weirder are cases in which one form of crystal is replaced by another. Take xylitol, a sugar alcohol used in chewing gum. In 1942, this was successfully crystallised, with a melting point of 61°C. A few years later, though, xylitol began crystallising in an entirely different way, gaining a new melting point of 94°C. Thereafter, the older form of crystallised xylitol disappeared, never to be seen again. It seems Nature's memory had forgotten it! One explanation is that crystal 'seeds', small fragments of newly formed crystals, find their way into laboratories after their first appearance, and contaminate the area, displacing the previous crystal forms. Anecdotally, wandering chemists are supposed to spread these seeds through their bushy, unwashed beards; but what if new crystal forms also instantly displace the old ones in isolated labs where itinerant hairy scientists are never seen? Sheldrake tries to account for this phenomenon by suggesting that, just as in the organic world, 'chemistry is not timeless; it is historical and evolutionary, like biology'. Maybe 'the new forms are more stable thermodynamically, and ... in competition with each other, the new forms win'.[92]

Applying evolutionary theory to inorganic matter sounds strange, but Sheldrake has pointed out a genuine anomaly. He isn't, like Haeckel, trying to say that crystals are living beings, only somehow *analogous* to them – and perhaps they are. The English philosopher A. N. Whitehead (1861–1947), who held that the entire universe was best thought of as a kind of organism, rather than a lifeless clockwork mechanism, once summed up his own thinking on such matters thus: 'Biology is the study of the larger organisms, whereas physics is study of the smaller organisms.' Did he mean by this that the atoms which make up living creatures are alive? Not precisely. Whitehead was a proponent of 'Organism', a school of thought which viewed Nature as prone to form wholes which are more than the sum of their parts; just as a blood cell is more than the atoms and chemicals which constitute it, so a man is more than the blood cells and chemicals which constitute him. Organism treats Nature *in effect* as alive, but not literally so.[93] The idea is not as weird as it seems. Think of the English environmental scientist James Lovelock's (b. 1919) celebrated 'Gaia Hypothesis', which also treats planet Earth as if it is alive; by this, Lovelock means that it is a kind of self-organising, self-regulating system that appears to make 'decisions' which ensure its well-being, not simply a big lump of inert minerals floating in space. Many people think similarly, saying things like 'If mankind keeps on abusing the Earth, then it will kill us all off through global warming'. Lovelock does not literally bow down and pray to an Earth goddess every evening; it's just a useful way to conceptualise the operations of the biosphere.

World in Motion

Can inanimate matter ever really be alive? By definition, no; if it's alive, then it isn't inanimate! The idea of a living universe, however, is as old as civilisation itself – if we take it that civilisation began with the Greeks. Much early Greek thought held that the universe was inherently animate. Things that moved were alive, as with animals, and so it was said by figures like Plato (*c.* 428 BC–*c.* 347 BC) that planets, stars and other astronomical bodies were alive too.[94] Anaxagoras (*c.* 500 BC–428 BC) thought that a layer of mind, or *nous*, was present in the universe, causing all motion. It led to a sort of rotation which spread gradually across the globe, causing heavy things to fall towards its centre and light things towards its circumference, a naive early conception of gravity infused with a sort of quasi-animation.[95] For Empedocles, meanwhile, forces called 'Love' and 'Strife' controlled all change and motion in Creation; Love brought substances together, whereas Strife separated them. For those less given to animism, these forces might be rechristened 'Growth' and 'Decay'.[96] Perhaps the use of such words was metaphorical, perhaps not. It is hard to tell; these early Greek thinkers were poets as much as philosophers. At the time some called them *physiologi*, which has since been translated as 'physicists', but really signifies something like 'students of Nature'.[97]

The Nature which they studied, however, was very different from our own. Heraclitus (*c.* 535 BC–*c.* 475 BC) believed in a sort of endless flux, governed by a force called 'Justice', which ensured that no one element was ever allowed to prevail over any other: 'Fire lives the death of air, and air lives the death of fire; water lives the death of earth, earth that of water.'[98] Anaximander felt something similar, his own concept of justice involving the idea that all things were ultimately one:

> Into that from which things take their rise they pass away once more, as is ordained, for they make reparation and satisfaction to one another for their Injustice according to the ordering of time.[99]

The early Greek religious concept of the cosmic order of things, in which everything and everyone had their correct place, taught that to overstep these divinely ordained boundaries was an act of hubris which would bring down inevitable punishment from eternal laws that governed even the lives of the gods. Thus, we can see how the primitive religious concept of Justice may well have fed into the later concept of justice in ancient Greek physics.[100] As modern science developed during the lead-up to the Enlightenment, however, such ideas of semi-animate forces within the universe began to lose their lustre, being increasingly replaced with a purely mechanistic outlook. What the Greeks were practising was not truly science, but a form of pre-scientific reasoning.[101]

It's All Greek to Them

Nonetheless, this pre-scientific philosophy has still made the occasional comeback well after Classical Greece had faded away. One man who would have made an excellent Greek was the German philosopher, psychologist and professor of physics Gustav Fechner (1801–87). After performing a series of ill-advised experiments in staring directly at the sun while formulating a theory of vision, Fechner began to suffer severe eye problems, followed by a spell of debilitating illness and mental breakdown. Three years later, when at last he was able to go out into the sunshine of his garden without a bandage over his eyes, Fechner saw the world anew and had a kind of mystical revelation in which he saw all at once that the universe around him was alive, down to the very last atom. In his book *Atomenlehre* (*Atom-Life*) Fechner argued that atoms are vortices of living energy which, when placed in combination, went towards making up a larger universe filled with living matter. Like Plato, he regarded the stars and planets as alive; the Earth itself was a kind of 'angel', he said, and all Creation was connected and animated by living flows of force like light and gravity.[102]

Another honorary Greek was James Burnett, better known as Lord Monboddo (1714–99), a celebrated Scottish judge and philosopher. In love with the classical world, he adopted a distinctly ancient view of physics, proposing that the whole of Creation had a substratum of psychic consciousness present within it, with each object in existence occupying a position at some point along the Great Chain of Being. Plants, for example, had a kind of 'vegetable mind' which made their leaves turn towards the sun and their roots seek out water. Monboddo also felt that magnetism had a mind; not a conscious one exactly, but a mind nonetheless, which allowed it to attract metals. Gravity was also a form of mind; Monboddo disliked the materialistic implications of Newton's thought, seeing the idea of impersonal forces causing planets to perform their movements as a temptation towards atheism. Instead, suggested Monboddo, why not say that planets had mineral minds, and orbited in cycles because this was the way they *wanted* to move, not because gravity was *forcing* them to do so? Likewise, when a stone fell to earth, or rolled down an incline, it was not due to gravity, but a motion initiated by the mind which dwelled within the pebble. Even Newton's famous apple did not really fall because of gravity; its vegetable mind made it seek out the ground beneath when the appropriate moment came. Monboddo knew his theories to be accurate, as in 1778 he said that a beautiful female ghost had appeared before him and delivered a detailed lecture in French, confirming his every word.[103]

Maybe this same spirit lady had also made an appearance before the French philosopher Pierre Louis Moreau de Maupertuis (1698–1759), who held similar ideas; he also argued that intelligence appeared in matter in a series of increasing stages, from stones to plants, then up through animals and human beings. Further, Maupertuis ascribed some degree of sentience to the individual particles which constituted any living being. After all, he asked, how would it be possible for the creatures made up from such particles to be intelligent if these atoms were not themselves intelligent too?[104] Another Frenchman, the *Encylopèdiste* Denis Diderot (1713–84), felt something similar, viewing each organ inside a body as being in itself a kind of 'animal'. A happy union of 'living molecules' within an organism were arranged together in such a way as to make a greater whole which was in itself alive, a process he compared to individual bees combining to make a greater living swarm. Diderot fantasised that if such a thing were true, it really ought to be possible to resurrect the dead, as they could be cloned by recombining their individual living atoms after death to make them live again through a new kind of palingenesis. People could be 'split up into an infinity of atomised men' which could then be carefully stored 'between sheets of paper' acting as cocoons, and allowed to grow back to life, leading to 'an entire region [being] populated by the fragments of a single individual'. Maybe their atoms could even be rearranged somehow to make them into super-men; 'who knows what new race could result some day from such a huge heap of sensitive and living points?' he asked.[105]

Even when Robert Brown first observed pollen grains moving about by themselves in a dish of water in 1827, he saw this as evidence not of the truth of atomic theory as Einstein later did, but as a suggestion that microscopic particles of organic matter were full of 'vital force' that caused them to jump around endlessly. When he found that bits of mineral dust behaved in the same way, Brown modified his views somewhat, but he need not necessarily have done. Brown accepted the theories of the great French naturalist Comte Buffon that all living things were made of some substance termed *matière vive*, whereas all inanimate things were formed of *matière brut*. However, seeing as this latter substance was, to Buffon, merely what the former became after death, the distinction between what was alive and what was dead was not entirely clear-cut.[106]

Chemical Clairvoyance

For one man, this kind of thinking raised a very interesting possibility; if all matter contained some element of mind buried somewhere within it, then might it be feasible to communicate with it? The sensible answer would be 'no' but, as we saw earlier on with his attempts to create evolutionary trees for fairy folk, the Theosophist C. W. Leadbeater was not a very sensible man. He was, though, highly clairvoyant – or so he said – and so co-wrote a truly ridiculous book, *Occult Chemistry*, detailing his attempts to peer into the souls of the chemical elements. The genesis of the book was revealed in its expanded 1919 edition, where one of Leadbeater's Theosophical colleagues, A. P. Sinnett (1840–1921), explained how the great man was staying at his house one night and amazing everyone present with his incredible psychic powers. Impressed, Sinnett asked Leadbeater 'if he thought he could actually *see* a molecule of physical matter' via paranormal means. Leadbeater said he'd try, and established psychic contact with a molecule of gold. Its structure proved far too difficult for him to explain, however, so he made a second attempt, this time with hydrogen. This proved less complex, a diagram was produced, and Leadbeater had found himself a new hobby.[107] So, what did hydrogen look like? Apparently:

> On looking carefully at it, it was seen to consist of six small bodies contained in an egg-like form. It rotated with great rapidity on its own axis, vibrating at the same time, and the internal bodies performed similar gyrations. The whole atom spins and quivers, and has to be steadied before exact observation is possible.[108]

The next element subjected to intense Theosophical scrutiny was oxygen, 'a far more complicated and puzzling body'. Apparently, it was full of tiny coloured snakes, made of 'small bead-like bodies' living within another 'ovoid body'. Sometimes these snakes broke into brilliant spots of light, 'reminding one of fire-flies stimulated to wild gyrations'.[109] Even better, Leadbeater's researches revealed to him the existence of several new chemical elements, utterly unknown to science, then or now; he called them occultum, kalon and meta-kalon. Some of the more standard elements, meanwhile, proved to be structured after the fashion of ancient esoteric symbols. Titanium, for example, surprisingly resembled the old Rosicrucian symbol of the Rosy Cross, whereas helium had 'an attractively airy appearance, as of a fairy-element' – no surprise, seeing as a race of 'tiny, ingenious builders' called elementals were apparently responsible for its construction. In addition, Leadbeater managed to observe some entirely new states of matter, such as the cumbersomely named 'hyper-meta-proto-elemental', the first state of molecular structure above that of the pure atom.[110]

What, though, *was* the pure atom? Nothing less than a part of God Himself, 'dying in matter', the world we see around us being the 'perpetual sacrifice' of His own life.[111] Looking at them closely, Leadbeater deduced that all atoms were either male or female, and were really heart-shaped bodies of pure, holy life force, as opposed to the tiny spherical billiard balls of the popular imagination. According to Leadbeater:

> The atom can scarcely be said to be a 'thing', though it is the material out of which all things are composed. It is formed by the flow of the life force and vanishes with its ebb. When this force arises in space ... atoms appear; if this be artificially stopped for a single atom, the atom disappears; there is nothing left. Presumably, were that flow checked but for an instant, the whole physical world would vanish, as the cloud melts away in the empyrean.[112]

Even weirder, according to Leadbeater's equally mad 1913 book *The Hidden Side of Things*, all inanimate objects that come into contact with human consciousness, like machines and building materials, subsequently become slightly sentient themselves:

> Just as the driver or rider becomes accustomed to his machine and learns to know exactly what it will do, and to humour its various little tricks, so the machine in its turn becomes used to the driver and will do more for him in various ways than for a stranger ... Iron which has formed part of a machine ... may be thought of as somewhat more [mentally] developed than iron which has not yet been used in the building of a self-contained system ... It is more awake than other iron ... Another interesting point with regard to this curious composite consciousness is that after a certain time it gets tired ... Metals show plainly that they are subject to fatigue ... A barber often finds his razor refusing to take a keen edge, and it is quite customary for him to say that it is 'getting tired' and to put it aside to rest. Some days later that same razor will be in perfect order, keen and sharp as ever ...[As regards] our houses ... it obviously follows that stone which has been used for building is never afterwards in the same condition as the stone which is as yet unquarried. It has been permeated, probably for years in succession, with [mental] influences ... We are therefore actually assisting in the evolution of the mineral kingdom when we use these various materials for our buildings ... the different influences which we put into them react upon us; so that just as a church radiates devotion and a prison radiates gloom, so each house in the business part of a city radiates anxiety and effort, too often coupled also with weariness and despair.[113]

Needless to say, all these atomic visions existed nowhere but within Leadbeater's own head. For the really skilled alchemist, it appears, the actual laboratory is a mere optional extra; imagining your own fictional subatomic world is a far less messy way of eliminating the space between spirit and matter than going to all that fuss of trying to physically transmute lead into gold using furnaces and acids. Why live in the world when you can live in your head?

Penetrating Insights

Another person who lived primarily within his own head was Alfred W. Lawson (1869–1954), a cult leader, economic theorist, aerospace pioneer and scientific genius who was, undoubtedly, the saviour of mankind. We know this because in 1940 Lawson published a book, *Songs That Will Be Sung Forever*, which included poorly scanning ditties in his own praise such as the following:

> We cannot change a wicked past but we can all agree
> To follow Alfred Lawson on to save humanity!
> We asked for guidance and expect an answer to our call,
> So recognise this man God sent to teach and lead us all ...
> Lawson will save us, yes save us, yes save us forever,
> His name forever all mankind will remember,
> Mankind will honour, his name will live on,
> Telling posterity what he had done![114]

And what, precisely, had Alfred W. Lawson done? It would be easier to ask what he hadn't done. According to Lawson's book *Lawsonian Religion*, 'there he stood, throughout the age of extreme falsity, like an immovable rock of righteousness, for the improvement of man before God'.[115] Born in London in 1869, baby Lawson was taken to Canada by his parents, the family finally settling in America during 1872. Aged nineteen, Lawson became a low-ranking professional baseball player. Quickly realising his true talents lay elsewhere, Lawson set up his own rival baseball league, clearing land to create a new stadium in Pennsylvania by dynamiting trees into the air instead of chopping them down. More successfully, Lawson later entered the aviation industry, coining the word 'aircraft' and operating the world's first airliner. Regrettably, Lawson insisted upon helping to fly this machine himself and kept on crash-landing it in fields. Thankfully, his insane proposal to transfer passengers from flight to flight in mid-air by dropping them down through big tubes never came to pass. He did, however, claim to have invented the aircraft carrier, proposing a big line of ships with flattened runways be strung across the ocean to allow planes to hop over to Europe.[116] According to Lawson, he was even directly responsible for America's entry into the First World War; he credited his obscure 1916 short story 'The Death of Liberty: The Outstanding Tale of All Time' with finally persuading President Woodrow Wilson to commit troops to Europe. He subsequently offered his services to Wilson as 'Air Generalissimo' of the US Air Force, but nothing ever came of it.[117] Because of having performed such wonders, when Lawson finally met his God after death, he greatly anticipated that the deity would 'grant him a special audience in which HE will say – ALFRED LAWSON, YOUR WORK HAS BEEN WELL DONE'.[118]

Even more noteworthy of divine attention were Lawson's achievements in the field of science. During the three years prior to Lawson's birth, his father had put immense mental effort into trying to invent a perpetual motion machine. These patterns of thought, said Lawson, had passed from his father's mind into his sperm, resulting in Lawson being born with special insights into physics. According to him, his initial moment of scientific enlightenment occurred at the tender age of four when, confined to his room with measles, he noticed specks of dust swirling around in mid-air.[119] Then came baby Lawson's first groundbreaking discovery:

> By blowing out his breath against these particles he learned that he could push them away and scatter them apart by PRESSURE and that by drawing in his breath he could pull them in and hold them together by SUCTION.[120]

Building upon this incredible insight, aged twelve Lawson ran away from home and, so he said, began clinging to the front of speeding freight trains, thereby 'studying atmospheric resistance to moving bodies' at first hand.[121] On 19 September 1922 the adult Lawson held a press conference in Washington DC to announce the most important scientific theory of all time to the world – but, disgracefully, only three people bothered to turn up. Nonetheless, here he explained for the first time his doctrine of Lawsonomy, something he felt mankind's previous 'new teacher with advanced intellectual equipment', Jesus Christ, had been at serious disadvantage for not knowing about.[122] The basic idea of Lawsonomy was that there is no such thing as energy (the mere creation of 'some fanciful mind'). Nor was there such a thing as empty space. Instead, there were only substances of heavy and light density, with substances of heavy density moving towards those of lesser density through the forces of suction and pressure, and all under the control of an overarching 'Law of Penetrability'. Everything was explicable in these terms: light and sound entered the eyes and ears via suction, gravity was attributable to suction, and food was drawn into the stomach through suction, then expelled again out the other end through pressure.[123]

Such forces govern everything for Lawson, meaning the entire universe is alive. The Earth, for example, apparently has a digestive system. The North Pole is its mouth, which sucks in bits of space debris, transforming them into a gaseous substance called lesether, which fills the planet and allows it to float in orbit. The Northern Lights are really space gas being eaten before it passes through hidden arteries beneath the Earth's crust, this gas ultimately being expelled again through a secret planetary anus at the South Pole. That's right – the Earth has a bum, and it farts! Earthquakes, meanwhile, were the result of indigestion, and an unwelcome sign that one day the whole delicate ecosystem would collapse and our planet die in an orgy of wanton flatulence.[124] Lawsonomy also implied that the cosmos was nothing but one gigantic and eternal sex act: the Law of Penetrability applied even to sexual penetration, in which a male force of pressure and density was drawn into a female suctional void. Whenever a dense substance penetrated a less dense one, it was having a kind of sex with it; if a stone sunk through water, then the stone was male and the water female, and the process basically rape.[125]

Later, Lawson added a further element of folly to his thinking in the form of 'menorgs' and 'disorgs' ('mental organisers' and 'disorganisers'), whose ceaseless activities were hymned in a classic song:

> Menorgs are wondrous builders all,
> Builders of the great and small,
> All of life they permeate,
> All formations they create.
> Disorgs tear down eternally
> While menorgs build faithfully.[126]

These mental organisms were the tiniest elements of all living matter, nothing less than a breed of conscious atoms. Menorgs were 'masters of physics, economics, metaphysics, mechanics and chemistry', given authority by God to create the world around us and interfere in the evolution of life. Some menorgs were more intellectually gifted than others, however, and were credited with inventing various new life forms; Lawson acclaimed the menorg who designed blades of grass as an absolute genius, for example, ditto the mastermind who first made a plant walk, thus inventing animals. Eventually, God decided to take things to another level by ordering the menorgs to create humans, which they did by joining together in their billions, with our entire physiological structure being made up of continual interaction between these hordes; if 'those unknown numbers of menorgs' should suddenly flee from your body, said Lawson, then it 'would fall to the ground like a dead chunk of putty'. Hidden away somewhere within 'a tiny cell of man's brain', meanwhile, was a supreme commander of all the body's menorgs, a 'thinking creature' who made sure our mental and physiological systems worked properly. (Thomas Edison thought something similar; according to him, millions of microscopic 'little peoples' lived inside our brains, controlled by tiny 'master entities' who told them what to do.[127]) Disorgs, meanwhile, were less benevolent living atoms, and when an excess managed to penetrate a human brain or body, this led to all kinds of diseases, mental, moral and physical. Together, these tiny beings operated a karmic wheel, with those who allowed themselves to be led astray by disorgs made to suffer from Lawson's 'Law of Manoeuvrability'. Souls were indestructible, said Lawson, and after death the menorgs created you a new body for your next incarnation which appropriately reflected your sins during life; Lawson's bizarre example is of a man who hates goats later being reincarnated as one to teach him a lesson. Presumably when Lawson himself eventually returned back to life it would have been as some kind of gigantic nut ...[128]

Sane Orgone Mad?

Another pseudoscientific theoriser often dismissed as a huge nut is Wilhelm Reich (1897–1957), a follower of Sigmund Freud who later moved on from psychoanalysis to bizarre speculations about biology. For Reich, sexuality was the key to understanding the entire world, and he drew upon the work of two German biologists, Max Hartmann (1876–1962) and Ludwig Rhumbler (1864–1939), to prove it. Hartmann and Rhumbler subjected amoebas to a variety of stimuli, finding they either moved towards them or fled. Reich felt this was a parallel with the psychological impulse of his patients to either release their mental energy in a positive way through sexual pleasure, or bottle it up through anxiety and sexual repression. Proposing that an orgasm was really a release of electricity from within the body, Reich claimed that the processes of ejaculation and organic cell division were closely linked.[129]

This idea led to Reich performing experiments in microscopy, following advice that the best way to get hold of amoebas was to place some grass blades in water and wait for a fortnight. Amoebas would then spawn, being infused via tiny spores from the air. Observing the samples closely, Reich became fascinated by how cells would gradually detach themselves from the edge of the grass blades and form vesicles which then clumped together into amoeba-like shapes. Reich concluded that this was how amoebas actually developed, rather than through spores, and termed these vesicles 'bions', calling them a transitional form between inanimate and animate matter. Soon, Reich began placing various substances into water and examining them, mixing up foodstuffs and staring at the resultant broth beneath a lens to see if more bions would appear. He said they did. According to Reich, these bions moved around, 'ate' other vesicles, and divided and grew, thus appearing to reproduce. If really true, then this was an amazing discovery, but Reich's critics have suggested that his samples were actually contaminated by bacteria from the air, or else that he was simply misinterpreting natural phenomena like Brownian motion. Reich denied this, however, proclaiming he had discovered how life itself began.[130]

His researches unearthed several different types of bion, some of which were blue and seemed to emit radiation. Continuing his researches in America from 1939 onwards, Reich created a special box through which to safely observe these radioactive bions. Surprisingly, however, he found he could see blue glows in these boxes even when no bions had been introduced. From this, he concluded that this blue radiance was a kind of natural energy which permeated the entire universe, and christened it 'orgone energy', a combination of the words 'orgasm' and 'organism'. As this implies, Reich felt that orgone was basically sexual in nature, and began seeing it everywhere, even through a telescope while looking at the night sky – probably due to an optical illusion. He even speculated that UFOs were powered by orgone, on account of their reputedly often having a blue glow to them; waste radiation from their engines might be having harmful effects upon the planet below, he warned. So excited was he by these findings that he contacted Einstein, who agreed to meet with him. Initially Einstein seemed impressed, but ultimately he concluded Reich was deluding himself.[131] A fan of Henri Bergson, for Reich orgone became a kind of new *élan vital*, a life force which made the sky and ocean blue, caused the shimmering air you see during heatwaves, and created natural electrical discharges like lightning. This electrical life force was concentrated in human genitalia during sex and released throughout the whole body at orgasm, helping destroy cancer cells. The whole universe was analogous to such a sex act, with orgone energy units moving in spiral paths which embrace and twist around one another, making life; small units of entwined orgone could create cells, whereas larger units could create galaxies. In this way, microcosm and macrocosm were reunited once more, with man and the universe revealed as twin manifestations of orgasmic energy. So pervasive was this force, said Reich, it was even responsible for the functioning of gravity![132]

Believing he could control the weather, Reich then began offering his services as a 'cloud-burster', using a series of long, gun-like pipes to influence the amount of orgone in the sky and create rain for farmers in drought-stricken areas. These pipes were connected to wells, and Reich thought that the water which flowed through them helped dissipate negative-energy clouds filled with harmful orgone radiation, or else caused ordinary clouds to spill their load. On the same principle, Reich began applying water-filled pipes to people's bodies to draw off harmful orgone radiation, or introduced pipes filled with beneficent orgone to enhance their health. According to him, inserting a glass tube filled with orgone-soaked steel wool into a vagina could sterilise all its bacteria within a single minute.[133] Reich liked to play at being a doctor, claiming that cancer-cells were caused by the sexual orgone energies of patients not being released properly due to their neuroses; he said he had examined such cells under the microscope and found they had fish-tails and swam around. According to Reich, cancer-bions were constantly forming out of the neurotic patient's tissue just as they spawned from grass placed in water, and, were it not for the fact that they died, cancer patients would eventually disintegrate into big piles of pure protozoa. Fortunately, certain special bions could be introduced into the patient to eat these cells, or else people could sit inside big boxes called 'orgone energy accumulators', and have their symptoms alleviated by bathing in positive orgone.[134] Reich described his basic sexo-medical thinking thus:

> In a running brook, the water changes constantly. This makes possible the self-purification of the water ... In stagnant water, on the other hand, processes of putrefaction are not only not eliminated, but furthered. Amoebae and other protozoa grow poorly or not at all in running water but copiously in stagnant water.[135]

It was the same with sexual orgone energy – it you weren't dispelling enough of it through an active and fulfilling sex life, then you were giving yourself cancer. It was an unlikely theory, and America's Food and Drug Administration deemed it to be nonsense, forbidding Reich to continue offering his apparently worthless therapy. Reich did not entirely obey, however, with the end result that he was jailed for two years, during which time he suffered a heart attack and died. Had he been allowed a few more conjugal visits, perhaps his health would not have declined so rapidly.

The final word on these subjects, I think, should go to Goethe, who summed up the desire to put man back into matter, as the alchemists and spontaneous generation enthusiasts did, thus:

> Man himself, inasmuch as he makes use of his healthy senses, is the greatest and most exact physical apparatus; and that is just the greatest evil of modern physics – that one has, as it were, detached the experiment from man and wishes to gain knowledge of Nature merely through that which artificial instruments show.[136]

The aim to reverse this unfortunate process is, in itself, a noble one; but what if, like Reich and Lawson, the human experimenter's senses are *not* healthy ones? Subjective science is all very well for those private individuals who choose to pursue its path for their own personal enjoyment and enlightenment. For society in a wider sense, however, it is probably best if science continues to follow the less romantic, though far more practical, route of objectivity. Practise alchemy as a private hobby if you want, by all means – but please, if you are employed as a genuine scientist, then not during normal working hours.

Conclusion: Delusions on a Cosmic Scale

Dead of all countries, unite!

Cosmist Manifesto, 1920[1]

As everybody knows, Soviet Russia was full of five-year plans; a five-year plan to industrialise the country, a five-year plan to transform agriculture, a five-year plan to build up the army, and so on forever. Fewer people, however, are aware of the fact that for many at the time this was not enough. The slogan 'The Five Year Plan in Four Years' was also advanced, and the slogan '2 + 2 = 5' began to appear all over the place – in newspapers, on billboards, written out in foot-tall letters in electric lights on buildings.[2] The slogan seemed to defy all logic; but that was exactly the point ...

Such utterly irrational notions were entirely in keeping with the delusions held by many Soviet thinkers about the prospect of transforming Russia into a kind of People's Paradise. We all know where that particular social experiment ended up, of course – with the gulag, mass starvation, huge environmental degradation and ultimate economic collapse. However, Communism itself was not the only inspiration behind such events. Some of the earliest utopian theorising in the Soviet lands was also directly influenced by a kind of quasi-Rosicrucian religio-scientific idealism, in the shape of a bizarre utopian philosophy now known as 'Russian Cosmism' – that same movement whose members we encountered in this book's introduction, trying to resurrect Lenin. Communism and Cosmism were not identical. Lenin and Stalin were not Cosmists themselves, and the movement was directly suppressed under Soviet rule, with many of its leading lights being banished or executed, and their writings banned. Still, however, their desire to transform society lived on, with some distinctly Cosmist schemes being put into direct application under the disguised banner of Communism.[3]

To many, the two movements seemed entirely compatible. Alexander Gorsky (1886–1943), for example, was a prominent Cosmist who initially approved of Soviet engineering projects in areas like drought alleviation as being excellent examples of Cosmism in action. He wrote letters to commissars, telling them of Cosmist ideas and asking they be implemented by the state; he even wrote to Stalin. For his efforts, Gorsky was ultimately arrested by the security services. Asked if he was a mystic, Gorsky replied that he was, but that 'mysticism ought to lead to a path of great scientific progress'. He told his interrogators that he had initially welcomed the revolution, because it apparently 'contained that which moved humanity forward, toward realisation of the dream of controlling all the forces of Nature', even up to 'the complete liquidation of Nature's idiocies – death and dying'. Executed as a traitor in 1943, however, Gorsky was ultimately forced to realise that Communism had not abolished death after all.[4]

So what was Cosmism? The title of a 1920s Cosmist journal, *Immortality*, might give us a clue. The journal and its followers had the slogan 'Immortalism and Interplanetarism', and issued a manifesto which declared the existence of two fundamental human rights: 'the right to exist forever, and the right to unimpeded movement through interplanetary space'. The basic idea was that, after vanquishing the bourgeoisie, Communism would ultimately conquer the forces of time, space and death too, through advanced science.[5] In essence, Russian Cosmism was a kind of twentieth-century Rosicrucianism of the Proletariat, a woefully unrealistic project in which, as one modern scholar has put it, mankind is 'called upon to overcome not only the outer world, but also his own inner nature'.[6] Under the brutalities of Soviet rule, it was the second part of that quest which was to prove the most unrealistic.

A Handful of Dust

While few of the main Cosmists much mentioned Rosicrucianism in their writings, its unspoken influence upon them seems clear enough.[7] Rosicrucian and Masonic movements were present in Russia's cities long prior to the emergence of Cosmism, greatly thanks to the efforts of Nikolai Novikov (1744–1818), a writer who pushed the idea of creating a new utopia in the land and ended up having his work suppressed by Catherine the Great (1729–96) as a destabilising influence due to its popularity. In this atmosphere, fabulous stories about scientific wonder-workers of the past began to spring up, with Jacob Bruce (1669–1735), a Scottish Freemason, alchemist and servant of the Russian Crown, being credited with having brewed up medicines to restore youth to old men and resurrect dogs from the dead. Another prominent Russian Freemason was Prince Ivan Alekseevich Gagarin (1771–1832), the grandfather of the founder of Russian Cosmism, the Moscow librarian and philosopher Nikolai Fedorov (1829–1903). Gagarin died when Fedorov was only three so can have hardly initiated him as a Mason, but the first Cosmist seems to have picked up the bug from somewhere, because his writings are absolutely suffused in Rosicrucian-type thought.[8] Though fêted by such literary giants as Leo Tolstoy (1828–1910) and Fyodor Dostoyevsky (1821–81) as a modern-day saint, Fedorov was virtually unpublished during his own lifetime. His weird philosophy was only laid out posthumously in a huge publication called *The Philosophy of the Common Task*, where he explained that mankind's overriding cosmic quest (the titular 'common task') was to do nothing less than resurrect the dead. And when Fedorov said 'the dead', he meant *all* the dead; every single one of them, all the way back to Adam and Eve.

All human problems, said Fedorov, had their roots in the fact of death and disintegration. According to the Book of Ecclesiastes, men 'all go unto one place; all are of the dust, and all turn to dust again',[9] but this seemed like little more than shameful defeatism. Yes, Fedorov had to admit that, left to itself, everything in Nature, even the tallest mountain, would ultimately crumble down into separate particles. But so what? Mankind's job, no matter how long it may take, was to treat the entire cosmos as a giant jigsaw puzzle and strive to reunite all these particles back into their original form; to Fedorov, the tale of Humpty Dumpty was a challenge, not a warning. Fedorov coined the term 'ancestral dust' to describe the disintegrated dead, and said that it was now every individual's sacred duty to set out to collect all the dust particles which had once constituted their parents, and reassemble them again. Then, your previously dead parents could resurrect their own dead parents, and so on *ad infinitum*, until we ended up with Adam stood there before us wearing a fig leaf and smiling – palingenesis on a grand scale. According to Fedorov, the particles in our own bodies and those of our ancestors' dust would vibrate in cosmic sympathy, allowing us to identify which bits of fluff and lint were our dead relatives and which mere strangers, thus making the task easier. An alternative Fedorovian path to immortality, though, would be to develop some kind of cloning technique, extracting genetic material from children and using this to create perfect copies of parental bodies. These clones would merely be perfect *physical* copies of the dead, though, without the dead person's memories or personality, so this would be only a temporary measure. In whatever way our scientists finally managed to resurrect the dead, said Fedorov, once Adam and Eve were back then the original Eden could at last be restored – the 'Great Instauration' so desired by Francis Bacon and his Rosicrucian-sympathising peers made real at last.

Fedorov was quite serious in this proposal. He didn't mean it all as some kind of spiritual metaphor, but as something which literally should be done in the real, physical world, using the ever-progressing methods of science. To make a start, he spent hours in his workplace, Moscow's Rumyantsev Library, combing through old documents and making lists of forgotten dead people to make sure they wouldn't be left behind when the moment of universal resurrection came (had he been reading Gogol's *Dead Souls*?). This may have seemed pointless, but he anticipated that progress towards immortality could only be made through a series of small steps – by bringing a person temporarily back to life for a few seconds or hours immediately after their demise, for instance, as with the *autojektors* we examined earlier. While his project of restoring Eden obviously had a distinct spiritual dimension to it, the committed Christian Fedorov was adamant his aims would be achieved through technology, not religion. The old prayer 'Give us this day our daily bread', for example, could be recast scientifically to mean 'Let us use science to improve our crop-yields'. Yes, by resurrecting the dead mankind would be following in the footsteps of Jesus and Lazarus, admitted Fedorov, but through scientific, not supernatural, means.

Fedorov was adamant that no problem was beyond science's future solution. For example, there was always the chance that some people's ancestral dust might by now have drifted off into space. How were dutiful offspring meant to get that back? Simple! If part of your dad was now orbiting somewhere beyond the moons of Jupiter, then you would just have to knuckle down and get on with transforming Earth into a gigantic moveable spaceship which could be flown throughout the solar system in search of his missing kneecap. 'Spaceship Earth', as Fedorov called it, would be powered by the erection of great electromagnetic cones at various points upon the planet's surface, allowing us to point it in the direction of the nearest space cloud of floating ancestor dust. As he explained, 'the Earth will be the first star [*sic*] in Heaven to be moved not by the blind force of gravity but by reason, which will have countered and prevented gravity and death'.[10]

In this way, Eden would be restored not only to Earth, but also to the moon, other planets and eventually the whole universe. In principle, all matter had once had its divinely ordained place during the original time of Adam's Paradise – and so, like T. H. Huxley working backwards, we would have to work out in what specific arrangement it had all lain back then, and restore things back to how they were before Eve ate that apple. In short, Fedorov wanted nothing less than total 'knowledge and control over all atoms and molecules of the world'. Once this common task had been performed successfully, all other aspects of Eden would inevitably be restored too. There was no war, competition, illness, death, sex, childbirth, work nor unhappiness in Eden, proclaimed the sage of Rumyantsev Library, and nor would there again be on some happy day in the future.

Eating People Is Wrong

That perfect future (or perfect past, depending on your perspective) was still some way off, though, with life in Fedorov's Russia being ultimately reducible down to only two quite terrible things – pornography and cannibalism. Modern life was nothing but a 'pornocracy', said Fedorov, with horrid physical desires ruling over us all. In Eden, though, prior to their tasting of the fruit of knowledge, Adam and Eve had not even noticed that they were naked, never mind wanting to go into the bushes and play hide-the-serpent with each other. Instead of this appalling porn-based civilisation, in which people went around performing such filthy acts as falling in love, having sex and giving birth to children, an Edenic 'psychocracy', in which spiritual matters were mankind's sole concern, had to be created. Sex and childbirth must be quickly abolished, and new ways of reproduction devised instead; rather than trying to create new life through sex, people should devote their time to trying to recreate *old* life, through resurrection. In this fashion, what Fedorov called 'the problem of cannibalism' would be solved, too – cannibalism, in his view, being the way that all young people preyed on the old, devouring their resources. 'Biologically, "progress" is the younger swallowing the older,' he wrote. Primitive cannibals, said Fedorov, should not be condemned for eating their parents in a literal sense, as at least they were honest about the fact that they chomped down on mum and dad; whereas us supposedly 'civilised' Europeans dishonestly made use of things like vegetables and items of clothing whose atoms once used to be our great-great-grandparents, and pretended there was nothing at all wrong with us doing this.

The current plague of 'covert people-eating', said Fedorov, worked thus. First of all, people die and return to dust; the nutrients from this corpse-dust nourish plants, and these plants also nourish animals, which we then either skin for clothing, or eat ourselves. So, when you see a man walking down Nevsky Prospekt dressed in a fur hat and eating a potato, say, then he's actually engaging in a disgraceful public act of incestuous cannibalism, whether he knows it or not. Worse, if we don't get our act together and do something about all this, said Fedorov, then one day *we* will be eaten by our own great-great-grandchildren in turn, too! Through this endless cycle of death and resurrection, mankind will find itself trapped upon this vile planet forever, being eternally resurrected in the form of fruit, vegetables and who knew what other sickening abominations – unless we finally managed to pull ourselves together and resurrect the dead. Once the dead had risen, of course, science would immediately progress in leaps and bounds, Fedorov implied, seeing as all people who had ever lived would all be present simultaneously, thus giving us access to a limitless storehouse of universal knowledge. Want to know how the pyramids were built? Just ask Pharaoh. Puzzled by some obscure aspect of mathematics? Set Pythagoras onto the task. Need a new law code? Napoleon's your man, with a little help from King Solomon. Nikolai Fedorov did not know it, but he had accidentally written the plot to *Bill and Ted's Excellent Adventure* a hundred years too early.

Fedorov also dreamed about recovering the original language of Adam and Eve, to replace our current Babel of babble. Once this ur-tongue had been successfully rediscovered, said Fedorov, everybody would recognise and understand it immediately as the perfect language of God, and use it in place of such inferior imitations as Russian and English. Therefore, Plato could understand Confucius with ease, and vice versa. With access to all knowledge, man would soon transform the world as it *is* into the world as it *should* be, through a process termed 'supramoralism'. This meant that all passive knowledge gained through observation would become a new active knowledge of transformation – so 'meteorology' would become 'meteorurgy', as mankind progressed from observing the weather to controlling it. Biology, too, would become 'biorurgy', and men would directly interfere in the course of their own evolution. Initially, mankind might give himself new body parts and physical powers, thus reducing our reliance upon machines, which Fedorov deemed mere 'prostheses'. You wouldn't need an aeroplane if you could already fly like Superman, nor a telescope if your eyes were super-powered enough to zoom out and see events on distant planets – a direct echo of Rosicrucian thought.

However, all this was only a stopgap. A state of 'autotrophy' was the ultimate evolutionary goal we should aim for; namely, the way in which certain plants and bacteria manage to nourish themselves only from the sun and the air, thus rendering the consumption of our vegetable ancestors wholly unnecessary. Gradually, mankind must try and shed each unnecessary body part, whether super-powered or not, until ultimately we become a race of asexual psychic plants, living in 'rich mind-fields'. This strange fate, said Fedorov, was mankind's only option. Otherwise we would just go on and on eating and being eaten forever, endlessly being squeezed from our offspring's bowels and onto the fields to fertilise the very same crops which we too would one day become. Only by 'introducing will and reason into Nature' and attempting to control it could we hope to escape the material planetary prison in which we were now trapped.[11] Fedorov's thinking could best be summed up in the following phrase: 'There is no purposefulness in

Nature – it is for man to introduce it.'[12] And, with such a project, so death and Darwin would at last be conquered!

It's Not Rocket Science

Obviously, this is a cartoonish discussion of Fedorov's philosophy. In reality, he must have cut a most impressive figure, though more for his saintly and ascetic ways than anything else. Fedorov wasn't overly keen on the idea of property, including intellectual property, believing like some modern-day Internet evangelist that all knowledge should be free. Unlike most contemporary Silicon Valley hypocrites, however, for a long period he refused any salary whatsoever for his toils in Rumyantsev Library, slept on a wooden chest with a book for a pillow and his tattered, rag-like overcoat for a blanket, drank only tea and ate only stale cheese and hard bread rolls, leavened with the occasional onion (maybe he was scared one might be his granddad). Living in cheap rented rooms the size of closets, Fedorov objected to furniture so much that he flatly refused to own any; once, when a friend gave him a bed, Fedorov simply donated it to somebody else. His greatest fear was that one day he would be found dead with a small amount of money in his pocket, which he had failed to give away to the poor. He did his best to practise autotrophy himself, then; Tolstoy said he was 'proud to have lived at the same time as such a man', and the more spiritual, religious side of his philosophy also had an influence on the famed mystics P. D. Ouspensky (1878–1947) and Vladimir Solovyov (1853–1900).[13]

But nobody ever took his actual scientific ideas seriously, did they? Amazingly, they did. One of Fedorov's proposed supramoralist branches of science was termed 'astronautics', the refinement of astronomy into a science which would one day catapult man beyond our atmosphere onto the moon and beyond – and one man who was particularly captivated by this idea of Fedorov's was Konstantin Tsiolkovsky (1857–1935), the father of Soviet rocket science. In 1873, Tsiolkovsky, aged only sixteen, had made his way to Moscow from the provinces and turned up at Rumyantsev Library in search of enlightenment. As Fedorov recognised, this was a noble cause, and he was soon playing the role of guru to the teenage savant, who went on to pen a number of science-fiction fantasies in which the influence of his mentor was apparent. Sci-fi for Tsiolkovsky was a kind of literary laboratory in which future ideas for actual experiments and technological advancements could be explored quickly and cheaply. 'First the idea must be conceived, almost like a fantasy,' he once wrote, 'then, with scientific work and calculation, ultimately the idea is crowned.' Perhaps this was why many of his stories had comically utilitarian titles like *Living Beings in Space* and *Changes in Relative*

Weight and read more like lightly disguised essays than adventures. Inspired during his own youth by the tales of the French father of science-fiction Jules Verne (1828–1905), Tsiolkovsky viewed sci-fi as a kind of propaganda, brainwashing the general public into believing that space travel might one day be possible, and encouraging the young to pursue careers in rocketry and aeronautics. He filled his yarns with then impractical ideas which he hoped future scientists would be motivated into making come true; his notion of 'space-cottages', for instance, was the first time anyone had suggested the possibility of creating an Earth-orbiting space station. Following 1917's Bolshevik revolution, Tsiolkovsky also tried to ingratiate himself with Russia's new rulers by penning tales in which it was shown how Soviet-style command economies which deliberately devoted huge resources towards massive engineering projects were the best way for science to advance. He also created the memorable character of 'The Gravity-Hater', a man obsessed with the idea that gravity was holding mankind back, and that if we wanted to evolve any further we had better get on with conquering it and building Communist colonies in outer space.[14]

However, eager to fulfil his mentor's ambition of freeing mankind from its earthly jail, Tsiolkovsky was not satisfied simply to pen pulp-fiction, and set about formulating various mathematical equations which successfully laid out the physics needed to fire rockets into outer space, something which gave the Soviets a head start in the space race decades later. Specifically, Tsiolkovsky calculated the amount of fuel which would be necessary for a rocket to escape the Earth's gravitational pull, devised gyroscopes for rocket stabilisation, conducted early wind-tunnel research in his home lab, helped develop methods of using liquid fuel, and was one of the first to theorise about the possibility of multi-stage rockets. Even today, NASA scientists still have to make use of some of the equations Tsiolkovsky formulated when planning their missions, as did Russia's cosmonauts – including Yuri Gagarin (1934–68), the first man in space, who was actually a member of the same old aristocratic family as Fedorov himself (the illegitimate son of a Gagarin princeling) was![15]

For Fedorov, one of mankind's most holy scientific tasks was to transform from a 'horizontal' way of thinking to a 'vertical' one; instead of taking death lying down, we should make corpses stand up on their own two feet again, and rather than settle for Alexander Sukhovo-Kobylin's definition of a bicycle as a form of 'horizontal flight', as mentioned earlier, we should try and send spaceships up into the atmosphere, flying vertically. In particular, Fedorov was obsessed with a report he had read about the experiments of a group of American artillerymen who had shot their cannons not horizontally at their enemies, but vertically up at clouds to try and make it rain during droughts.[16] This example is particularly interesting, as it carries direct echoes of a pair of Jules Verne novels, 1865's *From the Earth to the Moon*, and its 1889 sequel, *The Earth Turned Upside-Down*. These both involved a gang of ex-US Civil War artillerymen who formed an illegal-sounding organisation called The Gun Club devoted to the creation of exciting new items of extreme weaponry. Initially, these men manufacture a giant cannon to shoot some of their number off to the moon in a special capsule, while in the sequel they nearly destroy the entire planet by building a gun so huge its blast when fired will knock the Earth onto a new axis, altering its climate for the better (or so they think).[17] Did Fedorov read Verne? Possibly. His disciple certainly did; according to Tsiolkovsky, 'the first seeds of the idea' of space travel were sown in his mind by Verne's sci-fi, tales which 'awakened my mind in this direction'.[18] Scientifically speaking, *From the Earth to the Moon* was one of the most influential novels ever written, with several of the early pioneers in rocketry citing it as a direct motivation for their subsequent work. According to the US sci-fi novelist Ray Bradbury (1920–2012), Verne planted within these men the distinctly Fedorovian idea that 'if we reach the stars, one day we will be immortal', writing that 'his name never stops. At aerospace or NASA gatherings, Verne is the verb that moves us to space ... Without Verne there is a strong possibility we would never have romanced ourselves to the moon.'[19]

An Eccentric Orbit

So who was really responsible for turning Tsiolkovsky on to the idea of rocketry, Verne or Fedorov? Some commentators have tried to downplay Fedorov's influence, with the younger man's Western biographer James Andrews arguing that, after he left Moscow in 1877, Tsiolkovsky barely mentioned Fedorov in his writings, and that he was thus not really his disciple at all.[20] Others, though, have argued the influence was strong. In his 1928 pamphlet *The Will of the Universe: Unknown Rational Powers*, for instance, Tsiolkovsky speculated openly about the possible future attainment of Fedorovian autotrophy:

> The conquest of the air will be followed by the conquest of ethereal space. Will not the creature of the air turn into a creature of the ether? These creatures will be born citizens of the ether, of pure sunshine ... Thus, there is no end to life, to reason and to perfection of mankind. Its progress is eternal. And if that is so, one cannot doubt the attainment of immortality.[21]

Or, at least, immortality for some – humans who refuse to evolve, along with grossly physical plants and animals, will simply die out, or perhaps even be intentionally exterminated, something of which the peaceable Fedorov himself might not have approved.[22] Uniquely, Tsiolkovsky even believed that the Second Law of Thermodynamics (which states that energy is always dissipating outwards through entropy) might one day prove reversible, thus allowing us to avoid the ultimate heat death of the universe; the energetic equivalent of Fedorov's ideas about reversing the entropy of ancestral dust.[23] Fedorov's ancestral dust also finds an echo in Tsiolkovsky's idea of *atom-dukh*, or 'atom spirit', the now familiar notion that each and every particle of matter in the universe was alive. As Tsiolkovsky put it, 'I am not only a materialist, but also a panpsychist ... Since everything that is matter can, under favourable circumstances, convert to an organic state, we can conditionally say that inorganic matter is (in embryo) potentially living.'[24]

In Tsiolkovsky's scheme of things, lower forms of life consisted of matter with very little spirit, and higher forms spirit with very little matter until, eventually, even these advanced life forms shed their gross material skins and achieved a state of disembodied oneness with the universe. In sci-fi tales like *Biology of Dwarfs and Giants*, he laid out details of how, through a combination of strict exercise regimes and space travel, humanity would begin to develop bizarre new bodies, as indeed would our animals. The most advanced beings on Earth during his own day, said Tsiolkovsky, were the scientists, who should be allowed to govern over all inferior men in a kind of One-World Government – a belief not entirely incompatible with the Communist ideal of global revolution. After this, however, mankind should aim at union not only among all men, but also with alien life forms and the wider universe itself. He believed not only in life on other planets, but also in ethereal life inhabiting cosmic gas clouds, and the depths of interstellar space. These disembodied aliens, thought Tsiolkovsky, were mind readers and telepaths, with superior figures like poets, scientists and inventors already gaining unknowing inspiration from them via psychic means. Maybe it was the space spirits who actually sent him his ideas about rocketry, then? Possibly they were inviting humanity to join them floating around in outer space, where we would further evolve? This was Tsiolkovsky's fervent hope. In his view, both these space beings and our own backwards species were ultimately made up of the same atom spirits, so eventually, if we only allowed ourselves to be unquestioningly led by scientific geniuses and great spiritual leaders, in whom matters of the mind already predominated over the grosser demands of the body, we would climb up the Great Chain of Being to arrive at a state of incorporeal godly perfection.[25] Furthermore, seeing as we were all made up of atom spirits, this was proof of the reality of a kind of impersonal, Ernst Haeckel-like 'atomic reincarnation' process:

> We always have lived and always will live, but each time in a new form and, it goes without saying, with no memory of what came before ... In dying one says farewell forever to one's circumstances. After all, they are in the brain, but the brain is decomposing. New circumstances will arise when the atom spirit finds itself in another brain.[26]

It's the Sun What Done It

Reading such words, it seems obvious that, through Konstantin Tsiolkovsky's life and work, the influence of Nikolai Fedorov lived on, his atom spirit being reincarnated again and again within the new brains of later Soviet scientists. For example, Tsiolkovsky was friends with the heliobiologist Alexander Chizhevsky (1897–1964), who studied the alleged influence of solar cycles upon the behaviour of mankind, professing to have discovered correlations between coronal mass-ejections of energy from the sun and an increased instance of violent events like wars and revolutions here on Earth. While he did make several genuine contributions to science, being nominated for a Nobel Prize and helping develop technology for air-purification devices, Chizhevsky's work in the field of sunspots did not win official Soviet favour, with Stalin having him arrested in 1942 and labelled as an 'Enemy Under the Mask of a Scientist'. It turned out his notion of great historical events being driven by factors other than Marx, Lenin and the 'historical inevitability' of dialectical materialism did not go down too well with the authorities.

A skilled chemist, Chizhevsky proposed that human beings were really just walking, talking chemical processes which would naturally be affected by cosmic rays of all kinds; any idea of us as separate, individual actors from the universe around us was thus entirely false. As he said, our blood flows not only within our own veins but within the 'veins of the cosmos'; a real echo of the old alchemical idea of man representing the point of union between microcosm and macrocosm. Seeing as sunspots and suchlike tended to come in cycles, Chizhevsky proposed that human affairs followed a similar pattern, with wars and revolutions also following corresponding sequences. To prove this he combed through history, drawing deeply dubious correlations between history's greatest events (or what he arbitrarily deemed to be such ...) and the cycles of the sun and planets, hoping to unite 'the science of matter' with 'the science of human culture' into a grand overarching Theory of Everything.[27] Chizhevsky's idea has since been revived by Western scientists, with implausible ideas about the geomagnetic storms which occur with greater frequency during peaks of sunspot activity allegedly scrambling people's brains and causing them to start wars being put forward by several fringe researchers.[28] However, the idea seems a blatant pseudoscience about on a par with Prince Charles' mad 2015 claim that the Syrian Civil War had been caused largely by climate-change.[29]

For a time, the teenage Chizhevsky lived and exchanged ideas with his mentor Tsiolkovsky in a small town in central Russia named Kaluga, now home to the Tsiolkovsky-Chizhevsky Museum for Space and Cosmobiology. Once known as 'the Kaluga Eccentric' because of his bizarre ways – he was a schoolteacher in the town who went around lecturing his students in the open air with big, home-made tin funnels sticking out of his ears to aid his hearing, and used to stand on his roof lecturing on astronomy and reading out his sci-fi stories to anyone who would listen – Tsiolkovsky is now very much the local hero.[30] Due to his humble background, undoubted genius and willingness to pretend that his theories were inherently Marxist in nature, Tsiolkovsky was acclaimed by Stalin as the 'founding father' of space travel, and dubbed 'Grandfather Space' by the USSR's next dictator, Nikita Khrushchev (1894–1971). Sputnik, the first space satellite, was coincidentally launched in 1957, the centenary of his birth, leading to his name and life story being exploited for propaganda purposes. Seeing as during the 1940s and 1950s the Soviets were eager to make false claims that everything from lightbulbs to radio were really Russian inventions, the existence of a genuine national genius was exploited for all it was worth, making Tsiolkovsky a domestic celebrity.[31] In the years since, Kaluga has become a place of pilgrimage for many Russians – and not only those who love their space science. Nowadays, the very word 'Kaluga' carries implications and associations similar to those possessed by the word 'Glastonbury' to an Englishman. However, there is a major difference between the two places. Glastonbury is a centre of esotericism which functions essentially as a science-free zone; in Kaluga, the terms 'science' and 'mysticism' are seen as complementary in nature, not oppositional.[32]

Die, Death!

During the Soviet era, various prominent and influential scientists of a Cosmist bent were given positions of power from which they surely would have been barred in the West due to the inherently bizarre nature of some of their ideas. For example, take Vasily Kuprevich (1897–1969), a biologist who for many years was the highly respected President of the Belorussian Academy of Sciences. Botany was Kuprevich's particular speciality, and he found himself impressed by the way that certain plants and trees, like the giant redwood, were able to live for over 2,000 years. This got Kuprevich thinking. Maybe, just like Fedorov had once proposed, human beings might one day come somehow to resemble such vegetation? Why should we, too, not live for thousands of years – perhaps even forever? Maybe there were some animals on Earth which had secretly already mastered this skill, he mused; just possibly, the Loch Ness Monster and the Yeti were not mere legends after all, but prehistoric animals which had somehow managed to keep on regenerating themselves again and again over millions of years.

Eventually, Kuprevich developed an idea that death itself was merely a 'historical phenomenon' that had not always existed, and need not go on existing forever. Instead, he proposed, death was only a temporary solution which had been designed by Nature to allow evolution to take place. If old animals did not die, he said, then they would have no need to reproduce, and with no long lines of offspring changing over huge amounts of time into new species, there would be no chance for life to evolve any complexity to it – we would still be stuck as mere amoebas. Now, though, mankind had evolved to the point where death was no longer really needed for evolution to occur. Marx, said Kuprevich, taught his disciples to question everything, so why should Soviet scientists not also question the necessity of death? In a piece of thinking straight out of Fedorov, Kuprevich proclaimed that 'death is against human nature' and speculated that pretty soon we would be able to manage our own artificial means of evolution, with scientists discovering how to renew our cells indefinitely, thus lending us all immortal youth. If the human liver and fingernails could keep on renewing themselves, then why could not whole bodies be made do the same? Kuprevich died in 1969, but perhaps one day, when his ancestral dust finally comes to be resurrected, he will be able to continue with his good work.[33]

Maybe Kuprevich was actually onto something; there are a select few species of animal in the world which genuinely don't seem to age over very long spans of time, such as the Blanding's Turtle (a North American species in which twenty-year-old and eighty-year-old specimens are basically indistinguishable), and nobody is sure precisely why age-related decay and death actually occur anyway. By studying such Methuselah-like creatures, it is not entirely impossible that some means of artificially extending the human life span might one day be found, though perhaps not to the extent envisaged by Kuprevich. Furthermore, some modern-day scientists are currently trying to stop, reverse or slow down aging by turning on and off genetic 'switches' which apparently help control the process, though their level of success is controversial – altering such genetic switches may stop cells from dying normally, but appears to have the side effect of also turning them cancerous. The dream of scientifically devised immortality, it seems, is one that will never die.[34]

Green Thinking

Fedorov's ideas were also echoed by an early environmental-type theoriser named Vladimir Vernadsky (1863–1945). In Russia, Vernadsky is thought to rank alongside Darwin and Newton as an all-time genius of science due to his theory of what he termed the 'noösphere' (from the Greek *nous*, meaning 'mind'). Essentially, Vernadsky thought that Earth had undergone three separate evolutionary stages. First of all was the 'geosphere', the period of initial planetary formation, when there was no life. Then followed the 'biosphere', that period from the emergence of the first life forms to now, when plants and animals helped shape the planetary environment. The third stage, which was just beginning, he termed the 'noösphere', a period when mankind would, via technological means, begin deliberately altering the face of the Earth and the life forms which inhabited it. Rather than viewing humanity as inherently separate from Earth, he viewed us as a sort of manifestation of it, with all life ultimately arising from inanimate matter, and returning back to it again after death. For example, calcium was originally an inert element found in rocks, but now also helps make up our bones; when we die, it will eventually find its way back into the rocks again. People, he said, were therefore merely 'one component in a cycle of physical and chemical interactions and transmutations', which together constituted our planet. Because of this, Vernadsky denied that the various sciences were truly separate from one another, coining terms like 'biogeochemistry' to describe new interfaces between previously distinct disciplines; you suspect August Strindberg would have wholeheartedly approved.

Thus, to view some human modification of the landscape like a reservoir or dam as being artificial was just a trick of perspective. Seeing as mankind was really a temporary living manifestation of the planet, such structures were simply aspects of the noösphere, which Vernadsky defined as 'thinking matter', or the storehouse of human knowledge. Claiming that all scientific data was 'noöspheric matter', he saw mankind's ability to alter the landscape through exploitation of this information as perfectly natural; just as trees might cover the land with forest as part of the biosphere, so mankind might cover the land with cities as part of the noösphere. To Vernadsky, humanity's scientific and social development – including its anticipated movement towards socialism – were less conscious decisions made by individual humans or civilisations, and more expressions of natural evolutionary laws which could hardly be avoided. The idea of human progress, both scientific and social, was thus embedded within the universe itself. Just as for Fedorov, the final result of the whole process, said Vernadsky, would be our deliberate technologically driven development of an autotrophic lifestyle, with us surviving on sunlight and air alone and living on a planet which had been placed under our complete control. As he put it, the development of autotrophy would be 'a crowning achievement of paleontological evolution, and will not be an act of free human will but a manifestation of a natural process'.[35]

Vernadsky didn't go so far as to rechristen our globe as 'Spaceship Earth', but the ultimate implications of his thinking were clear – that, eventually, our planet would become a kind of artificial (or *apparently* artificial) human construct, with Mother Nature herself being deliberately channelled to become how we want her to be, not how she naturally would be.[36]

This Is Planet Earth

This kind of thinking still finds its modern echo in Russia today in the writings of Svetlana Semenova, whom we met earlier directing her ire against 'eco-sophists' who protest about mankind's attempts to alter and improve his natural environment. Where modern environmental doom-mongers say that the melting of the ice caps will drown us all, some Cosmists made proposals to *deliberately* melt them in order to provide water supplies for drought-stricken areas of the USSR, and it is precisely this kind of 'can-do' attitude that Semenova finds so appealing about the movement.[37] As Fedorov said, blind Nature-worship was a sort of death wish for the species; we had to *change* Nature, not simply *accept* it.[38] Many modern-day Russian scientists would seem to agree. Grandiose Cosmist-style schemes to modify Nature did not come to an end with the collapse of Communism. Russian experiments in cloud-seeding to control the weather and plans to resurrect woolly mammoths from their frozen remains preserved within the Siberian permafrost are still reported in the media from time to time. Indeed, it would appear that Russia is the undisputed world leader in Fedorovian meteorurgy, with the government in Moscow altering the weather on a routine basis in order to guarantee clear skies for Putin's interminable military parades through Red Square.[39] During the 1990s, Russia even tried (and thankfully failed) to abolish night from its lands by virtue of a 20-metre-wide 'space-mirror' called Znamya-2 which was supposed to be angled so as to reflect sunlight down onto selected locations to transform darkness into light, a scheme which was truly Cosmist in its ambitions. The device actually worked, producing a 5-kilometre-wide bright spot which raced across the surface of Europe towards western Russia one night during February 1993. However, it then burned up in the atmosphere over Canada, and the project was ultimately abandoned.[40] Svetlana Semenova would surely have approved, though, had the plan come off.

Semenova is Cosmism's leading modern-day scholar and it is to her that we really owe the now generally accepted definition of

Cosmism as being a movement devoted above all to the notion of 'active evolution', and it is she who first comprehensively gathered together scientists and thinkers like Vernadsky and Kuprevich under the Cosmist umbrella. None of these people were direct followers of Fedorov, exactly – at least not in public – but they were clearly inspired by him, albeit often at second hand. They departed from his thought in various original ways, but several still took some initial inspiration from his work before pursuing their own individual direction – although others may never even have read him.[41] To a certain extent, then, the very idea we have of Cosmism today is a partial construct of Semenova's own creation; she is the movement's most eager promoter, running scholarly conferences, writing books, running a Fedorov museum and generally just spreading the Cosmist gospel. She has even helped create a kind of Cosmist calendar after the vague fashion of Auguste Comte, in which dates of spiritual and scientific significance are accorded the status of holy days; in April 2011, for instance, the two main red-letter days were Easter and the fiftieth anniversary of Yuri Gagarin being first man in space. As Cosmism's main serious historian in the West, George M. Young, put it in his excellent book *The Russian Cosmists* (from which I have adapted much of this information), these two holidays are directly related in Cosmist terms, being 'spiritual and scientific commemorations of human victory over gravity and death'.[42]

Thanks to Semenova, Cosmism is currently undergoing an upsurge in popularity across Russia, being embraced as a kind of cuddly, New-Age version of Communism without the gulags. According to Young, its long-term project of resurrecting the dead offers 'a more comprehensive thousand-year plan to replace the five-year plans that didn't work'.[43] Even Stalin, whose popularity is also making something of a contemporary comeback, is being retrospectively claimed as an accidental Fedorovian by modern Russian commentators. Such people say that society under Stalin was all directed towards one common task, just like Fedorov would have wanted – the creation of a technological 'Paradise' on Earth.[44] According to the Russian ultranationalist writer Alexander Prokhanov (b. 1938), for example, 'Communism … is the defeat of death. The whole pathos of Soviet futurology and Soviet technocratic thought was directed at creating an "elixir of immortality".'[45]

Such a statement, I suggest, would be simply incomprehensible to a reader without some kind of basic background knowledge of the Russian Cosmist movement and its main scientific adherents. It is hard, at first sight, to see ideologically motivated mass murder as the first step on the path towards eternal life – but, to some Russians, it was indeed so.

Would You Adam and Eve It?

As with the 'Great Instauration' of Francis Bacon, Fedorov's dreams were to a great degree a movement back towards an initial earthly Paradise, the main difference being that Fedorov gave Eden a specific geographical location – namely, Turkestan. Turkestan today, a vast desert wasteland lying between Russia and China, can be few people's idea of what Paradise might resemble. Fedorov, though, learned of local legends that the place was once highly fertile, and had in fact been the Garden of Eden; Adam's bones themselves were said to be buried somewhere beneath the region's mountains. Fedorov saw the desertification which had since occurred in Eden as a prime example of the disastrous consequences which followed whenever mankind abandoned his God-given duty to alter the natural environment for his own benefit, and proposed that, in symbolic defiance of this calamitous mindset, Turkestan should become the centre of Cosmist research and endeavour. Observatories should be built on the region's mountains so as to better track the whereabouts of lost ancestral dust in space, he said, and international science stations be established there with the aim of restoring fertility to Eden and life to the dead; including Adam, once his bones had been located.[46]

The later Bolshevik refashioning of this fantasy of returning to Eden, however, was a little darker than Fedorov might have liked. The Soviet novelist Andre Platonov (1899–1951), for example, believed that the Communist project could only really be understood in esoteric and Fedorovian terms, and in his *A Technical Novel* put his own enthusiastic youthful beliefs about the Party into the mouth of a commissar:

> We shall dig up all the dead, we'll find their boss Adam, set him on his feet and ask: 'Where did you come from, either God or Marx – tell me, old man! If he speaks the truth, we'll resurrect Eve.'[47]

People like Platonov belonged to a movement known as the 'God-builders', a number of Marxist intellectuals and soon-to-be Soviet officials whose main concern was devising plans for the development of a 'New Adam' fit to live in the 'New Eden' of the USSR. Many were active disciples of Fedorov, and argued strongly for the state to reshape the environment around them along Cosmist lines.[48] The God-builders recognised the sublimated religious impulses which lay behind the politics and science-worship of early Bolshevism, as inadvertently revealed by Lenin in a statement he made to Leonid Krasin in 1918:

> Electricity will take the place of God. Let the peasant pray to electricity; he's going to feel the power of the central authorities more than that of Heaven.[49]

Krasin, of course, was the man behind the failed plan to resurrect Lenin, but some of his fellow God-builders preferred the idea of mankind tasting eternity not simply in a personal sense, but through the immortality of the race; Anatoly Lunacharsky (1875–1933), for example, 'Commissar of Enlightenment' under Lenin, liked to think that 'God' was simply another term for 'the socialist humanity of the future', a God who was 'not yet born, but being built' within Russia's new technological Eden. A Theosophist who founded the Russian Committee for Psychical Research, Lunacharsky declared that the true goal of Communism was to cause mankind to mutate into a new species, to transform the 'human spirit' into an 'All-Spirit'.[50] The Russian neurologist and parapsychologist Vladimir Bekhterev (1857–1927) would have agreed, arguing that he had found scientific proof that the human soul 'lives on eternally as a particle of universal human creativity' which is endlessly reincarnated in future humans, just like Tsiolkovsky thought.[51]

Auguste Comte had once declared that 'there is only one Eternal Man, always learning',[52] by which he meant that all humanity was really one, and the advance of the race the most important thing of all; in Tsiolkovsky and Bekhterev's terms, Comte's Eternal Man was continually renewing itself, as each 'particle of universal human creativity' was reborn anew in the more advanced Soviet humans of the future. Tsiolkovsky's quest for interplanetary travel has been defined by his biographer James Andrews as 'trying to overcome death through conquering outer space', with mankind achieving immortality not in a personal sense, but through 'the ceaselessness of mankind as it conquered the outer limits of the universe'. As each individual man died, his atom spirit would later be reincarnated in a super-advanced space-Communist of the far-flung future.[53] In a way, this God-building idealism and unquestioning faith in progress all seems at first glance to be quite noble. The unfortunate (though logical) conclusion of the idea that Soviet man's true immortality was not individual but collective, however, was to devalue the worth of any individual Soviet life in and of itself.

A Living Death

Killing large numbers for the wider benefit of the Russian people as a whole was always considered more than acceptable to Soviet leaders. Indeed, to God-builders like Bekhterev you were really doing such people a favour by liquidating them, seeing as the energy particles which constituted their souls would eventually be reincarnated anew in a different, but much more advanced, fresh body living within a forthcoming Soviet utopia. Perhaps the best illustration of this philosophy in action was Stalin's notorious White Sea Canal project, in which around 125,000 political prisoners were made to dig a

140-mile canal in harsh and primitive conditions within the space of only twenty months. So many labourers – possibly as many as 25,000 – died that their corpses themselves were utilised as building materials; lacking iron, human bones were mixed with the concrete to reinforce it. Stalin's purpose in constructing the canal was not remotely esoteric; he just wanted a new canal. Various Soviet writers inclined towards God-building were given tours of the slave-labour project, however, and began layering strange Cosmist-like meanings upon it. For the novelist Maxim Gorky, whom we met earlier on praising famine as an excellent means of weeding out the weakest elements from society, the White Sea Canal scheme was a form of practical alchemy, being carried out upon the human body. Through their toil, the labourers were to 'reforge' themselves, being transmuted from the base metal of enemies of the state and into the pure gold of useful servants of Communism. As such, the wretched prisoners should really have been grateful for their treatment, as they were helping make their race immortal by dying themselves. Gorky wrote a book about the topic, *The Canal Named Stalin*, which contained the phrase 'In changing Nature, man changes himself' – precisely the same logic as lay behind Lysenkoism. The White Sea Canal itself, incidentally, ended up being barely used, but at least lots and lots of people died making it, so it wasn't a complete waste of time and effort.[54]

During their first four years in office, the Bolsheviks killed more people than the Russian tsars had done during their entire three centuries of power, often through methods which would make even Islamic State think twice. Nobody knows how many died in Russia between the revolution of 1917 and the Nazi invasion of 1941; some say twenty million, others sixty.[55] Of course, most of this was due to simple cruelty and a paranoid desire to eliminate all potential enemies. Another analysis, though, as espoused by this book's favourite philosopher John Gray, holds that 'Bolshevism was a variant of Gnosticism', an ancient mystical view of life which deemed the material world of the Earth to be a deeply imperfect realm, and a 'prison of souls'. Traditionally, the best way for an individual Gnostic to flee the fleshly prison of his own Earth-bound body was through spiritual exercises and contemplation, allowing his soul to escape into the realms of pure spirit. The Soviets, though, were fanatical materialists who dynamited churches and executed priests, so their type of Gnosticism was much more materialist in nature. As Gray put it, 'materialism in practice meant the dematerialisation of the physical world' for the Bolsheviks; 'unnumbered humans had to die, so that a new humanity could be free of death'.[56]

This was where Fedorov's fantasies of autotrophy finally led; the liberation of man from the prison of his own body through the direct intervention of a bullet. Consider the case of Maxim Gorky, whose commitment towards man as a purely materialist phenomenon ironically led on to his prediction of a purely immaterial future for the human race. His views about the human mind being essentially electrical in nature, rather than spiritual, might have found favour with Nikola Tesla:

> Every year more and more thought-energy accumulates in the world, and I am convinced that this energy – which, while possibly related to light or electricity, has its own unique inherent qualities – will one day be able to effect things we cannot even imagine today.[57]

Gorky's materialist philosophy, though, would then inevitably lead on to mankind's complete escape from materiality and away into a kind of unadulterated noösphere:

> Personally, I prefer to imagine man as a machine, which transmutes in itself so-called 'dead matter' into a psychical energy, and will, in some far-away future, transform the whole world into a purely psychical one ... At that time nothing will exist except thought. Everything will disappear, being transmuted into pure thought, which will alone exist, incarnating the entire mind of humanity.[58]

Stalin had spoken of writers as 'engineers of souls', a paradoxical description which could certainly have applied to Gorky (to whom the phrase is often erroneously attributed).[59] Treating something immaterial as if it is a material engineering project could only ever end both badly and bizarrely, however; and you do have to ask, if such Fedorovian dreams of autotrophy had ever been achieved, then would the disembodied Bolshevik mind-souls inhabiting space really have been human at all? As John Gray once perceptively put it, the search for immortality is really 'a programme for human extinction' as humans, by definition, are *not* immortal![60]

Captain Planet

It would be easy for us to mock the Russian Cosmists, but this would be to ignore the fact that several strains of Cosmist-like thought have been present in Western science and politics down the years, too. For example, Vernadsky's noösphere has been recently recast in negative terms as the 'Anthropocene Age', an idea which proposes that, since the start of the Industrial Revolution, mankind has been altering its environment to such an extent that, during the next century or two, our resources will be exhausted, our flora and fauna become extinct, and climate-change eventually render everyone's favourite oblate spheroid virtually uninhabitable. James Lovelock, meanwhile, has also effectively reformulated Vernadsky's ideas in his famous 'Gaia Hypothesis', which, as we saw earlier, posited that the Earth and its inhabitants could be usefully conceived as one gigantic living 'animal', or self-regulating system – but a system which, thanks to mankind's foolishness, was about to be thrown disastrously out of balance. Appropriately enough for our environmentally conscious age, even Fedorov's phrase 'Spaceship Earth' has since been accidentally recycled, with the American architect, inventor and ecologist R. Buckminster Fuller (1895–1983) also using the term in an attempt to convince mankind of the need to lead a more sustainable lifestyle. Fuller's 1968 book *Operating Manual for Spaceship Earth* sounds like exactly the kind of thing which Fedorov would like to have seen stocked in his Rumyantsev Library.[61] The Russian Cosmist movement is exceedingly obscure in the West, and I would not claim that any of these theories were directly inspired by it, but the very fact of their occurrence does speak of a certain unrecognised parallel existing between Russian and Western modes of scientific thought.

Or is it really spiritual thought in 'scientific' disguise? At first sight, the notion of the Anthropocene is simply a direct pessimistic reworking of Vernadsky's noösphere – but, upon closer examination, there is a get-out clause. Through a programme of extreme self-denial of which Fedorov himself would surely have been proud, giving up our addiction to fossil fuels and excessive consumption, we can all still save ourselves, we are constantly being told. Rather than cannibalising our Earth's resources, we can instead become practitioners of technological autotrophy – imbibing energy from the sun and wind to power our lifestyles. If Spaceship Earth is currently in severe danger of breakdown, then at least it can, through such means, potentially be repaired, our current mindset goes. Perhaps we in the modern West are actually a little more like the Russian Cosmists than we would care to realise. The myth of eternal progress is not simply a possession of naive Communists – at the present moment it would appear to be the common deluded property of all mankind.

Another parallel to the thinking of the Cosmists can also be found in the work of Matt Ridley. Ridley, the climate-change lukewarmer whom we met earlier, and author of the tellingly titled tome *The Rational Optimist*, is in the habit of making various well-argued points to the effect that mankind's reliance upon fossil fuels has been, so far, an overwhelmingly good thing, and will continue to be so for quite some time. Citing the work of such researchers as Alex Epstein, author of what Ridley calls 'a bravely unfashionable book', *The Moral Case for Fossil Fuels*, Ridley quite happily talks in the newspapers about how fossils fuels have, in his opinion, 'contributed so dramatically to the world's prosperity and progress'. In purely environmental terms, he says, the use of coal instead of wood for fuel brought to a halt the deforestation of Europe and North America (until 'environmentally conscious' idiots recently began burning woodchips again, at least), the use of oil prevented the mass slaughter of whales and seals for their oily blubber, and the use of fertilisers manufactured using gas halved the amount of land needed to grow a given amount of crops on, thus allowing more people to be fed while also leaving more land free to return back to Nature. Meanwhile, the vast energy output derived from coal and gas since the Industrial Revolution is what has allowed us Westerners to develop our current technologically advanced lifestyles, he explains, increasing productivity, health, wealth and leisure time. As he says, 'to throw away these immense economic, environmental and moral benefits, you would have to have a very good reason'. Therefore, instead of kicking our addiction to fossil fuels to save humanity from an apocalypse which might not come, he recommends an insurance policy of directly altering our planet's biosphere through other means instead, such as 'fertilising oceans with nutrients that stimulate the growth of carbon-absorbing phytoplankton'.[62] Ridley probably doesn't know it, but such a proposed solution, being nothing less than meteorurgy in action, is deeply Fedorovian in nature!

Tesla's Trigger

Many people discussed throughout this book have had a distinctly Fedorovian aspect to their philosophies, from Francis Bacon to Wilhelm Reich. Perhaps unsurprisingly, Nikola Tesla was another one of our heroes to have been full to bursting with schemes that the Russian Cosmists would unquestionably have applauded. For example, he too wanted to control the weather, using something he termed 'Nature's Trigger'. Thinking that lightning flashes had the ability to activate rainfall, seeing as such phenomena often seemed to immediately precede downpours, Tesla formulated the plan of firing huge bolts of electricity into the sky in order to cause cloudbursts, thereby alleviating droughts, creating artificial lakes and reservoirs, and producing huge, never-ending torrents of water to drive hydroelectric perpetual motion generators. Sadly, however, Tesla's amazing eyes had for once deceived him; during a storm, rain actually comes first before any lightning strike appears, but raindrops take longer to reach the ground than an electricity bolt does, thereby giving the wrong impression to any observer on the ground. Nonetheless, Tesla continued to maintain that such a feat was possible, going on to hypothesise that weather control might already have been mastered by his imagined inhabitants of Mars.[63] He even speculated about the possibility of mankind one day developing an autotrophic, food-less existence, absorbing energy from an ethereal 'ambient medium' instead of through our gullets – highly appropriate for one with such a dainty diet. According to Tesla, 'there seems to be no philosophical necessity for food', something he thought was proven by the existence of crystals ('living beings', in his view). Arguing there was a great 'probability' of purely 'gaseous' beings existing on other planets, he proposed that such creatures might already be living among us here on Earth in secret, with us primitive humans being unable to perceive them due to their wholly incorporeal nature.[64]

Tesla also anticipated Russia's launch of the Znamya-2 space mirror by nearly a century, in 1914 conceiving the idea of flooding the atmosphere with electricity, thus transforming the sky into 'a giant lamp' and banishing the gloom of night time (and the beauty of stars and moonlight ...) forever.[65] Fedorov's long-held desire to abolish the pornocracy was also reflected in Tesla's longing for mankind to evolve into a race of sexless 'princess bees', with women running everything, laying eggs and consigning men to biological extinction. Only in such a way, said Tesla, could the 'exaggerated and perverse manifestations' of sexual desire which he alone among mankind had managed to escape, be banished from our fallen world forever. The 'perfect civilisation of the bee', filled with 'vast desexualised armies of workers', was, he said, one we should all desire to inhabit.[66] Most Fedorovian of all were these words of the great genius, taken from an unpublished article of his called *Man's Greatest Achievement*. If mankind could fully control Nature through science by one day gaining mastery over the primal substance or *protylē*, Tesla said, then

> he would have powers almost unlimited and supernatural. At his command, with but a slight effort on his part, old worlds would disappear and new ones of his planning would spring into being. He could fix, solidify and preserve the ethereal shapes of his imagining, the fleeting visions of his dreams. He could express all the creations of his mind on any scale, in forms concrete and imperishable. He could alter the size of this planet, control its seasons, guide it along any path he might choose through the depths of the universe. He could cause planets to collide and produce his suns and stars, his heat and light. He could originate and develop life in all its infinite forms. To create and to annihilate material substance, cause it to aggregate in forms according to his desire, would be the supreme manifestation of the power of Man's mind, his most complete triumph over the physical world, his crowning achievement, which would place him beside his Creator, make him fulfil his ultimate destiny.[67]

In other words, through science, man would effectively become one with God – the great desire of the Cosmists, Rosicrucians and alchemists alike. Tesla's biographer John J. O'Neill accuses his subject of having a 'Superman' complex.[68] If so, then he is not the only scientist to have suffered from such a pleasant delusion of hubris – remember these words of Tesla's in a few pages' time when we come to talk about something called 'foglets'.

Lawson's Laws

Another hitherto unacknowledged honorary Cosmist was none other than Alfred W. Lawson, who also conceived an independent idea of Spaceship Earth, albeit a profoundly insane one. For Lawson, the world's menorgs – those weird little conscious subatomic particles he spoke of – were constantly trying to encourage mankind to evolve, expanding our consciousness so as to come into contact with another race of menorgs which he claimed lived within the bowels of our shared planet. When this was finally achieved, Lawson said, mankind would develop a kind of collective planetary mind and become 'the Earth's captain and pilot'. By controlling the digestive gases lurking inside our globe, redirecting its suction and pressure currents, humanity could send it wherever we wished to go – the Earth would travel through space by the force of its own colossal farting.[69] He wrote:

> The interior of the world is populated by living beings [menorgs] who operate the machinery [digestive system] that draws sustenance into and then passes off the waste gases that would cause the Earth to explode if kept within it ... It will not take long, however, for an understanding to exist between them [and mankind] for the purpose of arranging a mutual working plan for the guidance of Earth's destinies.[70]

If we remember rightly, Lawson did not believe in energy. How, then, to account for new developments like the atomic bombs which were dropped over Japan in 1945? Simple. Menorgs – which were really what deluded nuclear physicists called 'atoms' – were also full of digestive gases. They ate, broke wind, and defecated, just like human beings did. Therefore, when scientists managed to split the atom, all they were really doing was forcing a menorg to squeeze out a massive poo or do a giant fart. Such an emission 'could produce enough force to demolish a mountain', apparently – so just imagine how powerfully all the Earth's menorgs breaking wind together out of our planet's secret South Pole bum-hole could propel us through the depths of interstellar space.[71] A cosmic symphony indeed!

The incurably utopian Lawson was also full of hot air, as could be seen in his 1904 novel *Born Again*, in which a sailor named John Convert comes across an undiscovered island, Sageland, where a super-evolved form of humanity known as the Sagemen have been living for millennia. According to Lawson 'many people' thought *Born Again* was 'the greatest novel ever written', though a quick glance at the plot might suggest otherwise. Sageland's inhabitants were so super-evolved they were seven feet tall, had prehensile toes and no teeth, were telepathic and could project their astral bodies. A Sagewoman named Arletta explains to Convert that he and his kind are merely ape-men who have failed to fully evolve, dismissing twentieth-century society as a form of primitive, organised selfishness. In an uncanny echo of Fedorov, Arletta tells him that any society which eats animal flesh is really a society of cannibals, due to an early version of Lawson's Law of Manoeuvrability, which held that human souls could easily be reincarnated within animals depending on their place upon the karmic wheel. The history of Sageland, however, was 'simply a record of Heaven on Earth', a place where the state owned and controlled all necessary enterprises, and in which each individual was assigned the particular job to which they were most suited. If anyone refused to do their share of work, they were put in a mental asylum – asylums which were now 'obsolete for want of inmates'. There was no crime, money, nor taxes in Sageland, and all citizens lived happy and healthy lives; indeed, to be unhealthy was a criminal offence! Occasionally a super-evolved Sageman appears among the ape-men and tries to tell them what they need to do to evolve, but, like Christ, he is generally crucified for his efforts. Now that he has been educated in this way, Arletta tells Convert (who is clearly Lawson himself) that he too is an honorary Sageman, and must return back to his so-called 'civilisation' and spread the message about the need for spiritual reform. Once this was done, ape-men would evolve into Sagemen and develop the ability to fly, run faster than speeding trains, and jump over tall buildings in a single bound.[72]

Lawson's biographer Lyell D. Henry Jr called his subject's creed of Lawsonomy the belief that 'humanity could eventually share in God's power through technology', and a 'modern gospel speaking to those who, finding that older faiths no longer seem relevant or adequate in a scientific age, confront a spiritual void'. It promised that, via a combination of advanced science and a kind of humanist spirituality in which man (in the guise of Lawson) is worshipped, we will all achieve salvation.[73] This became literally true when, in 1948, Lawson founded his Chapel of Lawsonomy, a bizarre church in which an impersonal God was worshipped by proxy through praise of the one person who had finally discovered his true laws – Alfred W. Lawson. Here, worshippers were read extracts from *Lawson's Mighty Sermons*, all of which ended with the phrase 'THUS SAYETH ALFRED LAWSON', and sang hymns like 'Lawsonomy Will Envelop the Earth' and 'We Give Thanks for Lawsonomy' in his honour.[74]

Declaring himself the 'First Knowledgian' and 'Wizard of Reason', in 1942 Lawson bought some abandoned university buildings in Iowa, and founded his University of Lawsonomy. Students signed up to spend *thirty years* studying for a degree, and to live in a weird, moneyless utopia. The basic aim was not only to study Lawson's many books, but to memorise them wholesale; texts on any other subject were banned. As a government inspector once accurately observed, the place was 'a university in name only' and 'a colony ... for the purpose of eulogising Alfred Lawson'. Disciples – sorry, students – had to renounce tobacco, alcohol, cosmetics, dancing, gambling and meat, and eat strange salads sprinkled with grass clippings from the college grounds. To maintain the perfect health necessary to evolve into Sagemen, students were encouraged to sleep nude, chew each mouthful of food fifty times before swallowing, and dunk their heads in cold water twice a day before repeatedly opening and closing their eyes to keep alert.[75] Lawson also instituted peculiar scientific research programmes in Iowa, including plans to create a 'solar coagulator', which would turn sunlight into glowing paint. Then, interior walls could be coated with it, producing free lighting without electricity, and allowing us all to live beneath the Earth's crust (should we want to ...). Other incredibly unrealistic engineering projects included wheel-less cars, smoke evaporators to bury pollution underground, and a 'solar engulferator', a weapon so fearsome it would eliminate war forever. Declaring the central purpose of his university was to 'lead the world in future development of machinery at present unknown to man', Lawson even proposed to build a solar-powered spaceship, and in 1948 contacted the US military requesting they send him 'atomic machinery' and radar systems with which to experiment! They did not comply.[76]

Lawson also directed his genius towards agriculture. His 1944 book *Gardening* was a demented call for mankind to stop fertilising crops with manure on the grounds that it was both 'filthy' and unnecessary; if plants really needed manure, Lawson asked, then how come they were able to grow successfully long before animals' bums were even invented by the menorgs? Because animals were so busy pooing all over the Earth twenty-four hours a day, said Lawson, mankind needed to take the matter in hand and eliminate them to clean our planet up. If we don't need their faeces for fertiliser, and eating them is soul-cannibalism, why not just exterminate them? In any case, eating greens instead of animal flesh will make mankind evolve into Sagemen more quickly, seeing as 'animals that subsist entirely upon the flesh of other animals are squatty by nature and grow upon the horizontal plan, while animals that subsist exclusively upon the substances emanating from plants are more upright and intelligent by nature and are built upon the perpendicular plan'.[77] Nikolai Fedorov wasn't the only man to have desired a transition from a horizontal way of life to a more vertical one, then.

Lighter than Air

Lawson's ideas about human evolution were particularly Cosmist. Essentially a Lamarckian, Lawson taught it was time to plant a 'seed' within society, from which a new humanity would then emerge, leaving the disgusting 'pig-bipeds' of the present far behind. The best way to do this would be to create a 'self-perpetuating social body' from which the new, 'unselfish animal' would arise. That perfect community was Lawson's university, where he decreed that nobody should marry and have children until the age of thirty when, after spending years studying Lawsonomy, this philosophy of the future would end up being transmitted towards any newborn babies in their very genes, much as his knowledge of physics had been implanted within him by the sperm of his perpetual motion-dabbling father. Lawson's cult, then, was the start of a giant Lysenkoist eugenics programme.[78] Gradually, the human race would then evolve to resemble the menorgs which made them up, joining together into one much larger whole in which every individual unit helped form a true unity. Even better, through Lawsonian science it might be possible one day to build a 'huge cosmic animal' made up of trillions of individual humans who, positioned correctly throughout the cosmos and communicating instantly via telepathy, would constitute a 'conscious workable machine' just like the individual human body was to the tiny menorgs – microcosm would at last be fully and harmoniously joined with macrocosm.[79]

Another obvious strain of Cosmist-like thought in Lawson's teaching came from his early years in aviation. Prior to launching the first airliner, Lawson had edited aeroplane magazines which he filled with bizarre prophecies about a future in which not only would planes have no wings and fly using 'certain natural forces ... as yet unknown', but where mankind would have evolved into an incorporeal 'bunch of intelligence without eyes, nose, ears, mouth, legs, or body', living via autotrophy. (In *Born Again*, Lawson had already prophesied Sagemen would lose their teeth due to eating all food in 'gaseous form'.[80]) By the year 3000, meanwhile, it would have been discovered that existing at high altitudes had the power to cure 'almost any human ailment', causing giant 'floating aerial hospitals' to be built.

Eventually, mankind would develop the power to (as Lawson put it) 'GET OFF THE EARTH', leading to us evolving ever further in space. Flying high in the sky would also liberate man from the strong gravitational pull of the Earth's surface, making his mind free to soar away towards new mental heights.[81] Ultimately, this would lead to the development of two separate breeds of humanity: the alti-men and the crab-men. By the year 10,000, Lawson said:

> Two distinct types of human being will inhabit this Earthly sphere – the alti-man and the ground-man [or crab-man]. The alti-man will be born and live his whole life at the very top of the atmosphere and will never go below a certain depth while the ground-man will live upon the crust of the Earth at the bottom of the atmospheric sea like a crab or an oyster, and will never go above a certain height.

The job of the crab-men – who would still be 'infinitely superior in intelligence to our present people' – would be to become the subordinate but willing slaves of the alti-men, who would use their position above the clouds to gain a fuller perspective upon life, and thus become capable of acts of Fedorov-like meteorurgy:

> He will cultivate the air in such a way that a distribution of the elements will be arranged to infinitely greater advantage than the present primitive method that Nature has adopted and spasmodic rain and snow storms will be done away with entirely. He will be as successful in air-culture as he has been and will be in horticulture.

Furthermore, the longer alti-man lived way up high, the more and more ethereal he would become; he would shed his need for artificial oxygen and invent 'a machine for turning all solids into gases'. By interacting with a purely gaseous atmospheric world, said Lawson, 'all danger of refuse falling upon the heads of the human crabs below will be eliminated'.[82] It seems that, during the first twenty years of the last century, many people were imagining the huge benefits to the human condition, from evolutionary advance to universal brotherhood, which would arise from the new age of flight.[83] Neither Lawson nor Fedorov were alone in dreaming that once freed from the bounds of gravity, mankind would inevitably float on up, up and away into the Paradise of scientific Heaven.

The Space Race

The key Fedorovian notion that mankind would evolve into a new immortal super-race in outer space, and that space travel was thus a fundamentally holy and sacred task, has also been echoed by numerous Western men of science. It is not greatly recognised, but a large number of the early pioneers of rocketry and spaceflight were, essentially, religious fundamentalists, or at least many of those who worked at NASA were. Some of them even helped develop satellites with the specific end in mind of using them to spread the Gospel. Several astronauts, like *Apollo 14*'s Edgar Mitchell (1930–2016) – who conducted his own personal experiments in telepathy when walking on the moon – took Bibles with them into outer space, and left them on the lunar surface. *Apollo 15* moonwalker Jim Irwin (1930–91), meanwhile, returned from our satellite only to lead expeditions to Mount Ararat in search of Noah's Ark, and later set up his own Baptist ministry 'in order to tell all men everywhere that God is alive, not only on Earth but also on the moon'.[84]

David F. Noble's 1999 book *The Religion of Technology* contains a very detailed chapter laying out the religious beliefs of prominent NASA employees and astronauts, and it makes for some most unexpected reading. For example, we may recall the Rosicrucians' love of the prophecy from the Book of Daniel reading 'many shall run to and fro, and knowledge shall be increased' come the end times, which was interpreted as meaning that, the more science and exploration progressed, the closer we were to Doomsday and the subsequent resurrection of the righteous in a state of Heaven on Earth. For Jerry Klumas, a long-time NASA systems engineer and founder of a religious institution called 'the NASA Church of the Nazarene', by helping bring about the era of space travel he was simultaneously helping bring about the this prophecy's fulfilment, too. Even stranger, Klumas speculated that the way in which Einstein's Law of Relativity worked meant that, as speed-of-light space travel came ever closer, mankind grew nearer to the brink of achieving immortality. To simplify, the faster you travel, the slower time goes and the slower you are meant to age; this phenomenon is known as 'time dilation'. We are all familiar with cartoons in which an astronaut zooms across space for a few years at light speed then returns to Earth to find half a century has passed for us ordinary mortals and that his son is now older than he is. By such means, theorised Klumas, astronauts might one day be made to go so fast in their spaceships that the process of aging stops altogether, an idea which Fedorov and Tsiolkovsky would surely have loved.[85]

The NASA aerospace engineer Tom Henderson, meanwhile, who helped train the early American astronauts for their missions, was also an evangelical Christian who claimed that 'science as a whole points to God', and that its onwards march helped mankind 'climb the hill of knowledge' back to that perfect state of all-pervading wisdom Adam had once enjoyed in Eden. Henderson's own personal vision of his fate at the end of time sounds like something straight out of Fedorov, but with autotrophic salvation achieved through Christ, not technology:

> When Christ returns to rule for a thousand years, the Earth will return to its pre-Flood state [i.e. Eden] ... Either when I die or when the Rapture of the Church occurs ... I will return to Earth with Christ; with a new immortal body I will live on Earth but not as a man; I will be able to travel in space without a spaceship; I will meet with Robert Boyle and Isaac Newton.[86]

And with Nikolai Fedorov too, probably. Even the leading light of NASA's early days, the Nazi rocket scientist Wernher von Braun (1912–77), who developed the V-2 missiles which blitzed London during the final months of the Second World War, became a born-again Christian after being taken over to America to help them win the space race – a rather unlikely development for a man who had previously been happy to make use of Jewish slave labour and rain down death upon innocent civilians from above. According to von Braun, space flight was necessary so that the word of God could be spread to other planets, with astronauts being remade as new Adams, sent out to populate new Edens all across the universe, thus assuring the race's immortality in a way which would have been highly familiar to Konstantin Tsiolkovsky.[87] According to von Braun, 'if man is Alpha and Omega, then it is profoundly important for religious reasons that he travel to other worlds, other galaxies; for it may be man's destiny to ensure immortality, not only of his race but even of the life spark itself'.[88]

The idea of man's ascent into space being sacred was also amusingly reflected in the American evangelist Emil Gaverluk's 1974 book *Did Genesis Man Conquer Space?*, an exercise in combing through the Bible and interpreting its stories as references to space travel and super-advanced technology. Inadvertently sounding all Rosicrucian, Gaverluk claimed that the famously long lives of Biblical figures like the 969-year-old Methuselah allowed them to develop laser beams, telecommunication devices, space travel and nuclear power (although 'some perhaps were killed as they playfully brought together the right combination of uranium'). The most advanced man in Gaverluk's view was the first man of all, Adam, who 'had a computer-like mind ... with computer memory-circuits unaffected by decay, breakdown, aging, deterioration, [or] death'. After the Fall, however – when 'a perfect brain ... made a wrong decision' – mankind slowly began to descend to the low level it is at today. Come the Rapture, though, says Gaverluk, the righteous portion of mankind will be resurrected in outer space, crowned as God's chosen people in a galaxy far, far away, and then be returned to Earth in 'the biggest spaceship' made of gold and crystal, where he will 'evolve into a superman'.[89] Another accidental Fedorov!

Your Body or Your Life

Fedorov's notion of mankind's future state of autotrophy has also been inadvertently echoed several times in the West. For St Augustine (AD 345–AD 430), for example, come the restoration of Paradise on Earth, the Chosen will develop 'a body utterly subject to our spirit and one so kept alive by spirit that there will be no need of any other food'.[90] A more scientifically coloured variant of this same myth was advanced by the British pioneer of x-ray crystallography J. D. Bernal (1901–71), in his 1929 book *The World, the Flesh and the Devil*. Here, Bernal outlined a hubristic fantasy in which man would first of all conquer death, by implanting his brain in mechanical bodies, and then conquer space, his new immortal metallic frame allowing him to do so. Seeing as it is only 'the brain that counts', said Bernal, 'sooner or later man will be forced to decide whether to abandon his body or his life'. Because 'normal man is an evolutionary dead-end', we will have no choice but to take control of our own evolution – until our individual human consciousness itself may 'end or vanish in a humanity that has become completely etherealised, losing the close-knit organism, becoming masses of atoms in space communicating by radiation, and ultimately perhaps resolving itself into light'.[91]

Bernal did allow that some 'reactionaries', trapped in 'primitive obscurantism', might object to relinquishing their bodies and becoming Cybermen and thence pure light, but proposed that these idiots could be left behind on Earth to wallow in their own ignorance, while the 'aristocracy' of disembodied scientists travelled off to populate space. 'The scientists would emerge as a new species,' he said, 'and leave humanity behind.' Bernal did admit that even some scientists might find this hard to do at first, but that ultimately, through an act of sheer willpower, they would manage to conquer their base bodily desires like a band of medieval ascetic monks, and float off towards the stars where eternal happiness awaited ... although if happiness it truly was, then how would they be able to feel it?[92]

Such petty considerations are apparently beneath a new breed of technological evangelists known as the 'Transhumanist' movement. Transhumanists are a growing band of idealists who wish to improve mankind's lot, first of all by splicing our bodies with advanced technology to give us super-powers and perfect health, and eventually by uploading our minds into all-powerful computers to lend us both immortality and a state of silicon-based autotrophy. Their slogan is 'h+', the 'h' standing for 'humanity', although I prefer to think it signifies 'hubris'. Probably Transhumanism's most prominent current prophet is the American computer scientist and inventor Ray Kurzweil (b. 1948), who predicts that sometime around 2045 computers will become more intelligent than their human creators, an event he terms the 'Singularity'. At this point, allegedly, we will be able to fuse our minds with these super-machines, and thus ascend to the next evolutionary level. 'When scientists become a million times more intelligent and operate a million times faster, an hour would result in a century's progress,' he says, which sounds exhausting. Ultimately, claims Kurzweil, humans will cease inhabiting their current fleshly bodies and acquire new ones made up from the interactions between millions of tiny nano-engineered robots called 'foglets' which sound uncannily like a contemporary version of Alfred Lawson's menorgs. According to Kurzweil, these foglet bodies will enable us to shift shape at will, and to increasingly inhabit virtual-reality environments where we will be able to live forever, doing whatever it is we want to do. Eventually, he forecasts, our disembodied virtual-reality minds will continue to expand until the point where 'the entire universe will become saturated with our intelligence', something Kurzweil considers to be the 'destiny' of Creation itself. In anticipation of such future wonders, Kurzweil has gone to extreme lengths to ensure he too will be one of that lucky breed who will live forever; not only does he follow a very extreme health regime of the sort which would make most people wish they were already dead, he has also arranged for his corpse to be cryogenically frozen should he not quite manage to live to see the Singularity for which he so longs.[93]

How likely is it that any of this will actually happen, though? Opinions vary, but we must remember that many of the things Kurzweil blithely talks about as future inventions simply do not exist. Take the case of foglets. These were first 'invented' (read: imagined) by the American nanotech pioneer John Storrs Hall as a solution for the alleged problem of having to wear a seatbelt. Instead of using such a restrictive device, he thought, would it not be better to simply seed the air in front of any car passenger with an invisible 'fog' made up from millions of microscopic robots with extended arms which would simply lock into position as a makeshift restraining barrier whenever they felt a collision occur? Yes, thought Hall, that would indeed be marvellous – but why stop there? Once manufactured, these tiny seatbelt robots could also be used to fill the entire atmosphere of the planet. Then, whenever you wanted anything – anything at all, from an umbrella to a skyscraper – you could just click your fingers and tell the nearby foglets to link arms and assume that shape. They could also pick you up and fly you to wherever you wanted to go, Hall proposed, all of which makes them sound less like robots and more like genies of the lamp.[94] Never before has the late sci-fi writer Arthur C. Clarke's (1917–2008) oft-repeated opinion that 'any sufficiently advanced technology is indistinguishable from magic'[95] seemed so appropriate – and yet Ray Kurzweil seems sincerely to believe in all this sorcery.

So, we might recall, did Nikola Tesla. So do many others. Kurzweil has many followers, particularly among Silicon Valley types, and is often treated as if he is a genuinely original thinker, although if you have been paying any attention at all throughout this book then you will quickly realise that he is not. He is a Gnostic, a Fedorovian, a Rosicrucian, a Cosmist and a Lawsonian – or, to put it another way, he believes in universal progress and mankind's ability to transcend his own fallen nature. It might not come naturally to associate computing with the religious impulse but, as we shall see, this unexpected nexus was there right from the discipline's very beginnings.

Gods in the Machine

All computing is based ultimately upon things called 'Boolean Algebra' and 'Boolean Logic-Gates', concepts devised by the English mathematician George Boole (1815–64). Simply put, the way that all machine code is ultimately reducible down to strings of zeros and ones has its basis in the work of Boole. Boole's big insight came to him one day in 1833 when, walking through a field, he suddenly underwent what he called a 'mystical' experience, in which it was revealed to him that the logical rules of man's patterns of thought could (in his view, anyway) ultimately be reduced down to algebraic formulas. An exceedingly religious man, who had once aimed to join the clergy, Boole was convinced that mankind's patterns of thought were modelled after those of God, and that to explain the one was thus to help explain the other.[96]

Of course Boole, being a Victorian, didn't expect or intend his algebraic formulas to be used to drive computer-programs – how could he? However, in the 1940s the American electrical engineer Claude Shannon (1916–2001), working within the labs of the Massachusetts Institute of Technology, realised that Boole's logic gates could be used to mathematically describe the way that the electronic computer circuits he was developing worked. To Shannon, this was a most significant discovery; if Boole's laws could describe both the logic of human thought and the logic of computer 'thought', then might this not mean that each was substantially similar? Like Emil Gaverluk in *Did Genesis Man Conquer Space?*, Shannon began to conceive of the human mind as a biological super-computer – a simplistic notion maybe, but some variant of this same idea seems to lie behind the Transhumanist belief that our minds will one day be uploaded into some future version of the Cloud.[97]

However, storing our minds on computers is not the end of it. Ray Kurzweil, like Fedorov and Tsiolkovsky before him, fantasises that one day these fabulous autotrophic computer minds of ours will merge with the entire universe around us. But how could this possibly occur? How can a computer mind suddenly just fuse with inanimate matter? One answer would be to speculate that perhaps the universe is all only one big computer program in any case, an idea which has been proposed down the years not only by inventive sci-fi writers like Philip K. Dick (1928–82), but also some genuine scientists, such as the early American AI pioneer Edward Fredkin (b. 1935). Fredkin was terrified by the idea of nuclear holocaust, and bought his own Caribbean island which he fortified with fall-out bunkers to allow him to avoid the anticipated apocalypse. An alternative method of escape from Soviet warheads, however, was for Fredkin to deny that the world really existed at all. He founded a new field of science called 'digital physics', which held that the whole universe was nothing more than a gigantic computer simulation, with God as the genius programmer behind it – a school of thought also known as 'pancomputationalism'.

Not only the cosmos, but all the humans and animals within it were essentially informational in nature, said Fredkin; one simple summary of such thought has it that 'biology reduces to chemistry, reduces to physics, reduces to computation of information.' Some of Fredkin's followers have since speculated that, like Boolean-derived computer code, all subatomic particles act effectively as zeros and ones, and that if a computer capable of processing 10^{90} bits was ever built, it would be able to simulate our entire universe as it currently is; so how do we know we aren't already living in just such a simulation? Wishing to spread the word of this self-penned Silicon Gospel, Fredkin began teaching courses at US universities in the new field of 'Saving the World'. Here, he informed undergraduates that the planet was 'a great computer' and that it was the task of the next generation of mathematicians and physicists to write a 'global algorithm', a piece of super-Boolean logic which, 'if methodically executed ... would lead to peace and harmony'. Considering the universe to be the programming work of God, for Fredkin it followed that, in imitating Him by programming our own holy algorithms, we would be resolving the distance between God's mind and our own, thus merging with Him and becoming gods ourselves.[98]

Even more optimistic in his views about what special features a future edition of Microsoft Windows might one day contain is the American physicist Frank Tipler (b. 1947), who honestly predicts that, some day in the distant future, our computers will become so advanced that they will become God and facilitate our resurrection from the dead. Essentially, Tipler has appropriated the thinking of the French

Jesuit scholar and mystic Teilhard de Chardin (1881–1955), who tried to make a kind of religion out of evolutionary theory by proposing the existence of something called the 'Omega Point'. Chardin was a noted palaeontologist, who – like so many of the people in this book – set out to try and reconcile God with Darwin. To this end, he made use of Vernadsky's concept of the world's passage from geosphere to biosphere and then on to noösphere, speculating that, as our species became more and more advanced, mankind was somehow merging with the world around him, frequently via technological means. Basically, Chardin argued that the universe itself was constantly evolving; firstly from inanimate particles into low forms of life like bacteria, and then into higher forms like plants, animals and then humans. Eventually, said Chardin, humans would themselves evolve into something higher, merging with one another and the environment around them into a kind of new, collective species – in which, paradoxically, we would all be more individualised in our personalities than ever before. However, this evolution was not occurring in a random, directionless way, as orthodox Darwinism would have it, but was being drawn onwards by the Omega Point. But what was this Omega Point? According to Chardin, it was the point at which mankind fused with the entire universe around him, thus becoming one with God and facilitating a reunion with Christ. Believing the common Catholic idea of 'the Body of Christ' to be a description of the cosmos itself, he referred to our evolutionary journey on towards the Omega Point as a process of 'Christogenesis'. In essence, then, the Omega Point is a religious version of the Singularity – which is highly appropriate, seeing as the Singularity is nothing more than a secular version of the Omega Point.[99]

Undoubtedly the man who has done the most to try and reconcile science and religion in this field, however, is Frank Tipler. Tipler is a genuine scientist who holds professorships in Mathematics and Physics at Louisiana's Tulane University. In 1994, he published a deeply controversial book, *The Physics of Immortality: Modern Cosmology, God and the Resurrection of the Dead*, which spun a complex tale arguing that, as humanity's computing power progressed, intelligent life would begin to merge with and colonise inanimate matter. At some point, our actual physical brains would be replaced with advanced computer emulations of them, Tipler said, whose capabilities would go on increasing all the time. However, our universe is at some point due to shrink and collapse in on itself, apparently. As this happens, the 'cost' of our computing will become smaller, as the space between processing points scattered throughout the universe shrinks, thus triggering a final expansion of our computer minds' potential to infinite levels. At this point – the Omega Point – we will become, in effect, God, who is really a cosmic super-computer.

With infinite processing power, our minds will be able to simulate any and all possible realities, and inhabit them as if they were real, predicts Tipler, in a way which will seem as if it lasts forever. Then, not only will we be able to resurrect the dead by running computer programmes in which they are imagined as still being alive, we will also be able to 'pre-surrect' (if that's even a word ...) all those people who haven't even been born yet, in a future which, paradoxically, is also experienced as being the present. Furthermore, we could even incarnate all those people who were never actually born at all, but might have been, had the world turned out differently to the way it did. This state, apparently, will be what Jesus meant when he talked about Heaven. Tipler presents all this as being both inevitable, and based upon genuine scientific laws. Among the devout Catholic Tipler's other 'scientific' beliefs, though, are that Jesus could fire neutrino-beams from his feet to enable him to walk on water, that there are genes in every human body which carry original sin, and that he knows how to build a time machine. An alternative name given to his theory by Tipler is that of the 'Final Anthropic Principle', or FAP – the critic Martin Gardner preferred to refer to it as the 'Completely Ridiculous Anthropic Principle', or CRAP, however, and I have to say I am minded to agree with him.[100]

Resurrecting the Rose

Many of today's Transhumanists, like those discussed above, make America their home. However, while California's Silicon Valley is, naturally enough, the world's number one Transhumanist centre, Nikolai Fedorov's Mother Russia is another, with a young man named Danila Medvedev (b. 1980) being one particular leading light. Medvedev firmly believes that one day our brains will be uploaded to computers, but accepts this is some way off, so is currently involved in the niche business of freezing people's brains and severed heads in big vats in his shed so they will be ready to be transferred to silicon come the glorious day. Medvedev views our brains as being essentially software in any case, so is more than willing to abandon all sense of being human when it finally becomes possible for him to evolve into an incorporeal computer-man. According to his own disturbing vision:

> There is no single human value that I expect to keep [after being uploaded to silicon]. Books are of no use to superintelligence. Love and sex are something many of us already want to get rid of, a dark vestige of our evolutionary past ... Work will no longer be needed, thought alone would accomplish everything I might ever need. And the food I need would be measured in ergs [energy units] not calories.[101]

It sounds to me that Mr Medvedev doesn't really want to be human at all; it sounds as if he wants to be dead, and there are already far easier ways of achieving that particular aim than becoming a computer. As John Gray has observed, 'The virtual afterlife is a high-tech variant of the Spiritualist Summerland' (their quaint old word for 'Heaven').[102] Maybe so – except most of us here in the modern-day West no longer believe in the existence of either the nineteenth-century Spiritualist Summerland, nor the even older myth of the Christian Heaven. We do, however, still believe in the myth of eternal progress, and so it is within cyberspace that we now locate our new, technological Heaven, a realm for which we evidently still have some psychological longing. The Transhumanists do not realise that their own dream of salvation will one day come to seem as silly as that of the Cosmists, just as the Cosmists never realised that their own dreams would one day seem as unlikely as those of the Rosicrucians. This is because, at root, all these dreams are really one and the same – the dream of the conquest of death, and the consequent creation of Heaven on Earth – clothed in various culturally appropriate disguises, whether religious or scientific in nature. We opened this book with a discussion of the quest to surmount death, and so we end it; because that, it would seem, is the ultimate aim of our science. Few sane readers can expect the ghost of Nikola Tesla one day to literally descend from his Theosophist Heaven and save us all from the manifest imperfections of our world like Margaret Storm did – but many of us, whether consciously or not, do still expect science as a whole one day to do so, even if not during our own lifetimes. Gnostic Rosicrucianism, I think, lives on still, its revered symbol no longer the mystic intertwining of the cross and the rose, but the equally inseparable interlacing of our new twin gods, the holy yet secular union of science and progress.

Bibliography

Aldersey-Williams, Hugh, *The Adventures of Sir Thomas Browne in the 21st Century* (London: Granta, 2015)
Al-Khalili, Jim, *Pathfinders: The Golden Age of Arabic Science* (London: Penguin, 2010)
Andrews, James T., *Red Cosmos: K. E. Tsiolvoksvii, Grandfather of Soviet Rocketry* (Texas: Texas A&M University Press, 2009)
Appleyard, Brian, *Aliens: Why They Are Here* (London: Scribner, 2006)
Bacon, Francis, *The Essays* (London: Penguin Classics, 1985)
Ball, Philip, *Unnatural: The Heretical Idea of Making People* (London: Bodley Head, 2011)
Ball, Philip, *Serving the Reich: The Struggle for the Soul of Physics Under Hitler* (London: Vintage, 2014)
Barrett, David V., *Secret Societies* (London: Godsfield Press, 2008)
Bartlett, Rosamund, *Tolstoy: A Russian Life* (London: Profile, 2010)
Birstein, Vadim J. *The Perversion of Knowledge: The True Story of Soviet Science* (Massachusetts: Westview Press, 2004)
Brooks, Michael, *13 Things That Don't Make Sense: The Most Intriguing Scientific Mysteries of Our Times* (London: Profile, 2009)
Campos, Luis A., *Radium and the Secret of Life* (Chicago: University of Chicago Press, 2015)
Dawkins, Richard, *The God Delusion* (London: Black Swan, 2006)
Dougan, Andy, *Raising the Dead: The Men Who Created Frankenstein* (Edinburgh: Birlinn, 2008)
Ekirch, A. Roger, *At Day's Close* (London: Weidenfeld & Nicolson, 2005)
Evans-Wentz, W.Y., *The Fairy Faith in Celtic Countries* (South Carolina: BiblioBazaar, 2008)
Flammarion, Camille, *Haunted Houses* (London: T. Fisher Unwin, 1924)
Fort, Charles, *The Book of the Damned* (London: John Brown, 1995)
Frayn, Michael, *The Human Touch: Our Part in the Creation of a Universe* (London: Faber & Faber, 2006)
Freeman, Richard, *Dragons: More Than a Myth?* (Exeter: CFZ Press, 2005)

Gardner, Martin, *Fads & Fallacies in the Name of Science* (New York: Dover Books, 1957)
Goodrick-Clarke, Nicholas, *The Occult Roots of Nazism* (London: IB Tauris, 2009)
Gray, John, *Straw Dogs: Thoughts On Humans and Other Animals* (London: Granta, 2003)
Gray, John, *The Immortalisation Commission: The Strange Quest to Cheat Death* (London: Penguin, 2012)
Gray, John, *The Silence of Animals: On Progress and Other Modern Myths* (London: Allen Lane, 2013)
Gray, Ronald Douglas, *Goethe the Alchemist* (Cambridge: Cambridge University Press, 2010)
Haeckel, Ernst, *Monism as Connecting Religion and Science* (London: A&C Black, 1895)
Hansen, George P., *The Trickster and the Paranormal* (Indiana: Xlibris, 2001)
Harpur, Patrick, *The Philosophers' Secret Fire: A History of the Imagination* (Victoria, Australia: Blue Angel Gallery, 2002)
Harpur, Patrick, *Mercurius: The Marriage of Heaven and Earth* (Glastonbury: Squeeze Press, 2008)
Harré, Rom, *Pavlov's Dogs and Schrödinger's Cat: Scenes from the Living Laboratory* (Oxford: OUP, 2009)
Henry, Lyell D., *Zig-Zag-and-Swirl: Alfred W. Lawson's Quest for Greatness* (Iowa: University of Iowa Press, 2009)
Hustvedt, Asti, *Medical Muses: Hysteria in 19th-Century Paris* (London: Bloomsbury, 2012)
Huxley, T. H., *Method and Results: Essays* (New York: D. Appleton & Co, 1899)
Jay, Mike, *The Atmosphere of Heaven: The Unnatural Experiments of Dr Beddoes and His Sons of Genius* (London: Yale University Press, 2009)
Jung, C. G., *Memories, Dreams, Reflections* (London: Harper Collins, 1995)
Jung, C. G., *Psychology and Alchemy: Revised Second Edition* (London: Routledge, 2008)
Jung, C. G., *Synchronicity: An Acausal Connecting Principle* (London: Routledge, 2010)
Kaku, Michio, *Physics of the Impossible* (London: Penguin, 2009)
Kershaw, Ian, *Hitler 1889–1936: Hubris* (London: Allen Lane, 1998)
Kirk, Robert, *The Secret Commonwealth* (New York: NYRB, 2007)
Koestler, Arthur, *The Sleepwalkers* (London: Penguin Modern Classics, 2014)
Kossy, Donna, *Kooks* (Oregon: Feral House, 1994)
Kossy, Donna, *Strange Creations: Aberrant Ideas of Human Origins from Ancient Astronauts to Aquatic Apes* (California: Feral House, 2001)
Kuhn, Thomas S., *The Structure of Scientific Revolutions* (Chicago: University of Chicago Press, 2012)

Lachmann, Gary, *The Dedalus Book of the Occult* (Cambridgeshire: Dedalus, 2003)
Leadbeater, C. W., *The Hidden Side of Things* (London: Theosophical Publishing House, 1913)
Leadbeater, C. W. & Besant, Annie, *Occult Chemistry: Clairvoyant Observations on the Chemical Elements* (London: Theosophical Publishing House, 1919)
Lebling, Robert, *Legends of the Fire Spirits: Jinn and Genies from Arabia to Zanzibar* (London: IB Tauris, 2010)
Lifton, Robert Jay, *The Nazi Doctors* (New York: Basic Books, 2000)
Long, David, *English Country House Eccentrics* (Stroud: The History Press, 2012)
Marx, Karl, *Das Kapital* (Oxford: Oxford World's Classics, 1999)
McGovern, Una & Rickard, Bob, *Chambers Dictionary of the Unexplained* (London: Chambers, 2007)
Melechi, Antonio, *Servants of the Supernatural: The Night-Side of the Victorian Mind* (London: Arrow, 2009)
Michell, John, *Simulacra* (London: Thames & Hudson, 1979)
Michell, John, *Eccentric Lives and Peculiar Notions* (Illinois: Adventures Unlimited, 1999)
Montefiore, Simon Sebag, *Stalin: 1939–1945* (London: Phoenix, 2004)
Mooney, Chris, *The Republican War on Science* (Cambridge, Massachusetts: Basic Books, 2005)
Moore, Steve (Ed.), *Fortean Studies Volume 1* (London: John Brown, 1994)
More, Sir Thomas, *Utopia* (London: Penguin Classics, 1965)
Noble, David F., *The Religion of Technology* (New York: Penguin, 1999)
O'Neill, John J., *Prodigal Genius: The Life of Nikola Tesla* (Tennessee: Bottom of the Hill, 2012)
Pauwels, Louis & Bergier, Jacques, *The Morning of the Magicians* (London: Souvenir Press, 2001)
Pilkington, Mark (Ed.), *Strange Attractor Journal Four* (London: Strange Attractor Press, 2011)
Price, Harry, *Poltergeist Over England* (London: Country Life, 1945)
Prideaux, Sue, *Strindberg: A Life* (New Haven & London: Yale University Press, 2013)
Pynchon, Thomas, *Gravity's Rainbow* (London: Vintage, 2000)
Randall, Lucien, *Disgusting Bliss: The Brass Eye of Chris Morris* (London: Simon & Schuster, 2010)
Rickard, Bob & Michell, John, *Unexplained Phenomena: A Rough Guide Special* (London: Rough Guides, 2000)
Robinson, Dave & Groves, Judy, *Introducing Philosophy* (Cambridge: Icon Books, 1998)
Rogers, Molly, *Delia's Tears: Race, Science and Photography in Nineteenth-Century America* (New Haven: Yale University Press, 2010)
Roud, Steve, *The Penguin Guide to the Superstitions of Britain and Ireland* (London: Penguin, 2006)
Roy, Archie E., *The Eager Dead* (Brighton: Book Guild, 2008)

Russell, Bertrand, *History of Western Philosophy* (London: Routledge, 2006)
Sagan, Carl, *The Demon-Haunted World: Science as a Candle in the Dark* (New York: Ballantine, 1997)
Screeton, Paul, *Mars Bar & Mushy Peas: Urban Legend and the Cult of Celebrity* (Loughborough: Heart of Albion, 2008)
Scull, Andrew, *Hysteria: The Disturbing History* (Oxford: OUP, 2011)
Sharaf, Myron, *Fury on Earth: A Biography of Wilhelm Reich* (Chicago: Da Capo Press, 1994)
Sheldrake, Rupert, *A New Science of Life* (London: Icon, 2006)
Sheldrake, Rupert, *The Science Delusion* (London: Coronet, 2013)
Shelley, Mary, *Frankenstein* (London: Penguin Classics, 2003)
Sitwell, Edith, *The English Eccentrics* (London: Pallas Athene, 2006)
Sokal, Alan & Bricmont, Jean, *Intellectual Impostures* (London: Profile, 2003)
Spengler, Oswald, *The Decline of the West* (New York: Oxford University Press, 1991)
Storm, Margaret, *Return of the Dove* (Maryland: Millennium Publications, 1957)
Strindberg, August, *Inferno* (London & New York: GP Putnams, 1913)
Swain, Frank, *How to Make a Zombie: The Real Life (and Death) Science of Reanimation and Mind-Control* (London: Oneworld, 2013)
Sweet, Matthew, *Inventing the Victorians* (London: Faber & Faber, 2001)
Tesla, Nikola, *My Inventions and Other Writings* (London: Penguin Classics, 2011)
Tesla, Nikola & Childress, David Hatcher, *The Fantastic Inventions of Nikola Tesla* (Illinois: Adventures Unlimited, 2009)
Thurston, Father Herbert, *Ghosts and Poltergeists* (Chicago: Henry Regnery Company, 1954)
Weinberg, Steven, *To Explain the World: The Discovery of Modern Science* (London: Allen Lane, 2015)
Wheen, Francis, *How Mumbo-Jumbo Conquered the World* (London: Harper Perennial, 2004)
Wilson, A. N., *God's Funeral* (London: Abacus, 2000)
Wilson, Bee, *The Hive: The Story of the Honeybee and Us* (London: John Murray, 2004)
Wilson, Colin, *The Occult* (London: Watkins, 2006a)
Wilson, Colin, *Mysteries* (London: Watkins, 2006b)
Wootton, David, *Bad Medicine: Doctors Doing Harm Since Hippocrates* (Oxford: OUP, 2007)
Yates, Frances, *The Rosicrucian Enlightenment* (London: Routledge Classics, 2002)
Young, George M., *The Russian Cosmists* (New York: OUP, 2012)
Frequently-cited article:
'Esoteric Elements in Russian Cosmism' by George M. Young in the *Rose + Croix Journal* volume 8, 2011, online at http://www.rosecroixjournal.org/issues/2011/articles/vol8-124-139-young.pdf

Notes

(All websites accessed between July and December 2015)

Frontispiece

1. Cited in Ball, 2011, p. 316

Introduction: Today's Science, Tomorrow's Superstition

1. Cited in Kuhn, 2012, p. 18
2. Wootton, 2007, pp. 94–5; Harré, 2009, pp. 92–3; Dougan, 2008, pp. 86–7
3. Wootton, 2007, p. 95
4. http://bja.oxfordjournals.org/content/early/2012/11/14/bja.aes388.full
5. Harré, 2009, pp. 92–100
6. Harré, 2009, pp. 108–110
7. Harré, 2009, pp. 101–2
8. Wootton, 2007, p. 107; http://publicdomainreview.org/collections/arthur-cogas-blood-transfusion-1667/
9. Wootton, 2007, p. 107
10. Swain, 2013, pp. 41–2
11. Swain, 2013, p. 201
12. Harré, 2009, p. 151; http://www.livescience.com/28996-hole-in-stomach-revealed-digestion.html
13. Wootton, 2007, pp. 107–8
14. https://www.newscientist.com/blogs/culturelab/2011/03/murder-medicine-and-the-first-blood-transfusions.html
15. Harré, 2009, p. 111–12
16. Harré, 2009, pp. 106, 108
17. Cited in Harré, 2009, p. 105
18. Harré, 2009, p. 106
19. Wootton, 2007, p. 82
20. Wootton, 2008, p. 45; Dougan, 2008, p. 84
21. Wootton, 2007, pp. 51, 73–76, 83–5
22. Wootton, 2007, pp. 90–1
23. Wootton, 2007, pp. 76–8
24. Harré, 2009, p. 159
25. Harré, 2009, p. 151
26. Cited In Harré, 2009, p. 159
27. Harré, 2009, pp. 151–4
28. Birstein, 2004, p. 16
29. Swain, 2013, p. 218; footage online at https://www.youtube.com/watch?v=uvZThr3POIQ
30. Swain, 2013, pp. 39–41; footage of film online at https://www.youtube.com/watch?v=KDqh-r8TQgs
31. Swain, 2013, pp. 80–1
32. *Fortean Times* 106, p. 22
33. Dougan, 2008, pp. 151–8

34. Cited in Gray, 2012, p. 161
35. Gray, 2012, pp. 161–3
36. Gray, 2012, p. 167, 169
37. Gray, 2012, p. 162
38. Young, 2012, pp. 179–180
39. Swain, 2013, pp. 59–60
40. Swain, 2013, pp. 60–1
41. Dougan, 2008, p. 133
42. Dougan, 2008, pp. 95–101; Harré, 2009, pp. 22–31
43. Sarah Bakewell, 'It's Alive!' in *Fortean Times* 139, pp. 34–9
44. Bob Rickard, 'The Archaeology of Madness' in *Fortean Times* 331, pp. 59–60
45. Dougan, 2008, pp. 103–4
46. Dougan, 2008, pp. 105–6
47. Swain, 2013, pp. 50–1
48. Dougan, 2008, pp. 109–11
49. Swain, 2013, p. 48–9
50. Jay, 2009, pp. 32–3, 67
51. Dougan, 2008, pp. 183–5
52. Dougan, 2008, pp. 182–3
53. Cited in Dougan, 2008, p. 141
54. Dougan, 2008, pp. 138–45
55. Dougan, 2008, pp. 66–9
56. Dougan, 2008, pp. 146–9
57. Dougan, 2008, pp. 136–7
58. Sitwell, 2006, pp. 88–91; Dougan, 2008, pp. 114–5; http://hoaxes.org/archive/permalink/grahams_celestial_bed
59. For the source of this misunderstanding, see http://users.dickinson.edu/~nicholsa/Romnat/frankmis.htm
60. Shelley, 2003, pp. 8, 11
61. Dougan, 2008, pp. 90–1
62. Shelley, 2003, xxv; Dougan, 2008, pp. 116–20
63. Dougan, 2008, p. 137
64. Swain, 2013, pp. 61–73; Dougan, 2008, pp. 190–1; http://www.madscientistblog.ca/mad-scientist-17-robert-cornish/#sthash.aXynrOmW.dpbs
65. Birstein, 2004, pp. 113–17
66. Gray, 2012, p. 147
67. Birstein, 2004, pp. 120
68. http://skeptoid.com/episodes/4219; http://www.scotsman.com/news/world/stalin-s-half-man-half-ape-super-warriors-1-686693
69. Lifton, 2000, pp. 289, 293–5, 301–2
70. Cited in Lifton, 2000, p. 292; quote slightly reordered
71. Lifton, 2000, pp. 292–3
72. Lifton, 2000, pp. 361–3
73. Cited in Lifton, 2000, p. 371
74. All taken from 'Sauerbruch's Sick Surgeon Syndrome' in *Fortean Times* 139, p. 32
75. Lifton, 2000, p. 301
76. http://www.madscientistblog.ca/mad-scientist-4-sigmund-rascher/#sthash.CpGFDwQj.dpbs; www.valas.fr/IMG/pdf/z-military_ethics.pdf; oethics.as.nyu.edu/docs/IO/30171/Steinberg.HumanResearch.pdf
77. Koestler, 2014, xvi
78. *The Times*, 8 September 2015, p. 32
79. http://www.madscientistblog.ca/mad-scientist-1314-vladimir-demikhov-and-robert-white/#sthash.NcoeJnyD.dpuf; footage of White at work online at https://www.youtube.com/watch?v=TGpmTf2kOc0 & https://www.youtube.com/watch?v=eW2RVq5ufgw
80. http://www.wired.com/2000/01/transplants/
81. *Daily Mail*, 21 August 2015, p. 15
82. Harré, 2009, pp. 124–9

83. *Private Eye* magazine issue 1400
84. Aldersey-Williams, 2015, p. x
85. *The Times*, 14 Jan 2015
86. Kuhn, 2012, p. 104
87. *Fortean Times* 32, p. 17
88. Wootton, 2007, p. 46
89. *Fortean Times* 32, p. 17
90. http://bestiary.ca/beasts/beast171.htm
91. Aldersey-Williams, 2015, p. xvii
92. Aldersey-Williams, 2015, p. 104
93. Aldersey-Williams, 2015, pp. 89–90
94. Aldersey-Williams, 2015, p. 164
95. Aldersey-Williams, 2015, pp. 98
96. Aldersey-Williams, 2015, pp. 85–6; http://penelope.uchicago.edu/psuedodoxia/pseudo322.html
97. Aldersey-Williams, 2015, p. 150
98. Aldersey-Williams, 2015, pp. 156–7
99. Aldersey-Williams, 2015, pp. 10, 27
100. Aldersey-Williams, 2015, p. 158
101. www.improbable.com/ig/iwinners
102. Aldersey-Williams, pp. xix, 20

Science Fictions

1. Cited in Sharaf, 1994, p. 291
2. Jay, 2009, p. 28
3. Jay, 2009, p. 145
4. Jay, 2009, pp. 150, 202–3
5. Long, 2012, pp. 103–4; http://www.gutenberg.us/articles/helena,_comtesse_d_noailles
6. Cited at http://www.katherinemansfieldsociety.org/today/
7. Jay, 2009, pp. 145–6
8. Jay, 2009, pp. 164, 233–4
9. Jay, 2009, pp. 60, 65
10. Bacon, 1985, p. 132
11. Bacon, 1985, p. 132
12. Sagan, 1997, p. 38
13. Sagan, 1997, p. 424–5
14. Cited in Sagan, 1997, pp. 425
15. Sagan, 1997, pp. 424, 405–6
16. Sagan, 1997, p. 426
17. Sagan, 1997, p. 38
18. Ball, 2014, pp. 263–7
19. Cited in Sagan, 1997, p. 202
20. Sagan, 1997, p. 325
21. Lebling, 2010, pp. 261–2
22. https://www.newscientist.com/article/dn4174-plasma-blobs-hint-at-new-form-of-life/
23. Paul Devereux, 'The Fourth State' in Pilkington, 2011, pp. 137–40
24. www.archives.gov/exhibits/charters/declaration_transcript.html
25. Jay, 2009, p. 59
26. Jay, 2009, p. 205
27. Jay, 2009, pp. 49–50
28. *Daily Mail*, 18 December 2015, p. 9
29. Kossy, 2001, pp. 108–109
30. Kossy, 2001, pp. 108–10
31. Cited at http://en.wikipedia.org/wiki/Frances_Cress_Welsing
32. Kossy, 2001, pp. 111–12
33. Cited in Ball, 2014, p. 93
34. Ball, 2014, p. 72
35. Cited in Ball, 2014, p. 72
36. Ball, 2014, pp. 70–2
37. Ball, 2014, pp. 83–90
38. Sagan, 1997, p. 261
39. Ball, 2014, pp. 91–2
40. Ball, 2014, p. 93
41. Ball, 2014, pp. 104–106
42. Ball, 2014, p. 265

43. Spengler, 1991, p. 412
44. Spengler, 1991, p. 413
45. Ball, 2014, p. 30
46. Ball, 2014, pp. 30–3
47. Sagan, 1997, p. 248
48. Cited in Sokal & Bricmont, 2003, pp. 100–101
49. *The Times*, 23 October 2015, pp. 1–2
50. Sokal & Bricmont, 2003, pp. 100–101, 110–12
51. Cited in Sokal & Bricmont, 2003, p. 113
52. Harré, 2009, pp. 254–5
53. *Daily Mail*, 7 Aug 2015, p. 8
54. http://www.todayifoundout.com/index.php/2014/04/cow-farts-really-significantly-contribute-global-warming/
55. *The Sport* 18 Feb 1998, cited in Screeton, 2008, p. 154
56. *The Times* 'T2' supplement, 19 January 2015, pp. 4–5; all subsequent quotes taken from this article
57. *The Times*, 27 October 2015, p. 24
58. Mooney, 2005, p. 36
59. *The Times*, 19 Oct 2015, p. 25; *The Times*, 21 Oct 2015, p. 32; http://www.theguardian.com/world/2015/nov/02/french-weatherman-fired-for-promoting-book-sceptical-of-climate-change
60. Gray, 2003, p. 19
61. http://news.bbc.co.uk/1/hi/programmes/more_or_less/7067003.stm
62. All quotes taken from the programme in question
63. Cited in Randall, 2010, p. 184
64. Sheldrake, 2013, pp. 166–70
65. Cited in Sheldrake, 2013, p. 168
66. Sheldrake, 2013, p. 172
67. Sheldrake, 2013, p. 173
68. Cited in Sheldrake, 2013, p. 17
69. http://www.theguardian.com/science/2013/nov/10/what-is-heisenbergs-uncertainty-principle
70. Frayn, 2006, pp. 72–3
71. All quotes cited in Sheldrake, 2006, pp. 20–1
72. Cited in Sheldrake, 2006, p. 22
73. Sheldrake, 2006, pp. 16, 247–98 summarises some of these tests and their findings
74. Sheldrake, 2006, pp. 286–7
75. Sheldrake, 2006, pp. 292–3
76. Sheldrake, 2006, pp. 274–5
77. Cited in Gould, 1920, p. 36
78. Compiled from Wilson, 2000, pp. 59–62; Gould, 1920, throughout; Noble, 1999, pp. 85–6
79. Wilson, 2000, p. 62
80. Noble, 1999, p. 83
81. Gould, 1920, pp. 10–11, 18
82. Gould, 1920, p. 19
83. Noble, 1999, pp. 79–80
84. Noble, 1999, pp. 92–3
85. Noble, 1999, pp. 17–18
86. Yates, 2002, p. 164; Sheldrake, 2013, p. 14; Noble, 1999, p. 173
87. McGovern & Rickard, 2007, p. 588; Barrett, 2008, pp. 62–7
88. Barrett, 2008, p. 80–3
89. Yates, 2002, pp. 164–8
90. Yates, 2002, p. 235
91. Yates, 2002, pp. 236–7
92. Barrett, 2007, pp. 68–9
93. Yates, 2002, pp. 242–3
94. Yates, 2002, pp. 200–19
95. Yates, 2002, p. 244
96. Yates, 2002, p. 210
97. 2:19–20
98. Sheldrake, 2013, pp. 13–14; Yates 2002, pp. 157–8
99. Noble, 1999, pp. 16–17

100. Noble, 1999, pp. 33–4
101. 12:4
102. Noble, 1999, pp. 47–8, 52, 115–17
103. Noble, 1999, p. 67
104. Marina Warner, 'Introduction' in Kirk, 2007, p. xviii
105. Cited in Noble, 1999, p. 61
106. Cited in Yates, 2002, pp. 210–11
107. Noble, 1999, p. 53–5
108. Yates, 2002, pp. 231–2
109. Yates, 2002, pp. 156–7
110. Yates, 2002, pp. 228–9
111. More, 1965, p. 8
112. Yates, 2002, pp. 230–2
113. Cited in Yates, 2002, p. 213
114. www.quotationspage.com/quote/26032.html
115. Cited in Yates, 2002, p. 211

Evolutionary Dead-Ends

1. Cited in Kossy, 2001, p. 162
2. Gray,2002, p. 31
3. Young, 2011, p. 136; Young, 2012, pp. 241–2
4. www.bbc.co.uk/news/science-environment-16554357
5. Robinson & Groves, 1998, pp. 77–81; Russell, 2006, 661–74
6. Young, 2011, pp. 132–33; Young, 2012, pp. 17–20
7. John Michell, 'The Myth of Darwinism' in *Fortean Times* 35, pp. 2–5
8. David McLellan, 'Introduction' to Marx, 1999, pp. xxiv-xxv
9. Wilson, 2000, pp. 195–222
10. Sheldrake, 2013, p. 174; Sheldrake, 2006, pp. 10–11; Gardner, 1957, p. 141
11. Gray, 2012, p. 146
12. Harré, 2009, p. 244–6; Birstein, 2004, p. 47
13. Harré, 2009, pp. 248–9; Birstein, 2004, p. 47; Gardner, 1957, p. 147
14. Gardner, 1957, pp. 144–5; Harré, 2009, p. 252; Sagan, 1997, p. 262
15. Birstein, 2004, p. 408
16. Birstein, 2004, pp. 264–5
17. Montefiore, 2004, p. 287
18. Harré, 2009, p. 252; Birstein, 2004, p. 49
19. Birstein, 2004, pp. 47–8
20. Gardner, 1957, p. 144, 146; Harré, 2009, p. 45; Birstein, 2004, pp. 45–6
21. Birstein, 2004, pp. 46, 48; Kossy, 2001, p. 59
22. Gardner, 1957, pp. 147–9
23. Birstein, 2004, pp. 49–50; Gardner, 1957, p. 334
24. Presuming he did really say this; see http://www.csmonitor.com/USA/Politics/2011/0603/Political-misquotes-The-10-most-famous-things-never-actually-said
25. Cited in Gray, 2012, p. 148
26. Both quotes cited in Gardner, 1957, p. 143
27. See 'Paul Kammerer and the Law of Seriality' by John Townley and Robert Schmidt in Moore, 1994, pp. 251–60
28. Townley & Schmidt, 1994, p. 251; Gardner, 1957, p. 143
29. *Fortean Times* 88, p. 21
30. Harré, 2009, pp. 255–264, Gardner, 1957, pp. 143–4
31. Sheldrake, 2006, p. 161; Harré, 2008, p. 258
32. Antoni Melechi, 'Evolution's Maze' in *Fortean Times* 318, pp. 48–52
33. *The Times*, 22 Aug 2015, p. 17
34. Wilson, 2000, pp. 195–222
35. Wilson, 2000, p. 235

36. Russell, 2006, p. 36
37. Russell, 2006, p. 61
38. Kossy, 2001, pp. 32–3
39. Gardner, 1957, pp. 153–4
40. Gardner, 1957, p. 155; Kossy, 2001, pp. 87–8
41. Goodrick-Clarke, 2009, pp. 90–9; Kossy, 2001, pp. 56, 84–8; Kershaw, 1998, pp. 49–52
42. Kossy, 2001, pp. 102–104
43. Kossy, 2001, pp. 227–30
44. Kossy, 2001, pp. 58–63
45. Goldberg, 2007, p. 40
46. Jay, 2009, pp. 1–8
47. Cited in Noble, 1999, p. 70
48. Noble, 1999, p. 195
49. Cited in Noble, 1999, p. 189
50. Russell, 2006, pp. 41–2; Robinson & Groves, 1998, p. 9
51. Russell, 2006, p. 167
52. Young, 2012, p. 131
53. Young, 2011, pp. 134–5; Young, 2012, pp. 119–33
54. Noble, 1999, p. 65
55. Harré, 2008, pp. 95–6; Noble, 1999, pp. 59–60, 64
56. Price, 1945, p. 43; Thurston, 1954, pp. 39–40
57. Price, 1945, p. 164
58. Roy, 2008, p. 81
59. Melechi, 2009, pp. 200–3
60. Sheldrake, 2013, p. 307
61. Sheldrake, 2013, p. 307
62. *Daily Mail*, 14 Oct 2015, pp. 30–1
63. Hansen, 2001, pp. 150–1
64. Hansen, 2001, p. 160
65. Sagan, 1997, p. 299
66. Hansen, 2001, p. 151
67. Hansen, 2001, p. 154
68. Hansen, 2001, pp. 291–306
69. Michell, 1979, p. 9
70. Hansen, 2001, pp. 319–20, 342
71. Hansen, 2001, pp. 326–7
72. Hansen, 2001, p. 327
73. Hansen, 2001, p. 328
74. Hansen, 2001, p. 329
75. Hansen, 2001, p. 336
76. Sheldrake, 2013, p. 303
77. Sagan, 1997, p. 272
78. Paul Chambers, 'Body and Soul' in *Fortean Times* 262, pp. 32–5
79. http://en.wikipedia.org/wiki/Scientism
80. http://www.stephenjaygould.org/library/gould_noma.html
81. http://en.wikipedia.org/wiki/Non-overlapping_magisteria
82. Dawkins, 2006, p. 77
83. http://www.telegraph.co.uk/news/science/science-news/10875912/Reading-fairy-stories-to-children-is-harmful-says-Richard-Dawkins.html; http://www.dailymail.co.uk/sciencetech/article-2648272/Should-ban-Christmas-Telling-children-Santa-Claus-damage-claims-Richard-Dawkins.html
84. Sheldrake, 2013, pp. 47–8
85. Sheldrake, 2013, pp. 163–5
86. Jeremy Harte, review in *Fortean Times* 329, p. 55
87. Cited in Wheen, 2004, p. 140
88. Sagan, 1997, pp. 174–5
89. Cited in Koestler, 2014, p. 405
90. Wilson, 2000, p. 242
91. Harré, 2009, pp. 217–19; Wilson, 2000, p. 242
92. Sagan, 1997, pp. 258–9
93. Roy, 2008, pp. 75–6
94. Roy, 2008, pp. 34–5
95. Melechi, 2009, p. 192; Roy, 2008, p. 73
96. Cited in Melechi, 2009, p. 193
97. Wilson, 2000, p. 238
98. Melechi, 2009, p. 193
99. Wilson, 2006b, pp. 189–90

100. Michell, 1979, p. 23
101. Michell, 1979, p. 23
102. Wilson, 2006b, pp. 190–2
103. Michell, 1979, pp. 22–4
104. Gardner, 1957, pp. 124–6
105. Kossy, 2001, pp. 186–9; https://en.wikipedia.org/wiki/Carl_Baugh
106. Sheldrake, 2013, pp. 105–7, 151; Russell, 2006, pp. 715–6
107. Wilson, 2006a, pp. 756–7
108. Evans-Wentz, 2008, p. 554
109. Evans-Wentz, 2008, p. 554
110. Evans-Wentz, 2008, p. 559
111. Evans-Wentz, 2008, pp. 555–6, 573
112. Evans-Wentz, 2008, pp. 532, 570
113. Evans-Wentz, 2008, pp. 532, 542, 554
114. Leadbeater, 1913, pp. 77–80
115. Leadbeater, 1913, p. 83
116. Leadbeater, 1913, pp. 83–4
117. Leadbeater, 1913, p. 86
118. Leadbeater, 1913, pp. 87–90
119. Leadbeater, 1913, p. 82
120. Wilson, 2006a, pp. 161–2
121. http://www.metanexus.net/essay/zoology-and-religion-work-alister-hardy
122. Wilson, 2006a, pp. 159–60
123. Jay, 2009, p. 46
124. Sheldrake, 2013, pp. 64–5
125. John Michell, 'When Feathers Fly' in *Fortean Times* 52, pp. 47–51
126. Cited in Appleyard, 2006, p. 222
127. Sheldrake, 2013, pp. 66–7
128. https://en.wikipedia.org/wiki/James_Ussher
129. Cited in Sheldrake, 2009, p. 3
130. Gray, 2003, p. 21
131. Kuhn, 2012, pp. 24, 68
132. Kuhn, 2012, pp. 36–7
133. Kuhn, 2012, p. 79
134. Kuhn, 2012, pp. x–xi
135. Kuhn, 2012, p. 90
136. Gardner, 1957, pp. 86–8
137. Cited in Gardner, 1957, p. 87
138. Kuhn, 2012, p. 152
139. Kuhn, 2012, p. 59
140. Rogers, 2010 accessed online at https://books.google.co.uk/books?isbn=0300163282; [no page numbers given]; Flammarion, 1924, p. 325 has an exaggerated version of this tale in which it is recommended Daguerre be committed to an asylum.
141. *Fortean Times* 74, p. 20
142. Cited in Flammarion, 1924, pp. 320–1
143. Flammarion, 1924, p. 322; Fort, 1995, pp. 17–18
144. Flammarion, 1924, p. 324; Fort, 1995, p. 18–19
145. Kuhn, 2012, p. 139
146. Kuhn, 2012, p. 137
147. Koestler, 2014, pp. 497–9
148. Kuhn, 2012, pp. 139–40
149. Kuhn, 2012, pp. 165–6
150. Kuhn, 2012, pp. 169–70
151. Sheldrake, 2013, p. 19
152. Kuhn, 2012, p. xxvi; Weinberg, 2015, p. 260
153. Cited in McGovern & Rickard, 2007, p. 31
154. McGovern & Rickard, 2007, pp. 235–7
155. William Corliss, 'Science Frontiers' in *Fortean Times* 46, pp. 46–7
156. McGovern & Rickard, 2007, p. 31
157. Sheldrake, 2013, p. 328

Two Little Boys

1. Cited in Gray, 2002, p. 20
2. Kaku, 2009, p. 260
3. Kaku, 2009, pp. 263–6; Sheldrake, 2013, pp. 69–70
4. Prideaux, 2013, 44–6, 153, 313

5. Prideaux, 2013, p. 45
6. Tesla, 2011, pp. 6–8
7. Tesla, 2011, p. 13
8. Tesla, 2011, p. 32
9. Tesla, 2011, pp. 13–14
10. Tesla, 2011, p. 25
11. Tesla, 2011, p. 25
12. Tesla, 2011, pp. 25–6, 31–2
13. Prideaux, 2013, p. 190
14. Prideaux, 2013, pp. 166–7
15. Prideaux, 2013, p. 166
16. Prideaux, 2013, p. 229
17. Prideaux, 2013, p. 215
18. Prideaux, 2013, p. 216
19. Cited in Prideaux, 2013, p. 193
20. Tesla, 2011, p. 35
21. O'Neill, 2012, pp. 84–5
22. O'Neill, 2012, p. 124
23. O'Neill, 2012, pp. 74–5
24. *The Times*, 15 Aug 2015, p. 3
25. Sheldrake, 2013, pp. 294, 299–300
26. Koestler, 2014, p. xvii
27. Gray, 2013, pp. 113–5
28. All quotes above taken from Haeckel, 1895; accessed online at http://www.gutenberg.org/files/44108/44108-h/44108-h.htm (no page numbers provided)
29. Prideaux, 2013, pp. 170–1
30. Prideaux, 2013, p. 236
31. Wilson, 2006a, p. 153
32. Strindberg, 1913, Ch.III
33. Prideaux, 2013, pp. 164–9
34. Prideaux, 2013, p. 212
35. Prideaux, 2013, pp. 187–8
36. Prideaux, 2013, p. 196
37. Strindberg, 1913, Ch.III
38. Cited in Sheldrake, 2013, p. 19
39. Huxley, 1899, pp. 224–5
40. Huxley, 1899, p. 223
41. Huxley, 1899, p. 242; my italics
42. Huxley, 1899, p. 244; my italics
43. Huxley, 1899, pp. 240–1
44. Huxley, 1899, p. 244
45. http://www.iep.utm.edu/epipheno/
46. Sheldrake, 2013, pp. 109–12
47. Swain, 2013, pp. 118–9
48. Huxley, 1899, p. 228
49. Cited in Huxley, 1899, pp. 247–9
50. Brooks, 2009, pp. 153–4
51. Sheldrake, 2013, p. 124
52. http://www.telegraph.co.uk/news/science/8058541/Neuroscience-free-will-and-determinism-Im-just-a-machine.html
53. *The Times*, 15 Oct 2015, p. 23; *Daily Mail*, 15 Oct 2015, p. 13
54. Melechi, 2009, pp. 43–4
55. Melechi, 2009, pp. 87–8
56. Cited in Melechi, 2009, pp. 88–9
57. Scull, 2009, p. 13; Wootton, 2007, p. 42
58. Scull, 2009, p. 129
59. Scull, 2009, pp. 106–11
60. Scull, 2009, p. 111
61. Hustvedt, 2012, pp. 50–3
62. Hustvedt, 2012, p. 52
63. Hustvedt, 2012, p. 54
64. Hustvedt, 2012, pp. 286–7
65. Hustvedt, 2012, p. 48–50
66. Scull, 2009, p. 111
67. Melechi, 2009, pp. 138–9
68. Hustvedt, 2012, pp. 116–9
69. Hustvedt, 2012, pp. 281–2
70. All quotes taken from Haeckel, 1895; accessed online at http://www.gutenberg.org/files/44108/44108-h/44108-h.htm (no page numbers provided)
71. Strindberg, 1913, Ch.IV
72. Strindberg, 1913, Ch.V

73. Strindberg, 1913, Ch.V
74. Prideaux, 2013, p. 225
75. Strindberg, 1913, Ch.XV
76. Strindberg, 1913, Ch.II
77. Strindberg, 1913, Ch.V
78. Strindberg, 1913, Ch.V
79. Prideaux, 2013, p. 216
80. Rickard & Michell, 2000, p. 203
81. Flammarion, 1924, pp. 136–8
82. Flammarion, 1924, p. 311
83. Flammarion, 1924, p. 316
84. Strindberg, 1913, Ch.V
85. Strindberg, 1913, Ch.V
86. Strindberg, 1913, Ch.I
87. Strindberg, 1913, Ch.V
88. Prideaux, 2013, p. 224
89. Strindberg, 1913, Ch.V
90. Strindberg, 1913, Ch.IX
91. Strindberg, 1913, Ch.IV & Ch.V; Prideaux, 2013, p. 224
92. Prideaux, 2013, pp. 224–5
93. Strindberg, 1913, Ch.V
94. Strindberg, 1913, Ch.VI
95. Strindberg, 1913, Ch.V
96. Strindberg, 1913, Ch.V
97. Strindberg, 1913, Ch.VII; Prideaux, 2013, p. 233
98. McGovern & Rickard, 2007, p. 658; Lachmann, 2003, pp. 16–19
99. Strindberg, 1913, Ch.VIII; Prideaux, 2013, p. 235
100. Strindberg, 1913, Ch.XIV
101. Strindberg, 1913, Ch.VIII
102. Prideaux, 2013, p. 239
103. Noble, 1999, pp. 156–7
104. Tesla, 2011, pp. 122–3; https://en.wikipedia.org/wiki/Nikola_Tesla
105. Tesla, 2011, p. 122–3
106. Tesla, 2011, p. 12
107. Tesla, 2011, pp. 38–9
108. Tesla, 2011, p. 104
109. Tesla, 2011, pp. 106–7
110. Tesla, 2011, p. 9
111. Tesla, 2011, p. 23
112. Tesla, 2011, p. 16
113. Tesla, 2011, p. 121
114. Tesla, 2011, p. 121
115. O'Neill, 2012, p. 140
116. O'Neill, 2012, pp. 143–4
117. O'Neill, 2012, p. 144
118. Cited in O'Neill, 2012, p. 142
119. O'Neill, 2012, p. 142
120. O'Neill, 2012, pp. 145–6
121. Flammarion, 1924, p. 316
122. O'Neill, 2012, p. 11, 151
123. Tesla, 2011, p. 100
124. Tesla, 2011, p. 99
125. Tesla, 2011, p. 99
126. Tesla, 2011, pp. 10–11, 43
127. Tesla, 2011, pp. 39–40; O'Neill, 2012, pp. 29–30
128. O'Neill, 2012, p. 153
129. Cited at http://www.madscientistblog.ca/mad-scientist-16-charles-babbage/#sthash.rjgMcNGK.dpbs; it is possible Babbage meant this as a joke, of course ...
130. Cited in Storm 1957, p. 236
131. Cited in Storm 1957, p. 137
132. Russell, 2006, pp. 61, 236–7
133. Jay, 2009, pp. 46–9
134. Canto IV, II.397–400
135. Gray, 2010, pp. 58–9
136. Gray, 2010, pp. 134–5
137. Gray, 2010, pp. 71–100
138. Gray, 2010, pp. 139–43
139. O'Neill, 2012, p. 49
140. Tesla & Childress, 2009, pp. 278–9; 282–3
141. Tesla & Childress, 2009, pp. 289–93
142. O'Neill, 2012, pp. 11, 155
143. O'Neill, 2012, pp. 39, 106
144. https://en.wikipedia.org/wiki/Otis_T._Carr
145. O'Neill, 2012, pp. 131–2
146. Storm, 1957, pp. 72, 173–4, 206
147. Tesla, 2011, p. xiv
148. Storm, 1957, p. 289

149. Storm, 1957, p. 55
150. Storm, 1957, pp. 141–2
151. Storm, 1957, pp. 47, 68–9
152. Storm, 1957, p. 23
153. Storm, 1957, pp. 89, 139–40
154. Storm, 1957, pp. 289–90
155. Storm, 1957, p. 54
156. Storm, 1957, pp. 249–52
157. Storm, 1957, p. 71
158. Storm, 1957, pp. 34–5, 62–3
159. Storm, 1957, p. 108
160. Storm, 1957, pp. 136–7
161. Storm, 1957, p. 148
162. Storm, 1957, p. 80
163. Storm, 1957, pp. 84–5
164. O'Neill, 2011, p. 168
165. Storm, 1957, pp. 143–4, 146
166. O'Neill, 2012, p. 140

The Birds, but Not the Bees

1. Michell, 1979, p. 9
2. https://en.wikipedia.org/wiki/Colony_collapse_disorder
3. Actually, he may have said no such thing; see http://quoteinvestigator.com/2013/08/27/einstein-bees
4. www.apinews.com/en/news/item/12780-china-hand-pollination
5. Wilson, 2004, pp. 66–8, 70
6. Judges 14; Wilson, 2004, pp. 73–4
7. Wilson, 2004, p. 71–4
8. Wilson, 2004, pp. 78–9
9. Wilson, 2004, pp. 75, 118
10. Wilson, 2004, pp. 78–9
11. Wilson, 2004, p. 104
12. Wilson, 2004, pp. 69, 99
13. Wilson, 2004, p. 95–100
14. Wilson, 2004, pp. 100–104
15. Wootton, 2008, pp. 116–17
16. Wootton, 2008, p. 82
17. Wootton, 2008, pp. 117–18; Sheldrake, 2013, pp. 160–1
18. Ball, 2011, p. 18; Wilson, 2004, pp. 76–7
19. Roud, 2006, p. 252
20. http://users.dickinson.edu/~nicholsa/Romnat/frankmis.htm
21. II.vii.25–6
22. Ball, 2011, p. 19
23. Wilson, 2004, pp. 74–5; Wootton, 2007, pp. 120–1
24. Wootton, 2007, p. 119
25. Wootton, 2007, pp. 121–3
26. Wootton, 2007, pp. 133–4
27. Sweet, 2001, pp. 36, 110
28. Sarah Bakewell, 'The Strange Story of Crosse's Acari' in *Fortean Times* 139, p. 38; Dougan, 2008, pp. 158–66; Ball, 2011, pp. 68–9; Gardner, 1957, pp. 116–17
29. Gardner, 1957, p. 117
30. Campos, 2015, pp. 57–99
31. Ball, 2011, pp. 128–9
32. Gardner, 1957, p. 117–19
33. Cited at http://www.rexreserach.com/morleymartin/morleymartin.htm
34. Gardner, 1957, pp. 119–22
35. Pauwels & Bergier, 2001, pp. 59–60; Jay, 2009, p. 38; Al-Khalili, 2012, pp. 64–5; Wilson, 2006b, pp. 405–6
36. Jung, 1995, p. 230
37. Cited in Harpur, 2008, p. 13
38. Harpur, 2002, p. 173
39. Jung, 2008, p. 317
40. Jung, 1995, pp. 226–49
41. Harpur, 2002, pp. 176–7; Jung, 2008, p. 293
42. Jung, 2008, p. 277
43. Cited in Jung, 2008, p. 269
44. Freeman, 2005, p. 28
45. Jung, 2008, p. 291–2; Freeman, 2005, p. 28
46. Freeman, 2005, p. 29
47. See Jung, 2008, p. 277 for one example
48. Jung, 2008, pp. 244–5

49. Cited in Jung, 2008, pp. 246–7
50. Jung, 2008, pp. 252–4
51. Jung, 1995, pp. 228–30
52. http://www.nytimes.com/1988/08/16/science/the-benzene-ring-dream-anlaysis.html
53. Jung, 2008, p. 413
54. Pynchon, 2000, pp. 487–91
55. http://www.steel.org/making-steel/how-its-made/processes/processes-info/coal-utilization-in-the-steel-industry.aspx
56. Pynchon, 2000, pp. 196–8
57. Harpur, pp. 174–5
58. Jay, 2009, pp. 157–8, 164–5, 210
59. Jay, 2009, p. 246
60. Jay, 2009, pp. 153, 194
61. Russell, 2006, pp. 641–3; Robinson & Groves, 1998, pp. 73–5
62. Jay, 2009, pp. 193, 198, 222, 252
63. Gray, 2010, pp. 60–1
64. Cited in Harpur, 2002, p. 187
65. Al-Khalili, 2012, pp. 51–2; Weinberg, pp. 6, 10
66. McGovern & Rickard, 2007, pp. 53, 452–3, 597, 694
67. Gray, 2010, pp. 9–10
68. David Hambling, 'The Alchemical Resurrection' in *Fortean Times* 330, p. 12
69. Gray, 2010, pp. 150–1
70. Ball, 2011, p. 43
71. Ball, 2011, pp. 39–41
72. Ball, 2011, p. 41
73. Al-Khalili, 2012, pp. 52, 55
74. Al-Khalili, 2012, pp. 61–2, 64
75. Ball, 2011, pp. 42–3
76. Cited in Gray, 2010, pp. 205–6
77. McGovern & Rickard, 2007, p. 318
78. Ball, 2011, p. 42
79. Ball, 2011, pp. 46–7
80. Cited in Gray, 2010, p. 207
81. Gray, 2010, pp. 206–8
82. Gray, 2010, pp. 209–12; Ball, 2011, pp. 16–18
83. Gray, 2010, p. 212
84. Noble, 1999, pp. 167–8
85. Ball, 2011, p. 134
86. Ball, 2011, p. 127
87. Ball, 2011, p. 135
88. Brooks, 2009, p. 72
89. Ball, 2011, p. 134
90. Gardner, 1957, p. 122; http://blogs.scientificamerican.com/history-of-geology/charles-darwin-and-the-early-search-for-extraterrestial-life/
91. Ball, 2011, p. 129
92. Sheldrake, 2013, pp. 101–4; Sheldrake, 2006, pp. 128–33
93. Sheldrake, 2013, pp. 48–9
94. Russell, 2006, p. 493–4
95. Russell, 2006, p. 69
96. Russell, 2006, p. 62
97. Weinberg, 2015, pp. 7, 12–14
98. Russell, 2006, p. 51
99. Russell, 2006, p. 35
100. Russell, 2006, p. 116
101. Weinberg, 2015, p. 27
102. Wilson, 2006b, pp. 594–6
103. Michell, 1999, pp. 100–1, 104
104. Ball, 2011, pp. 125
105. Ball, 2011, pp. 125–6
106. Ball, 2011, pp. 125, 130
107. Leadbeater & Besant, 1919, pp. 1–2
108. Leadbeater & Besant, 1919, p. 8
109. Leadbeater & Besant, pp. 10–12
110. Leadbeater & Besant, 1919, p. 19, 25, 33, 40, 116, 140
111. Leadbeater & Besant, 1919, p. 137
112. Leadbeater & Besant, 1919, p. 23
113. Leadbeater, 1913, pp. 514–20

114. Henry, 2009, pp. 162–3 (compiled from two songs)
115. Henry, 2009, p. 265
116. Henry, 2009, pp. 79, 83, 92–3, 104–5; Peter Brookesmith, 'Flights of Fancy' in *Fortean Times* 195, pp. 42–8
117. Henry, 2009, pp. 296–7
118. Henry, 2009, p. 268
119. Henry, 2009, pp. 4, 114–15
120. Cited in Henry, 2009, p. 115
121. Henry, 2009, p. 11
122. Henry, 2009, 184
123. Gardner, 1957, pp. 70–1; Henry, 2009, pp. 118–122
124. Henry, 2009, pp. 122–3
125. Henry, 2009, p. 124–5
126. Gardner, 1957, p. 76
127. Gardner, 1957, pp. 311–2
128. Henry, 2009, pp. 188–192, 213
129. Sharaf, 1994, pp. 208–10
130. Sharaf, 1994, pp. 218–23
131. Sharaf, 1994, pp. 276–280, 283–88, 413–14
132. Gardner, 1957, pp. 253–4, 260
133. Sharaf, 1994, pp. 378–82
134. Sharaf, 1994, pp. 299–308; Gardner, 1957, pp. 256–7
135. Cited in Sharaf, 1994, p. 299
136. Cited in Sharaf, 1994, p. 279

Conclusion: Delusions on a Cosmic Scale

1. Cited in Gray, 2012, p. 157
2. Gray, 2013, pp. 43–5
3. Young, 2012, pp. 5, 204
4. Young, 2012, pp. 201–7
5. Young, 2012, pp. 198–9
6. Svetlana Semenova cited in Young, 2012, p. 8
7. Young, 2012, p. 7
8. Young, 2012, p. 40–5
9. Ecclesiastes 3:20
10. Cited in Young, 2012, p. 81
11. All information compiled from Young, 2011, pp. 127–32; Young, 2012, pp. 46–91
12. Gray, 2012, p. 159
13. Young, 2011, p130–2; Young, 2012, p. 69; Bartlett, 2010, p. 297
14. Andrews, 2009, pp. 65, 68–73
15. Young, 2011, pp. 128, 132; Young, 2011, pp. 145–1; Andrews, 2009, pp. 37–40
16. Young, 2012, pp. 50, 60
17. Noble, 1999, pp. 117–19
18. Young, 2012, p. 147
19. Cited in Noble, 1999, p. 119
20. Andrews, 2009, p. 19
21. Cited in Gray, 2012, p. 151
22. Young, 2012, pp. 151–2
23. Andrews, 2009, p. 43
24. Cited in Young, 2012, p. 151
25. Young, 2012, pp. 151–3; Andrews, 2009, p. 76
26. Cited in Young, 2012, pp. 152–3
27. Young, 2012, pp. 163–71
28. *Fortean Times* 46, p. 67
29. www.bbc.co.uk/news/uk-34897796
30. Andrews, 2009, pp. 20, 77–8
31. Andrews, 2009, pp. 49, 91, 95
32. Young, 2011, p. 134
33. Young, 2011, p. 136; Young, 2012, pp. 171–6
34. Brooks, 2009, pp. 122–35
35. Cited in Young, 2012, p. 162
36. Young, 2011, p. 132; Young, 2012, pp. 155–62
37. Andrews, 2009, p. 115
38. Young, 2012, pp. 59–60
39. http://www.bbc.co.uk/news/technology-16068581; http://news.bbc.co.uk/1/hi/sci/tech/8587725.stm
40. Ekirch, 2005, p. 339; http://news.bbc.co.uk/1/hi/sci/tech/272103.stm; https://

en.wikipedia.org/wiki/Znamya_(satellite)
41. Young, 2012, pp. 220–1
42. Young, 2012, pp. 222
43. Young, 2012, pp. 236
44. Young, 2012, pp. 228–9
45. Cited in Young, 2012, p. 228
46. Young, 2012, pp. 71–3
47. Cited in Gray, 2012, p. 180
48. Young, 2012, pp. 180–1
49. Cited in Gray, 2012, p. 180
50. Gray, 2012, pp. 141–4
51. Gray, 2012, p. 144
52. Gould, 1920, p. 36; this is a paraphrase
53. Andrews, 2009, pp. 67, 80–1
54. Young, 2012, pp. 180–1; Gray, 2012, pp. 147–8; https://en.wikipedia.org/wiki/White_Sea%E2%80%93Baltic_Canal; the various figures relating to the Canal vary wildly from source to source
55. Gray, 2012, pp. 184–5; 182
56. Gray, 2012, p. 182
57. Cited in Gray, 2012, p. 140
58. Cited in Gray, 2012, p. 142
59. Gray, 2012, p. 152
60. Gray, 2012, p. 218
61. https://en.wikipedia.org/wiki/Buckminster_Fuller; https://en.wikipedia.org.uk/wiki/Spaceship_Earth
62. *Sunday Times News Review* supplement, 22 March 2015, p. 9
63. Tesla, 2011, pp. 58–9, 165–6; O'Neill, 2012, p. 20
64. Tesla, 2011, p. 111
65. O'Neill, 2012, pp. 79–80
66. O'Neill, 2012, pp. 162–3
67. Cited in O'Neill, 2012, pp. 135–6
68. O'Neill, 2012, p. 136
69. Henry, 2009, p. 125
70. Cited in Kossy, 1994, p. 152
71. Kossy, 1994, p. 153
72. Henry, 2009, pp. 40–6
73. Henry, 2009, pp. 53, 291
74. Henry, 2009, pp. 238–41
75. Henry, 2009, pp. 223–35
76. Henry, 2009, pp. 271–6
77. Cited in Henry, 2009, p. 266
78. Henry, 2009, pp. 198–9, 236
79. Henry, 2009, pp. 195–6
80. Henry, 2009, p. 46
81. Henry, 2009, pp. 69–74
82. All quotes cited in *Fortean Times* 195, p. 145
83. Henry, 2009, p. 77
84. Noble, 1999, p. 130, 133–4, 140–1
85. Noble, 1999, p. 131
86. Cited in Noble, 1999, p. 132
87. Noble, 1999, pp. 123–9
88. Cited in Noble, 1999, p. 126
89. Kossy, 2001, pp. 189–192
90. Noble, 1999, p. 12
91. Cited in Noble, 1999, p. 175
92. Noble, 1999, pp. 174–5
93. Gray, 2012, pp. 213–16
94. https://en.wikipedia.org/wiki/Utility_fog
95. https://en.wikipedia.org/wiki/Clarke%27s_three_laws
96. Noble, 1999, pp. 145–6
97. Noble, 1999, p. 149
98. Noble, 1999, pp. 153–4, 163–4; https://en.wikipedia.org/wiki/Edward_Fredkin; https://en.wikipedia.org/wiki/Digital_physics
99. https://enwikipedia.org/wiki/Pierr_Teilhard_de_Chardin; https://en,wikipedia.org/wiki/Omega_Point
100. http://www.csicop.org/si/show/the_strange_case_of_frank_jennings_tipler; https://en.wikipedia.org/wiki/Frank_J._Tipler; https://en.wikipedia.org/wiki/Tipler_cylinder
101. Cited in Young, 2012, p. 234
102. Gray, 2012, p. 216

Index

More books by S. D. Tucker

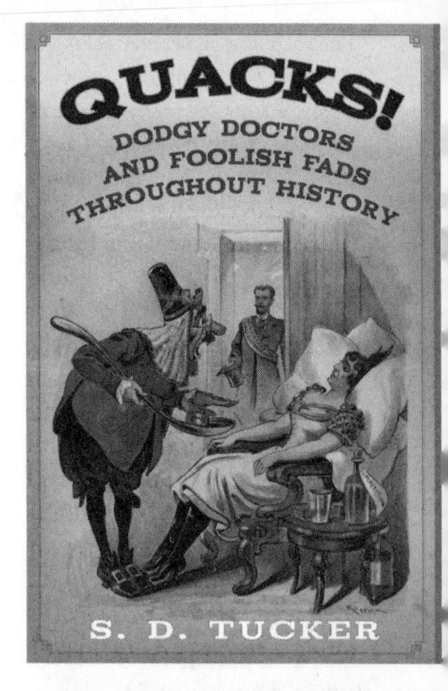

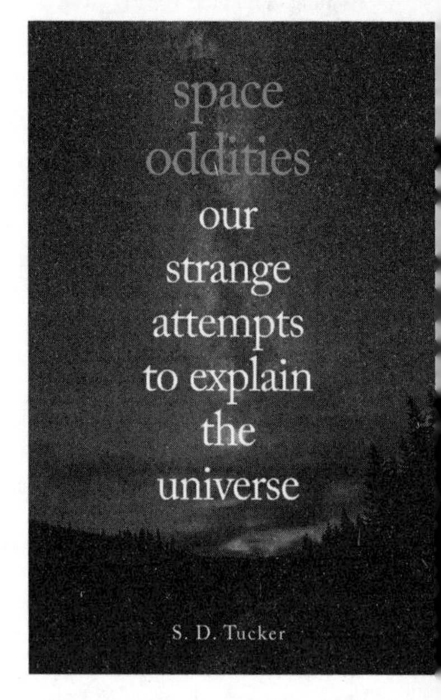